www.KnowledgePublications.com

DOE-HDBK-1011/1-92
JUNE 1992

DOE FUNDAMENTALS HANDBOOK

ELECTRICAL SCIENCE

Volume 1 of 4

U.S. Department of Energy **FSC-6910**

Washington, D.C. 20585

This document has been reproduced directly from the best available copy.

Available to DOE and DOE contractors from the Office of Scientific and Technical Information. P. O. Box 62, Oak Ridge, TN 37831; (615) 576-8401.

Available to the public from the National Technical Information Service, U.S. Department of Commerce, 5285 Port Royal Rd., Springfield, VA 22161.

Order No. DE92019785

ABSTRACT

The *Electrical Science Fundamentals Handbook* was developed to assist nuclear facility operating contractors provide operators, maintenance personnel, and the technical staff with the necessary fundamentals training to ensure a basic understanding of electrical theory, terminology, and application. The handbook includes information on alternating current (AC) and direct current (DC) theory, circuits, motors, and generators; AC power and reactive components; batteries; AC and DC voltage regulators; transformers; and electrical test instruments and measuring devices. This information will provide personnel with a foundation for understanding the basic operation of various types of DOE nuclear facility electrical equipment.

Key Words: Training Material, Magnetism, DC Theory, DC Circuits, Batteries, DC Generators, DC Motors, AC Theory, AC Power, AC Generators, Voltage Regulators, AC Motors, Transformers, Test Instruments, Electrical Distribution

FOREWORD

The *Department of Energy (DOE) Fundamentals Handbooks* consist of ten academic subjects, which include Mathematics; Classical Physics; Thermodynamics, Heat Transfer, and Fluid Flow; Instrumentation and Control; Electrical Science; Material Science; Mechanical Science; Chemistry; Engineering Symbology, Prints, and Drawings; and Nuclear Physics and Reactor Theory. The handbooks are provided as an aid to DOE nuclear facility contractors.

These handbooks were first published as Reactor Operator Fundamentals Manuals in 1985 for use by DOE category A reactors. The subject areas, subject matter content, and level of detail of the Reactor Operator Fundamentals Manuals were determined from several sources. DOE Category A reactor training managers determined which materials should be included, and served as a primary reference in the initial development phase. Training guidelines from the commercial nuclear power industry, results of job and task analyses, and independent input from contractors and operations-oriented personnel were all considered and included to some degree in developing the text material and learning objectives.

The *DOE Fundamentals Handbooks* represent the needs of various DOE nuclear facilities' fundamental training requirements. To increase their applicability to nonreactor nuclear facilities, the Reactor Operator Fundamentals Manual learning objectives were distributed to the Nuclear Facility Training Coordination Program Steering Committee for review and comment. To update their reactor-specific content, DOE Category A reactor training managers also reviewed and commented on the content. On the basis of feedback from these sources, information that applied to two or more DOE nuclear facilities was considered generic and was included. The final draft of each of the handbooks was then reviewed by these two groups. This approach has resulted in revised modular handbooks that contain sufficient detail such that each facility may adjust the content to fit their specific needs.

Each handbook contains an abstract, a foreword, an overview, learning objectives, and text material, and is divided into modules so that content and order may be modified by individual DOE contractors to suit their specific training needs. Each subject area is supported by a separate examination bank with an answer key.

The *DOE Fundamentals Handbooks* have been prepared for the Assistant Secretary for Nuclear Energy, Office of Nuclear Safety Policy and Standards, by the DOE Training Coordination Program. This program is managed by EG&G Idaho, Inc.

OVERVIEW

The *Department of Energy Fundamentals Handbook* entitled *Electrical Science* was prepared as an information resource for personnel who are responsible for the operation of the Department's nuclear facilities. A basic understanding of electricity and electrical systems is necessary for DOE nuclear facility operators, maintenance personnel, and the technical staff to safely operate and maintain the facility and facility support systems. The information in the handbook is presented to provide a foundation for applying engineering concepts to the job. This knowledge will help personnel more fully understand the impact that their actions may have on the safe and reliable operation of facility components and systems.

The *Electrical Science* handbook consists of fifteen modules that are contained in four volumes. The following is a brief description of the information presented in each module of the handbook.

Volume 1 of 4

Module 1 - Basic Electrical Theory

This module describes basic electrical concepts and introduces electrical terminology.

Module 2 - Basic DC Theory

This module describes the basic concepts of direct current (DC) electrical circuits and discusses the associated terminology.

Volume 2 of 4

Module 3 - DC Circuits

This module introduces the rules associated with the reactive components of inductance and capacitance and how they affect DC circuits.

Module 4 - Batteries

This module introduces batteries and describes the types of cells used, circuit arrangements, and associated hazards.

Module 5 - DC Generators

This module describes the types of DC generators and their application in terms of voltage production and load characteristics.

Module 6 - DC Motors

This module describes the types of DC motors and includes discussions of speed control, applications, and load characteristics.

Volume 3 of 4

Module 7 - Basic AC Theory

This module describes the basic concepts of alternating current (AC) electrical circuits and discusses the associated terminology.

Module 8 - AC Reactive Components

This module describes inductance and capacitance and their effects on AC circuits.

Module 9 - AC Power

This module presents power calculations for single-phase and three-phase AC circuits and includes the power triangle concept.

Module 10 - AC Generators

This module describes the operating characteristics of AC generators and includes terminology, methods of voltage production, and methods of paralleling AC generation sources.

Module 11 - Voltage Regulators

This module describes the basic operation and application of voltage regulators.

Volume 4 of 4

Module 12 - AC Motors

This module explains the theory of operation of AC motors and discusses the various types of AC motors and their application.

Module 13 - Transformers

This module introduces transformer theory and includes the types of transformers, voltage/current relationships, and application.

Module 14 - Test Instruments and Measuring Devices

This module describes electrical measuring and test equipment and includes the parameters measured and the principles of operation of common instruments.

Module 15 - Electrical Distribution Systems

This module describes basic electrical distribution systems and includes characteristics of system design to ensure personnel and equipment safety.

The information contained in this handbook is by no means all encompassing. An attempt to present the entire subject of electrical science would be impractical. However, the *Electrical Science* handbook does present enough information to provide the reader with a fundamental knowledge level sufficient to understand the advanced theoretical concepts presented in other subject areas, and to better understand basic system and equipment operations.

Department of Energy
Fundamentals Handbook

ELECTRICAL SCIENCE
Module 1
Basic Electrical Theory

TABLE OF CONTENTS

LIST OF FIGURES iv

LIST OF TABLES vi

REFERENCES vii

OBJECTIVES viii

ATOM AND ITS FORCES 1

The Atom 1
Electrostatic Force 2
The First Law of Electrostatics 3
Electrostatic Field 3
Potential Difference 5
Free Electrons 6
Summary 8

ELECTRICAL TERMINOLOGY 9

Conductors 9
Insulators 9
Resistors 9
Voltage 10
Current 10
Real and Ideal Sources 12
Summary 12

UNITS OF ELECTRICAL MEASUREMENT 13

System Internationale (SI) Metric System 13
Voltage 13
Current 14
Resistance 14
Ohm's Law 14
Conductance 16
Power 16

TABLE OF CONTENTS (Cont.)

Inductance 17
Capacitance 17
Summary 18

METHODS OF PRODUCING VOLTAGE (ELECTRICITY) 19

Electrochemistry 19
Static Electricity 20
Magnetic Induction 21
Piezoelectric Effect 21
Thermoelectricity 22
Photoelectric Effect 23
Thermionic Emission 24
Summary 26

MAGNETISM 27

Magnetism 27
Magnetic Flux 29
Magnetic Flux Density 29
Magnetic Materials 30
Electromagnetism 31
Polarity of a Single Conductor 31
Magnetic Field and Polarity of a Coil 32
Magnetomotive Force 33
Field Intensity 34
Reluctance 35
Summary 36

MAGNETIC CIRCUITS 37

Magnetic Circuits 37
BH Magnetization Curve 39
Hysteresis 40
Magnetic Induction 41
Faraday's Law of Induced Voltage 42
Lenz's Law 43
Summary 44

TABLE OF CONTENTS (Cont.)

ELECTRICAL SYMBOLS 45

Symbols 45
Summary 47

APPENDIX A Metric System and Powers of Ten A-1

LIST OF FIGURES

Figure 1 The Atom 1

Figure 2 The Carbon Atom 2

Figure 3 Electrostatic Force 2

Figure 4 Electrostatic Field 3

Figure 5 Electrostatic Field Between Two Charges of Opposite Polarity 4

Figure 6 Electrostatic Field Between Two Charges of Like Polarity 4

Figure 7 Potential Difference Between Two Charged Objects 5

Figure 8 Energy Shells and Electron Quota 6

Figure 9 Electron Flow Through a Copper Wire with a Potential Difference 11

Figure 10 Potential Difference Across a Conductor Causes a Current to Flow 11

Figure 11 Voltaic Chemical Cell 20

Figure 12 Static Electricity 20

Figure 13 Generator - Electromagnetic Induction 21

Figure 14 Pressure Applied to Certain Crystals Produce an Electric Charge 22

Figure 15 Heat Energy Causes Copper to Give up Electrons to Zinc 23

Figure 16 Producing Electricity from Light Using a Photovoltaic Cell 24

Figure 17 Vacuum Tube Diode 25

Figure 18 Electron Spinning Around Nucleus Produces Magnetic Field 27

Figure 19 Magnetic Domains 28

Figure 20 The Law of Magnetic Attraction and Repulsion 28

LIST OF FIGURES (Cont.)

Figure 21 The Magnetic Field Produced by Current in a Conductor 31

Figure 22 Left-hand Rule for Current Carrying Conductors 31

Figure 23 Left-hand Rule for Coils 32

Figure 24 Left-hand Rule to Find North Pole of an Electromagnet 33

Figure 25 Different Physical Forms of Electromagnets 35

Figure 26 Magnetic Current with Closed Iron Path 38

Figure 27 Typical BH Curve for Two Types of Soft Iron 39

Figure 28 Hysteresis Loop for Magnetic Materials 41

Figure 29 Induced EMF 42

Figure 30 Electrical Symbols 46

LIST OF TABLES

Table A-1 Base Units of the International Metric System A-1

Table A-2 Supplementary SI Units A-2

Table A-3 Derived SI Units A-3

Table A-4 Metric Prefixes Used in Electricity A-4

Table A-5 Powers of 10 . . . . A-5

Table A-6 Metric Prefixes Expressed as Powers of 10 . . . . A-8

REFERENCES

- Gussow, Milton, Schaum's Outline Series, Basic Electricity, McGraw-Hill.
- Academic Program for Nuclear Power Plant Personnel, Volume IV, Columbia, MD: General Physics Corporation, Library of Congress Card #A 326517, 1982.
- Sienko and Plane, Chemical Principles and Properties, 2nd Edition, McGraw-Hill.
- Academic Program for Nuclear Power Plant Personnel, Volume II, Columbia, MD: General Physics Corporation, Library of Congress Card #A 326517, 1982.
- Nasar and Unnewehr, Electromechanics and Electric Machines, John Wiley and Sons.
- Van Valkenburgh, Nooger, and Neville, Basic Electricity, Vol. 5, Hayden Book Company.
- Exide Industrial Marketing Division, The Storage Battery, Lead-Acid Type, The Electric Storage Battery Company.
- Lister, Eugene C., Electric Circuits and Machines, 5th Edition, McGraw-Hill.
- Croft, Carr, Watt, and Summers, American Electricians Handbook, 10th Edition, McGraw-Hill.
- Mason, C. Russel, The Art and Science of Protective Relaying, John Wiley and Sons.
- Mileaf, Harry, Electricity One - Seven, Revised 2nd Edition, Hayden Book Company.
- Buban and Schmitt, Understanding Electricity and Electronics, 3rd Edition, McGraw-Hill.
- Kidwell, Walter, Electrical Instruments and Measurements, McGraw-Hill.

TERMINAL OBJECTIVE

1.0 Given a simple electrical circuit, **APPLY** basic electrical theory fundamental principles to describe circuit operation.

ENABLING OBJECTIVES

1.1 **DESCRIBE** the following terms:
- a. Electrostatic force
- b. Electrostatic field
- c. Potential difference
- d. Electromotive force (EMF)
- e. Ion charge

1.2 **DEFINE** the following terms:
- a. Conductor
- b. Insulator
- c. Resistor
- d. Electron current flow
- e. Conventional current flow
- f. Direct current (DC)
- g. Alternating current (AC)
- h. Ideal source
- i. Real source

1.3 **DESCRIBE** the following electrical parameters, including the unit of measurement and the relationship to other parameters.
- a. Voltage
- b. Current
- c. Resistance
- d. Conductance
- e. Power
- f. Inductance
- g. Capacitance

1.4 Given any two of the three component values of Ohm's Law, **DETERMINE** the unknown component value.

ENABLING OBJECTIVES (Cont.)

1.5 **DESCRIBE** how the following methods produce a voltage:
- a. Electrochemistry
- b. Static electricity
- c. Magnetic Induction
- d. Piezoelectric effect
- e. Thermoelectricity
- f. Photoelectric effect
- g. Thermionic emission

1.6 **DEFINE** the following terms:
- a. Magnetic flux
- b. Magnetic flux density
- c. Weber
- d. Permeability
- e. Magnetomotive force (mmf)
- f. Ampere turns
- g. Field intensity
- h. Reluctance

1.7 **DESCRIBE** the following materials as they relate to permeability, including an example and an approximate relative permeability.
- a. Ferromagnetic materials
- b. Paramagnetic materials
- c. Diamagnetic materials

1.8 **EXPLAIN** the physical qualities of a simple magnetic circuit, including relationships of qualities and units of measurements.

1.9 Given the physical qualities of a simple magnetic circuit, **CALCULATE** the unknown values.

1.10 **DESCRIBE** the shape and components of a BH magnetization curve.

1.11 **EXPLAIN** the cause of hysteresis losses.

1.12 Given Faraday's Law of induced voltage:
- a. **DESCRIBE** how varying parameters affect induced voltage.
- b. **CALCULATE** voltage induced in a conductor moving through a magnetic field.

1.13 **STATE** Lenz's Law of induction.

ENABLING OBJECTIVES (Cont.)

1.14 Given a standard electrical symbol, **IDENTIFY** the component that the symbol represents. The symbols will be for the following components:

a. Resistor
b. Capacitor
c. Inductor
d. Relay
e. Contacts
f. Breaker
g. Switch
h. Transistor
i. Rheostat
j. Diode
k. Ground connections
l. Vacuum tube
m. Fuse
n. Junction
o. AC voltage source
p. Voltmeter
q. Ammeter
r. Wattmeter
s. Relay operated contacts
t. Potential transformer
u. Current transformer
v. Wye (Y) connection
w. Delta (Δ) connection
x. Light bulb
y. Battery

ATOM AND ITS FORCES

What is electricity? Electricity is defined as "the flow of electrons through simple materials and devices" or "that force which moves electrons." Scientists think electricity is produced by very tiny particles called electrons and protons. These particles are too small to be seen, but exist as subatomic particles in the atom. To understand how they exist, you must first understand the structure of the atom.

EO 1.1 DESCRIBE the following terms:
- **a. Electrostatic force**
- **b. Electrostatic field**
- **c. Potential difference**
- **d. Electromotive force (EMF)**
- **e. Ion charge**

The Atom

Elements are the basic building blocks of all matter. The atom is the smallest particle to which an element can be reduced while still keeping the properties of that element. An atom consists of a positively charged nucleus surrounded by negatively charged electrons, so that the atom as a whole is electrically neutral. The nucleus is composed of two kinds of subatomic particles, protons and neutrons, as shown in Figure 1. The proton carries a single unit positive charge equal in magnitude to the electron charge. The neutron is slighty heavier than the proton and is electrically neutral, as the name implies. These two particles exist in various combinations, depending upon the element involved. The electron is the fundamental negative charge (-) of electricity and revolves around the nucleus, or center, of the atom in concentric orbits, or shells.

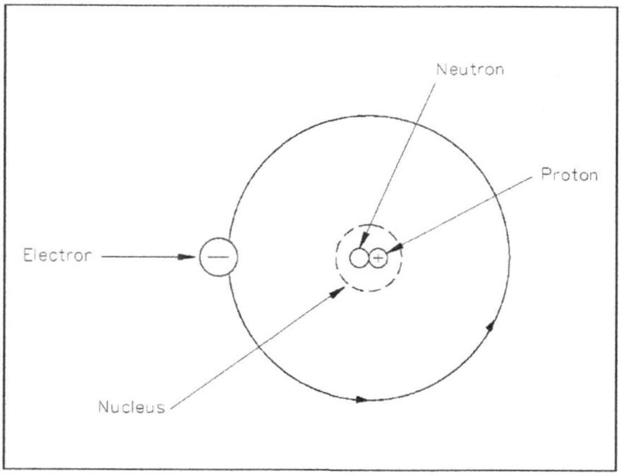

Figure 1 The Atom

The proton is the fundamental positive charge (+) of electricity and is located in the nucleus. The number of protons in the nucleus of any atom specifies the atomic number of that atom or of that element. For example, the carbon atom contains six protons in its nucleus; therefore, the atomic number for carbon is six, as shown in Figure 2.

In its natural state, an atom of any element contains an equal number of electrons and protons. The negative charge (-) of each electron is equal in magnitude to the positive charge (+) of each proton; therefore, the two opposite charges cancel, and the atom is said to be electrically neutral, or in balance.

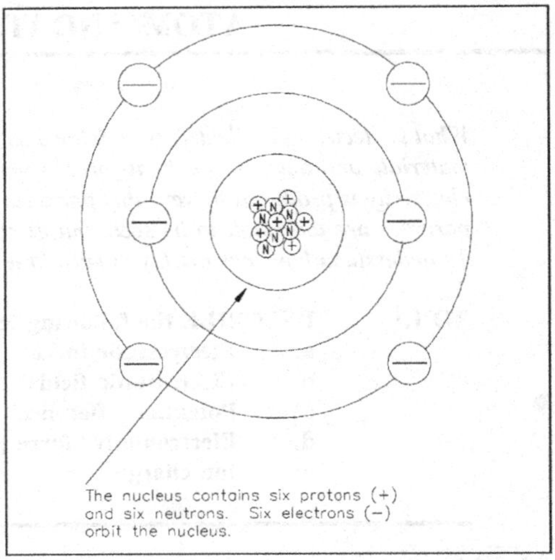

Figure 2 The Carbon Atom

Electrostatic Force

One of the mysteries of the atom is that the electron and the nucleus attract each other. This attraction is called *electrostatic force*, the force that holds the electron in orbit. This force may be illustrated with lines as shown in Figure 3.

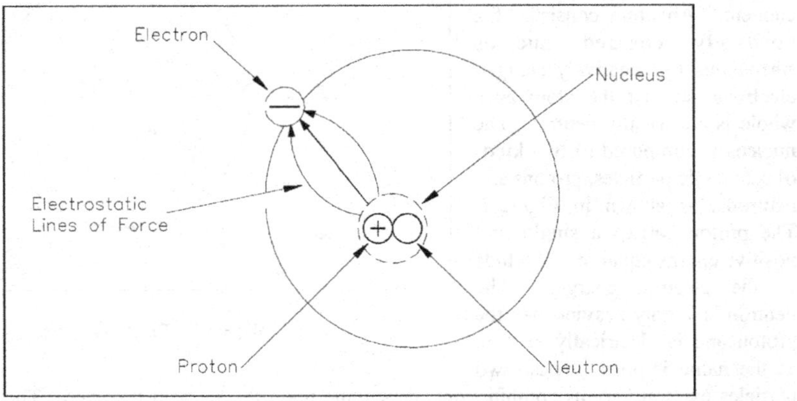

Figure 3 Electrostatic Force

Without this electrostatic force, the electron, which is traveling at high speed, could not stay in its orbit. Bodies that attract each other in this way are called charged bodies. As mentioned previously, the electron has a negative charge, and the nucleus (due to the proton) has a positive charge.

The First Law of Electrostatics

The negative charge of the electron is equal, but opposite to, the positive charge of the proton. These charges are referred to as electrostatic charges. In nature, unlike charges (like electrons and protons) attract each other, and like charges repel each other. These facts are known as the *First Law of Electrostatics* and are sometimes referred to as the law of electrical charges. This law should be remembered because it is one of the vital concepts in electricity.

Some atoms can lose electrons and others can gain electrons; thus, it is possible to transfer electrons from one object to another. When this occurs, the equal distribution of negative and positive charges no longer exists. One object will contain an excess of electrons and become negatively charged, and the other will become deficient in electrons and become positively charged. These objects, which can contain billions of atoms, will then follow the same law of electrostatics as the electron and proton example shown above. The electrons that can move around within an object are said to be free electrons and will be discussed in more detail in a later section. The greater the number of these free electrons an object contains, the greater its negative electric charge. Thus, the electric charge can be used as a measure of electrons.

Electrostatic Field

A special force is acting between the charged objects discussed above. Forces of this type are the result of an *electrostatic field* that exists around each charged particle or object. This electrostatic field, and the force it creates, can be illustrated with lines called "lines of force" as shown in Figure 4.

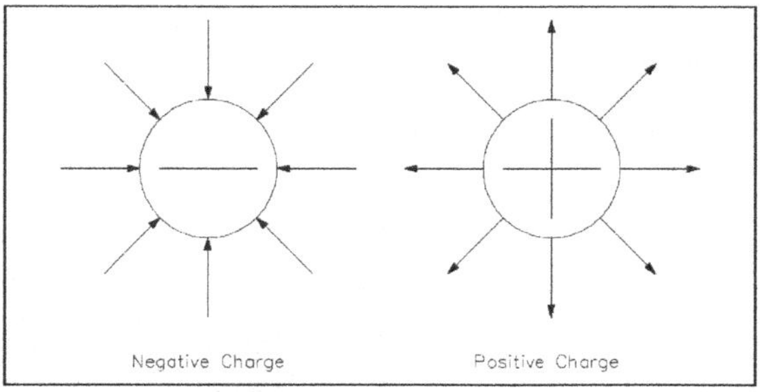

Figure 4 Electrostatic Field

Charged objects repel or attract each other because of the way these fields act together. This force is present with every charged object. When two objects of opposite charge are brought near one another, the electrostatic field is concentrated in the area between them, as shown in Figure 5. The direction of the small arrows shows the direction of the force as it would act upon an electron if it were released into the electric field.

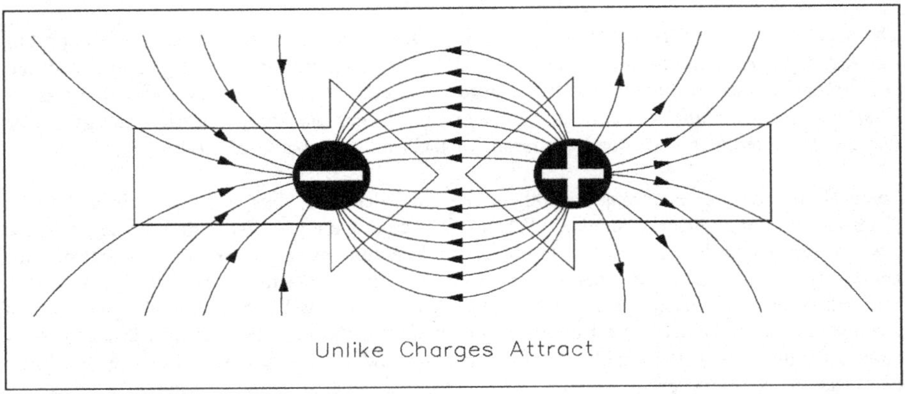

Figure 5 Electrostatic Field Between Two Charges of Opposite Polarity

When two objects of like charge are brought near one another, the lines of force repel each other, as shown in Figure 6.

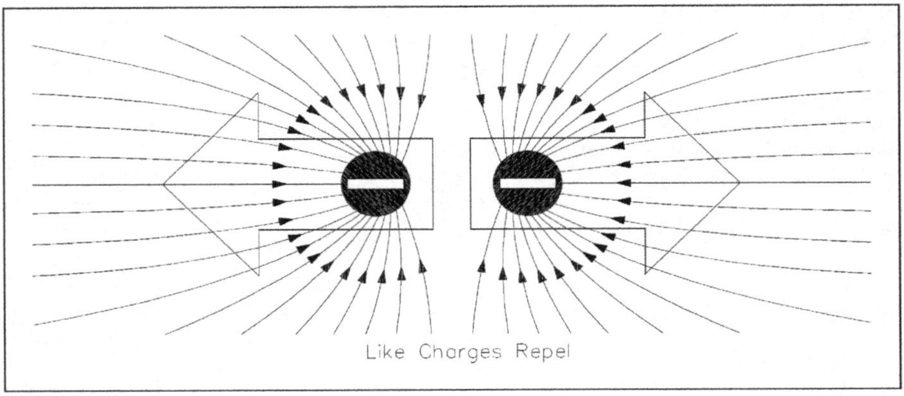

Figure 6 Electrostatic Field Between Two Charges of Like Polarity

The strength of the attraction or of the repulsion force depends upon two factors: (1) the amount of charge on each object, and (2) the distance between the objects. The greater the charge on the objects, the greater the electrostatic field. The greater the distance between the objects, the weaker the electrostatic field between them, and vice versa. This leads us to the law of electrostatic attraction, commonly referred to as Coulomb's Law of electrostatic charges, which states that the force of electrostatic attraction, or repulsion, is directly proportional to the product of the two charges and inversely proportional to the square of the distance between them as shown in Equation 1-1.

$$F = K\frac{q_1 - q_2}{d^2} \qquad (1\text{-}1)$$

where

F	= force of electrostatic attraction or prepulsion (Newtons)
K	= constant of proportionality (Coulomb ^{2}/N-m^2)
q_1	= charge of first particle (Coulombs)
q_2	= charge of second particle (Coulombs)
d	= distance between two particles (Meters)

If q_1 and q_2 are both either positively or negatively charged, the force is repulsive. If q_1 and q_2 are opposite polarity or charge, the force is attractive.

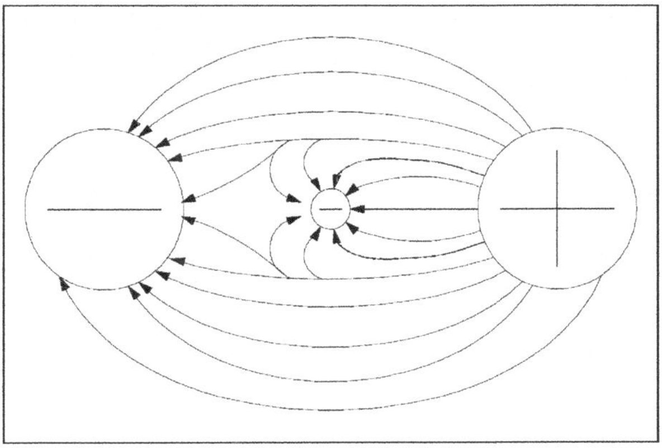

Figure 7 Potential Difference Between Two Charged Objects

Potential Difference

Potential difference is the term used to describe how large the electrostatic force is between two charged objects. If a charged body is placed between two objects with a potential difference, the charged body will try to move in one direction, depending upon the polarity of the object. If an electron is placed between a negatively-charged body and a positively-charged body, the action due to the potential difference is to push the electron toward the positively-charged object. The electron, being negatively charged, will be repelled from the negatively-charged object and attracted by the positively-charged object, as shown in Figure 7.

Due to the force of its electrostatic field, these electrical charges have the ability to do work by moving another charged particle by attraction and/or repulsion. This ability to do work is called "potential"; therefore, if one charge is different from another, there is a potential difference between them. The sum of the potential differences of all charged particles in the electrostatic field is referred to as *electromotive force* (EMF).

The basic unit of measure of potential difference is the "volt." The symbol for potential difference is "V," indicating the ability to do the work of forcing electrons to move. Because the volt unit is used, potential difference is also called "voltage." The unit volt will be covered in greater detail in the next chapter.

Free Electrons

Electrons are in rapid motion around the nucleus. While the electrostatic force is trying to pull the nucleus and the electron together, the electron is in motion and trying to pull away. These two effects balance, keeping the electron in orbit. The electrons in an atom exist in different energy levels. The energy level of an electron is proportional to its distance from the nucleus. Higher energy level electrons exist in orbits, or shells, that are farther away from the nucleus. These shells nest inside one another and surround the nucleus. The nucleus is the center of all the shells. The shells are lettered beginning with the shell nearest the nucleus: K, L, M, N, O, P, and Q. Each shell has a maximum number of electrons it can hold. For example, the K shell will hold a maximum of two electrons and the L shell will hold a maximum of eight electrons. As shown in Figure 8, each shell has a specific number of electrons that it will hold for a particular atom.

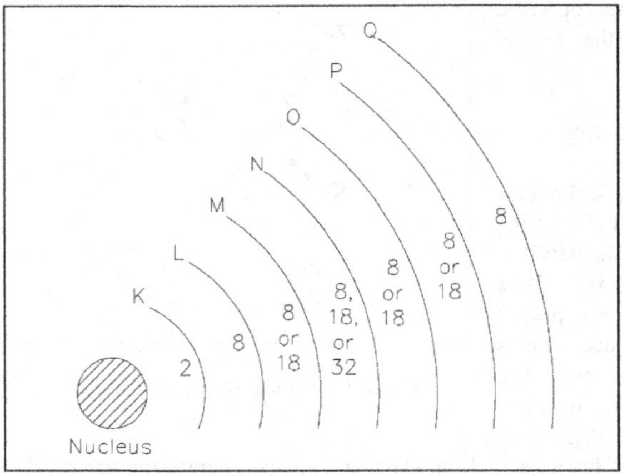

Figure 8 Energy Shells and Electron Quota

There are two simple rules concerning electron shells that make it possible to predict the electron distribution of any element:

1. The maximum number of electrons that can fit in the outermost shell of any atom is eight.

2. The maximum number of electrons that can fit in the next-to-outermost shell of any atom is 18.

An important point to remember is that when the outer shell of an atom contains eight electrons, the atom becomes very stable, or very resistant to changes in its structure. This also means that atoms with one or two electrons in their outer shell can lose electrons much more easily than atoms with full outer shells. The electrons in the outermost shell are called *valence electrons*. When external energy, such as heat, light, or electrical energy, is applied to certain materials, the electrons gain energy, become excited, and may move to a higher energy level. If enough energy is applied to the atom, some of the valence electrons will leave the atom. These electrons are called *free electrons*. It is the movement of free electrons that provides electric current in a metal conductor. An atom that has lost or gained one or more electrons is said to be *ionized* or to have an *ion change*. If the atom loses one or more electrons, it becomes positively charged and is referred to as a *positive ion*. If an atom gains one or more electrons, it becomes negatively charged and is referred to as a *negative ion*.

Summary

The important information contained in this chapter is summarized below.

Forces Around Atoms Summary

- Electrostatic Force - force that holds an electron in orbit around a nucleus
- Electrostatic Field - force acting between charged objects that causes them to repel or attract
- Potential Difference - measures how large the electrostatic force is between two charged objects. According to Coulomb's Law, charged bodies attract or repel each other with a force that is directly proportional to the product of their charges and is inversely proportional to the square of the distance between them.
- Electromotive Force (EMF) - sum of the potential differences of all charged particles in an electrostatic field
- Ion Charge - dependent on the loss or gain of free electrons (if an atom gains an electron - negative ion charge; if an atom loses an electron - positive ion charge)

ELECTRICAL TERMINOLOGY

Knowledge of key electrical terminology is necessary to fully understand principles in electrical science.

EO 1.2 DEFINE the following terms:

- **a. Conductor**
- **b. Insulator**
- **c. Resistor**
- **d. Electron current flow**
- **e. Conventional current flow**
- **f. Direct current (DC)**
- **g. Alternating current (AC)**
- **h. Ideal source**
- **i. Real source**

Conductors

Conductors are materials with electrons that are loosely bound to their atoms, or materials that permit free motion of a large number of electrons. Atoms with only one valence electron, such as copper, silver, and gold, are examples of good conductors. Most metals are good conductors.

Insulators

Insulators, or nonconductors, are materials with electrons that are tightly bound to their atoms and require large amounts of energy to free them from the influence of the nucleus. The atoms of good insulators have their valence shells filled with eight electrons, which means they are more than half filled. Any energy applied to such an atom will be distributed among a relatively large number of electrons. Examples of insulators are rubber, plastics, glass, and dry wood.

Resistors

Resistors are made of materials that conduct electricity, but offer opposition to current flow. These types of materials are also called *semiconductors* because they are neither good conductors nor good insulators. Semiconductors have more than one or two electrons in their valence shells, but less than seven or eight. Examples of semiconductors are carbon, silicon, germanium, tin, and lead. Each has four valence electrons.

Voltage

The basic unit of measure for potential difference is the *volt* (symbol V), and, because the volt unit is used, potential difference is called *voltage*. An object's electrical charge is determined by the number of electrons that the object has gained or lost. Because such a large number of electrons move, a unit called the "coulomb" is used to indicate the charge. One coulomb is equal to 6.28×10^{18} (billion, billion) electrons. For example, if an object gains one coulomb of negative charge, it has gained 6,280,000,000,000,000,000 extra electrons. A volt is defined as a difference of potential causing one coulomb of current to do one joule of work. A volt is also defined as that amount of force required to force one ampere of current through one ohm of resistance. The latter is the definition with which we will be most concerned in this module.

Current

The density of the atoms in copper wire is such that the valence orbits of the individual atoms overlap, causing the electrons to move easily from one atom to the next. Free electrons can drift from one orbit to another in a random direction. When a potential difference is applied, the direction of their movement is controlled. The strength of the potential difference applied at each end of the wire determines how many electrons change from a random motion to a more directional path through the wire. The movement or flow of these electrons is called *electron current flow* or just *current*.

To produce current, the electrons must be moved by a potential difference. The symbol for current is (I). The basic measurement for current is the ampere (A). One ampere of current is defined as the movement of one coulomb of charge past any given point of a conductor during one second of time.

If a copper wire is placed between two charged objects that have a potential difference, all of the negatively-charged free electrons will feel a force pushing them from the negative charge to the positive charge. This force opposite to the conventional direction of the electrostatic lines of force is shown in Figure 9.

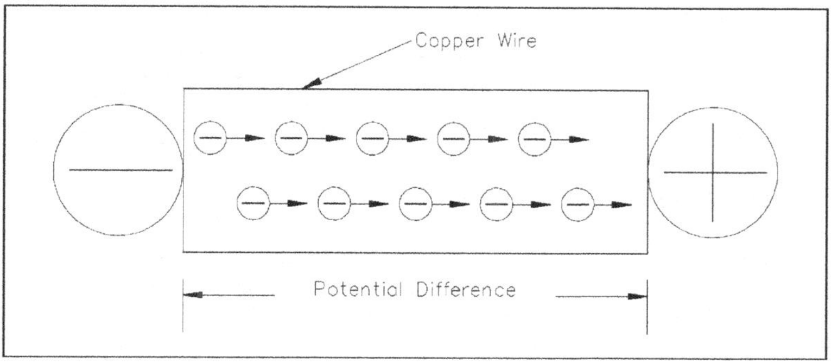

Figure 9 Electron Flow Through a Copper Wire with a Potential Difference

The direction of electron flow, shown in Figure 10, is from the negative (-) side of the battery, through the wire, and back to the positive (+) side of the battery. The direction of electron flow is from a point of negative potential to a point of positive potential. The solid arrow shown in Figure 10 indicates the direction of electron flow. As electrons vacate their atoms during electron current flow, positively charged atoms (holes) result. The flow of electrons in one direction causes a flow of positive charges. The direction of the positive charges is in the opposite direction of the electron flow. This flow of positive charges is known as *conventional current* and is shown in Figure 10 as a dashed arrow. All of the electrical effects of electron flow from negative to positive, or from a higher potential to a lower potential, are the same as those that would be created by a flow of positive charges in the opposite direction. Therefore, it is important to realize that both conventions are in use and that they are essentially equivalent; that is, all effects predicted are the same. In this text, we will be using electron flow in our discussions.

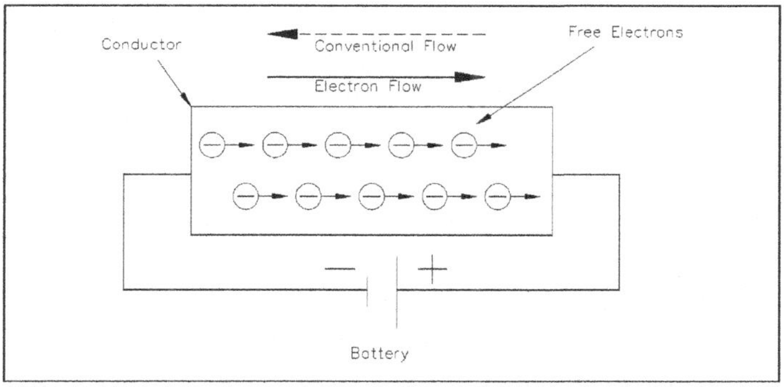

Figure 10 Potential Difference Across a Conductor Causes a Current to Flow

Generally, electric current flow can be classified as one of two general types: *Direct Current* (DC) or *Alternating Current* (AC). A direct current flows continuously in the same direction. An alternating current periodically reverses direction. We will be studying DC and AC current in more detail later in this text. An example of DC current is that current obtained from a battery. An example of AC current is common household current.

Real and Ideal Sources

An *ideal source* is a theoretical concept of an electric current or voltage supply (such as a battery) that has no losses and is a perfect voltage or current supply. Ideal sources are used for analytical purposes only since they cannot occur in nature.

A *real source* is a real life current or voltage supply that has some losses associated with it.

Summary

The important information contained in this chapter is summarized below.

Terminology Summary

- Conductor - material with electrons loosely bound to its atoms or that permits free motion of large number of electrons
- Insulator - material with electrons tightly bound to its atoms; requires large amounts of energy to free electrons from its nuclei
- Resistor - material that conducts electricity, but opposes current flow
- Electron Current Flow - current flow from negative to positive potentials
- Conventional Current Flow - current flow from positive to negative potentials
- Direct Current - current flow continuously in the same direction
- Alternating Current - current flow periodically reverses direction
- Ideal Source - theoretical current or voltage supply with no losses
- Real Source - actual current or voltage supply with losses

UNITS OF ELECTRICAL MEASUREMENT

Using Ohm's Law and the System Internationale (SI) Metric System, electrical measuring units can be derived.

EO 1.3 **DESCRIBE the following electrical parameters, including the unit of measurement and the relationship to other parameters.**

- **a. Voltage**
- **b. Current**
- **c. Resistance**
- **d. Conductance**
- **e. Power**
- **f. Inductance**
- **g. Capacitance**

EO 1.4 **Given any two of the three component values of Ohm's Law, DETERMINE the unknown component value.**

System Internationale (SI) Metric System

Electrical units of measurement are based on the International (metric) System, also known as the SI System. Units of electrical measurement include the following:

- Ampere
- Volt
- Ohm
- Siemens
- Watt
- Henry
- Farad

Appendix A provides more information concerning the metric system, metric prefixes, and powers of 10 that are used in electrical measuring units.

Voltage

Voltage, electromotive force (emf), or potential difference, is described as the pressure or force that causes electrons to move in a conductor. In electrical formulas and equations, you will see voltage symbolized with a capital E, while on laboratory equipment or schematic diagrams, the voltage is often represented with a capital V.

Current

Electron *current*, or amperage, is described as the movement of free electrons through a conductor. In electrical formulas, current is symbolized with a capital I, while in the laboratory or on schematic diagrams, it is common to use a capital A to indicate amps or amperage (amps).

Resistance

Now that we have discussed the concepts of voltage and current, we are ready to discuss a third key concept called resistance. *Resistance* is defined as the opposition to current flow. The amount of opposition to current flow produced by a material depends upon the amount of available free electrons it contains and the types of obstacles the electrons encounter as they attempt to move through the material. Resistance is measured in ohms and is represented by the symbol (R) in equations. One ohm is defined as that amount of resistance that will limit the current in a conductor to one ampere when the potential difference (voltage) applied to the conductor is one volt. The shorthand notation for ohm is the Greek letter capital omega (Ω). If a voltage is applied to a conductor, current flows. The amount of current flow depends upon the resistance of the conductor. The lower the resistance, the higher the current flow for a given amount of voltage. The higher the resistance, the lower the current flow.

Ohm's Law

In 1827, George Simon Ohm discovered that there was a definite relationship between voltage, current, and resistance in an electrical circuit. Ohm's Law defines this relationship and can be stated in three ways.

1. Applied voltage equals circuit current times the circuit resistance. Equation (1-2) is a mathematical respresentation of this concept.

$$E = I \times R \quad \text{or} \quad E = IR \tag{1-2}$$

2. Current is equal to the applied voltage divided by the circuit resistance. Equation (1-3) is a mathematical representation of this concept.

$$I = \frac{E}{R} \tag{1-3}$$

3. Resistance of a circuit is equal to the applied voltage divided by the circuit current. Equation (1-4) is a mathematical representation of this concept.

$$R \text{ (or } \Omega) = \frac{E}{I} \tag{1-4}$$

where

I = current (A)

E = voltage (V)

R = resistance (Ω)

If any two of the component values are known, the third can be calculated.

Example 1: Given that I = 2 A, E = 12 V, find the circuit resistance.

Solution:

Since applied voltage and circuit current are known, use Ohm's Law to solve for resistance.

$$R = \frac{E}{I}$$

$$R = \frac{12 \text{ V}}{2 \text{ A}} = 6 \ \Omega$$

Example 2: Given E = 260 V and R = 240 Ω, what current will flow through a circuit?

Solution:

Since applied voltage and resistance are known, use Ohm's Law to solve for current.

$$I = \frac{E}{R}$$

$$I = \frac{260 \text{ V}}{240 \ \Omega} = 1.08\overline{3} \text{ A}$$

Example 3: Find the applied voltage, when given circuit resistance of 100 Ω and circuit current of 0.5 amps.

Solution:

Since circuit resistance and circuit current are known, use Ohm's Law to solve for applied voltage.

$$E = IR$$

$$E = (0.5\ A)(100\ \Omega) = 50\ V$$

Conductance

The word "reciprocal" is sometimes used to mean "the opposite of." The opposite, or reciprocal, of resistance is called *conductance*. As described above, resistance is the opposition to current flow. Since resistance and conductance are opposites, conductance can be defined as the ability to conduct current. For example, if a wire has a high conductance, it will have low resistance, and vice-versa. Conductance is found by taking the reciprocal of the resistance. The unit used to specify conductance is called "mho," which is ohm spelled backwards. The symbol for "mho" is the Greek letter omega inverted (℧). The symbol for conductance when used in a formula is G. Equation (1-5) is the mathematical representation of conductance obtained by relating the definition of conductance (1/R) to Ohm's Law, Equation (1-4).

$$G = \frac{1}{\text{RESISTANCE}} = \frac{I}{E} \tag{1-5}$$

Example: If a resistor (R) has five ohms, what will its conductance (G) be in mhos?

Solution:

$$G \text{ (or } \mho) = \frac{1}{R} = \frac{1}{5} = 0.2\ \mho$$

Power

Electricity is generally used to do some sort of work, such as turning a motor or generating heat. Specifically, *power* is the rate at which work is done, or the rate at which heat is generated. The unit commonly used to specify electric power is the watt. In equations, you will find power abbreviated with the capital letter P, and watts, the units of measure for power, are abbreviated with the capital letter W. Power is also described as the current (I) in a circuit times the voltage (E) across the circuit. Equation (1-6) is a mathematical representation of this concept.

$$P = I \times E \quad \text{or} \quad P = IE \tag{1-6}$$

Using Ohm's Law for the value of voltage (E),

$$E = I \times R$$

and using substitution laws,

$$P = I \times (I \times R)$$

power can be described as the current (I) in a circuit squared times the resistance (R) of the circuit. Equation (1-7) is the mathematical representation of this concept.

$$P = I^2R \tag{1-7}$$

Inductance

Inductance is defined as the ability of a coil to store energy, induce a voltage in itself, and oppose changes in current flowing through it. The symbol used to indicate inductance in electrical formulas and equations is a capital L. The units of measurement are called henries. The unit henry is abbreviated by using the capital letter H. One henry is the amount of inductance (L) that permits one volt to be induced (V_L) when the current through the coil changes at a rate of one ampere per second. Equation (1-8) is the mathematical representation of the rate of change in current through a coil per unit time.

$$\left(\frac{\Delta I}{\Delta t}\right) \tag{1-8}$$

Equation (1-9) is the mathematical representation for the voltage V_L induced in a coil with inductance L. The negative sign indicates that voltage induced opposes the change in current through the coil per unit time ($\Delta I/\Delta t$).

$$V_L = -L\left(\frac{\Delta I}{\Delta t}\right) \tag{1-9}$$

Inductance will be studied in further detail later in this text.

Capacitance

Capacitance is defined as the ability to store an electric charge and is symbolized by the capital letter C. Capacitance (C), measured in farads, is equal to the amount of charge (Q) that can be stored in a device or capacitor divided by the voltage (E) applied across the device or capacitor plates when the charge was stored. Equation (1-10) is the mathematical representation for capacitance.

$$C = \frac{Q}{E} \tag{1-10}$$

Summary

The important information contained in this chapter is summarized below.

Electrical Units Summary

Parameter	Measuring Unit	Relationship
Voltage	volt (V or E)	$E = I \times R$
Current	amp (I)	$I = \frac{E}{R}$
Resistance	ohm (R or Ω)	$R = \frac{E}{I}$
Conductance	mho (G or ℧)	$G = \frac{I}{R} = \frac{I}{E}$
Power	watt (W)	$P = I \times E$ or $P = I^2R$
Inductance	henry (L or H)	$V_L = -L\left(\frac{\Delta I}{\Delta t}\right)$
Capacitance	farad (C)	$C = \frac{Q}{E}$ (Q = charge)

METHODS OF PRODUCING VOLTAGE (ELECTRICITY)

This section provides information on the following methods of producing electricity:

- *Electrochemistry*
- *Static (friction)*
- *Induction (magnetism)*
- *Piezoelectric (pressure)*
- *Thermal (heat)*
- *Light*
- *Thermionic emission*

EO 1.5 DESCRIBE how the following methods produce a voltage:

a. Electrochemistry
b. Static electricity
c. Magnetic induction
d. Piezoelectric effect
e. Thermoelectricity
f. Photoelectric effect
g. Thermionic emission

Electrochemistry

Chemicals can be combined with certain metals to cause a chemical reaction that will transfer electrons to produce electrical energy. This process works on the *electrochemistry* principle. One example of this principle is the voltaic chemical cell, shown in Figure 11. A chemical reaction produces and maintains opposite charges on two dissimilar metals that serve as the positive and negative terminals. The metals are in contact with an electrolyte solution. Connecting together more than one of these cells will produce a battery.

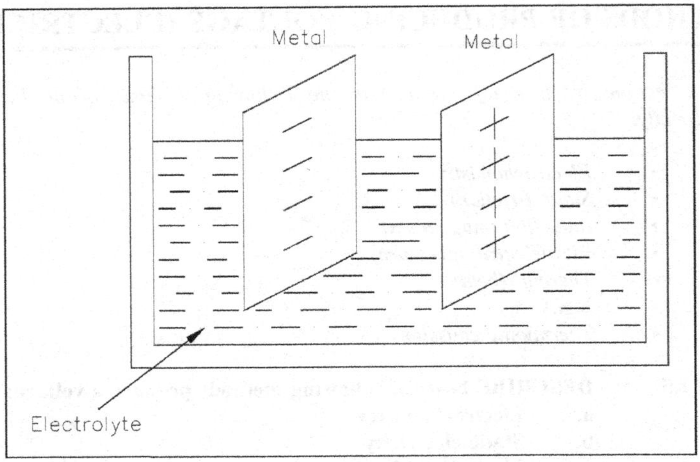

Figure 11 Voltaic Chemical Cell

Example: A battery can maintain a potential difference between its positive and negative terminals by chemical action. Various types of cells and batteries will be studied in more detail in Module 4, Batteries.

Static Electricity

Atoms with the proper number of electrons in orbit around them are in a neutral state, or have a "zero charge." A body of matter consisting of these atoms will neither attract nor repel other matter that is in its vicinity. If electrons are removed from the atoms in this body of matter, as happens due to friction when one rubs a glass rod with a silk cloth, it will become electrically positive as shown in Figure 12. If this body of matter (e.g., glass rod) comes near, but not in contact with, another body having a normal charge, an electric force is exerted between them because of their unequal charges. The existence of this force is referred to as *static electricity* or *electrostatic force*.

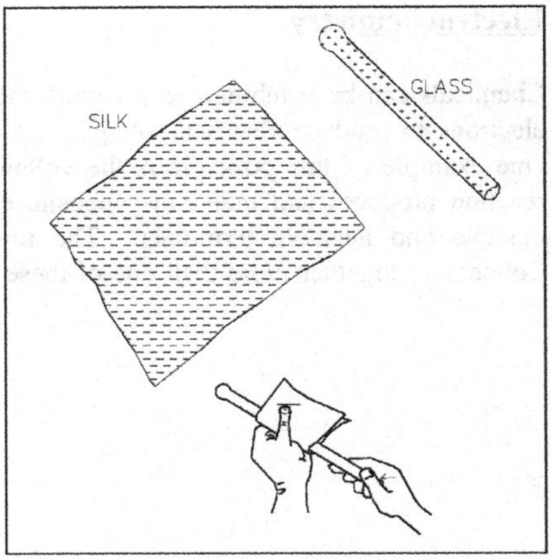

Figure 12 Static Electricity

Example: Have you ever walked across a carpet and received a shock when you touched a metal door knob? Your shoe soles built up a charge by rubbing on the carpet, and this charge was transferred to your body. Your body became positively charged and, when you touched the zero-charged door knob, electrons were transferred to your body until both you and the door knob had equal charges.

Magnetic Induction

A generator is a machine that converts mechanical energy into electrical energy by using the principle of *magnetic induction*. Magnetic induction is used to produce a voltage by rotating coils of wire through a stationary magnetic field, as shown in Figure 13, or by rotating a magnetic field through stationary coils of wire. This is one of the most useful and widely-employed applications of producing vast quantities of electric power. Magnetic induction will be studied in more detail in the next two chapters "Magnetism," and "Magnetic Circuits."

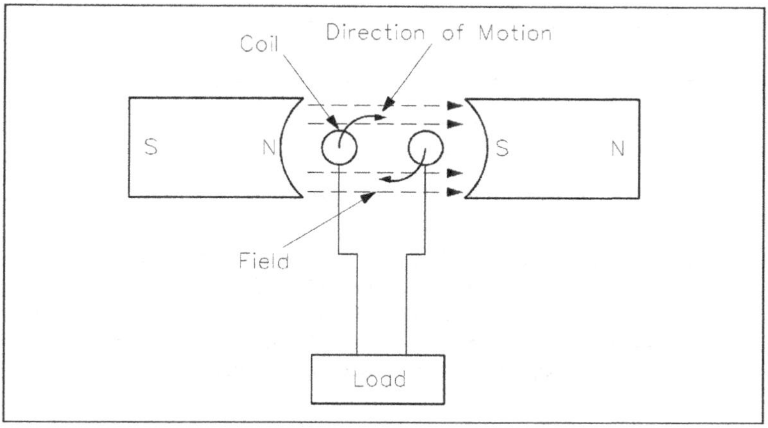

Figure 13 Generator - Electromagnetic Induction

Piezoelectric Effect

By applying pressure to certain crystals (such as quartz or Rochelle salts) or certain ceramics (like barium titanate), electrons can be driven out of orbit in the direction of the force. Electrons leave one side of the material and accumulate on the other side, building up positive and negative charges on opposite sides, as shown in Figure 14. When the pressure is released, the electrons return to their orbits. Some materials will react to bending pressure, while others will respond to twisting pressure. This generation of voltage is known as the *piezoelectric effect*. If external wires are connected while pressure and voltage are present, electrons will flow and current will be produced. If the pressure is held constant, the current will flow until the potential difference is equalized.

When the force is removed, the material is decompressed and immediately causes an electric force in the opposite direction. The power capacity of these materials is extremely small. However, these materials are very useful because of their extreme sensitivity to changes of mechanical force.

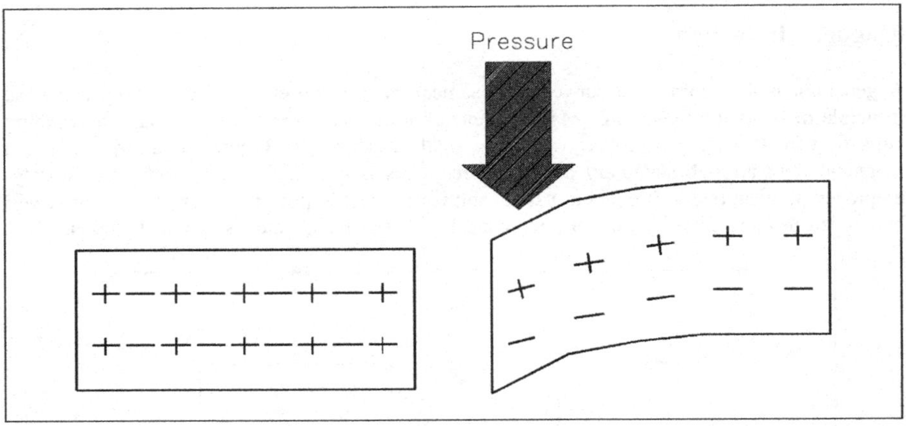

Figure 14 Pressure Applied to Certain Crystals Produces an Electric Charge

Example: One example is the crystal phonograph cartridge that contains a Rochelle salt crystal. A phonograph needle is attached to the crystal. As the needle moves in the grooves of a record, it swings from side to side, applying compression and decompression to the crystal. This mechanical motion applied to the crystal generates a voltage signal that is used to reproduce sound.

Thermoelectricity

Some materials readily give up their electrons and others readily accept electrons. For example, when two dissimilar metals like copper and zinc are joined together, a transfer of electrons can take place. Electrons will leave the copper atoms and enter the zinc atoms. The zinc gets a surplus of electrons and becomes negatively charged. The copper loses electrons and takes on a positive charge. This creates a voltage potential across the junction of the two metals. The heat energy of normal room temperature is enough to make them release and gain electrons, causing a measurable voltage potential. As more heat energy is applied to the junction, more electrons are released, and the voltage potential becomes greater, as shown in Figure 15. When heat is removed and the junction cools, the charges will dissipate and the voltage potential will decrease. This process is called *thermoelectricity*. A device like this is generally referred to as a "thermocouple."

The thermoelectric voltage in a thermocouple is dependent upon the heat energy applied to the junction of the two dissimilar metals. Thermocouples are widely used to measure temperature and as heat-sensing devices in automatic temperature controlled equipment.

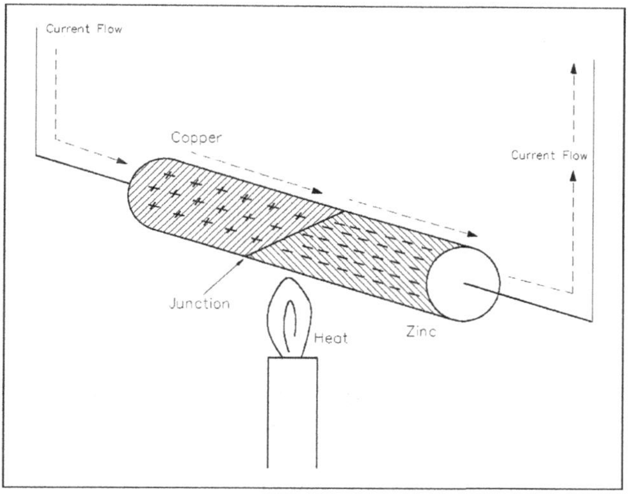

Figure 15 Heat Energy Causes Copper to Give up Electrons to Zinc

Thermocouple power capacities are very small compared to some other sources, but are somewhat greater than those of crystals.

Generally speaking, a thermocouple can be subjected to higher temperatures than ordinary mercury or alcohol thermometers.

Photoelectric Effect

Light is a form of energy and is considered by many scientists to consist of small particles of energy called photons. When the photons in a light beam strike the surface of a material, they release their energy and transfer it to the atomic electrons of the material. This energy transfer may dislodge electrons from their orbits around the surface of the substance. Upon losing electrons, the photosensitive (light sensitive) material becomes positively charged and an electric force is created, as shown in Figure 16.

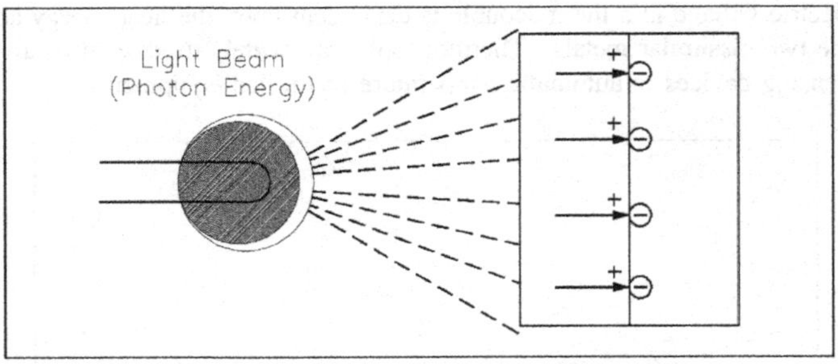

Figure 16 Producing Electricity from Light Using a Photovoltaic Cell

This phenomenon is called the *photoelectric effect* and has wide applications in electronics, such as photoelectric cells, photovoltaic cells, optical couplers, and television camera tubes. Three uses of the photoelectric effect are described below.

- Photovoltaic: The light energy in one of two plates that are joined together causes one plate to release electrons to the other. The plates build up opposite charges, like a battery (Figure 16).

- Photoemission: The photon energy from a beam of light could cause a surface to release electrons in a vacuum tube. A plate would then collect the electrons.

- Photoconduction: The light energy applied to some materials that are normally poor conductors causes free electrons to be produced in the materials so that they become better conductors.

Thermionic Emission

A thermionic energy converter is a device consisting of two electrodes placed near one another in a vacuum. One electrode is normally called the cathode, or emitter, and the other is called the anode, or plate. Ordinarily, electrons in the cathode are prevented from escaping from the surface by a potential-energy barrier. When an electron starts to move away from the surface, it induces a corresponding positive charge in the material, which tends to pull it back into the surface. To escape, the electron must somehow acquire enough energy to overcome this energy barrier. At ordinary temperatures, almost none of the electrons can acquire enough energy to escape. However, when the cathode is very hot, the electron energies are greatly increased by thermal motion. At sufficiently high temperatures, a considerable number of electrons are able to escape. The liberation of electrons from a hot surface is called *thermionic emission.*

The electrons that have escaped from the hot cathode form a cloud of negative charges near it called a space charge. If the plate is maintained positive with respect to the cathode by a battery, the electrons in the cloud are attracted to it. As long as the potential difference between the electrodes is maintained, there will be a steady current flow from the cathode to the plate.

The simplest example of a thermionic device is a vacuum tube diode in which the only electrodes are the cathode and plate, or anode, as shown in Figure 17. The diode can be used to convert alternating current (AC) flow to a pulsating direct current (DC) flow.

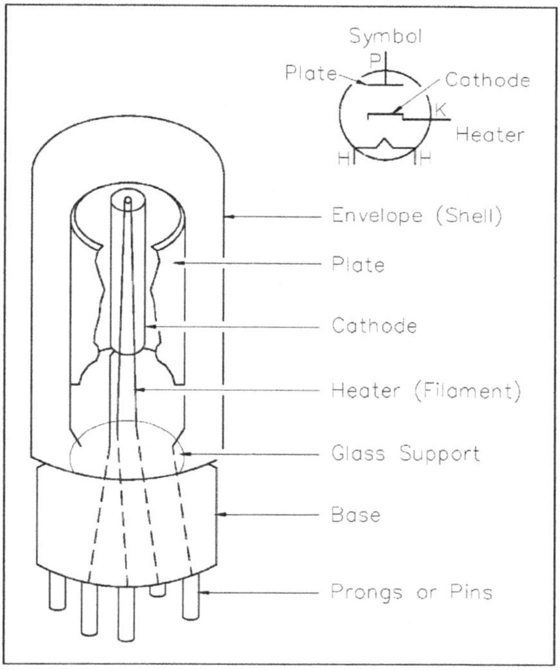

Figure 17 Vacuum Tube Diode

Summary

The important information contained in this chapter is summarized below.

Methods of Producing Electricity Summary

- Electrochemistry - Combining chemicals with certain metals causes a chemical reaction that transfers electrons.
- Static electricity - When an object with a normally neutral charge loses electrons, due to friction, and comes in contact with another object having a normal charge, an electric charge is exerted between the two objects.
- Magnetic induction - Rotating coils of wire through a stationary magnetic field or by rotating a magnetic field through a stationary coil of wire produces a potential.
- Piezoelectric effect - Bending or twisting certain materials will cause electrons to drive out of orbit in the direction of the force. When the force is released, the electrons return to their original orbit.
- Thermoelectricity - Heating two joined dissimilar materials will cause a transfer of electrons between the materials setting up a current flow.
- Photoelectric effect - Dislodging of electrons from their orbits by light beams creates positively-charged objects.
- Thermionic emission - Freeing electrons from a hot surface causes electrons to escape.

MAGNETISM

Certain metals and metallic oxides have the ability to attract other metals. This property is called magnetism, and the materials which have this property are called magnets. Some magnets are found naturally while others must be manufactured.

EO 1.6 DEFINE the following terms:
a. Magnetic flux
b. Magnetic flux density
c. Weber
d. Permeability
e. Magnetomotive force (mmf)
f. Ampere turns
g. Field intensity
h. Reluctance

EO 1.7 DESCRIBE the following materials as they relate to permeability, including an example and an approximate relative permeablity.
a. Ferromagnetic materials
b. Paramagnetic materials
c. Diamagnetic materials

Magnetism

Magnetism is a result of electrons spinning on their own axis around the nucleus (Figure 18).

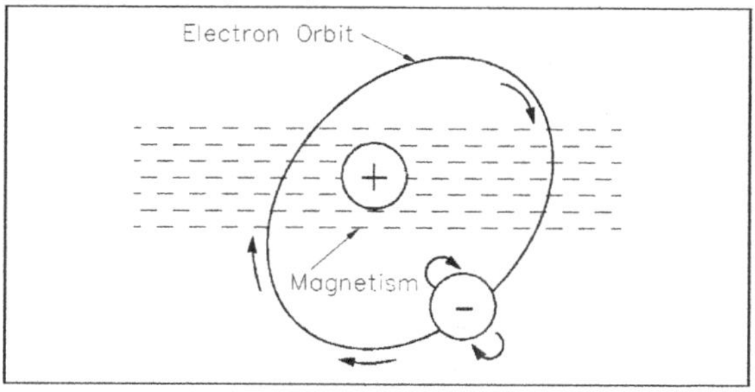

Figure 18 Electron Spinning Around Nucleus Produces Magnetic Field

In magnetic materials, the atoms have certain areas called domains. These domains are aligned such that their electrons tend to spin in the same direction (Figure 19).

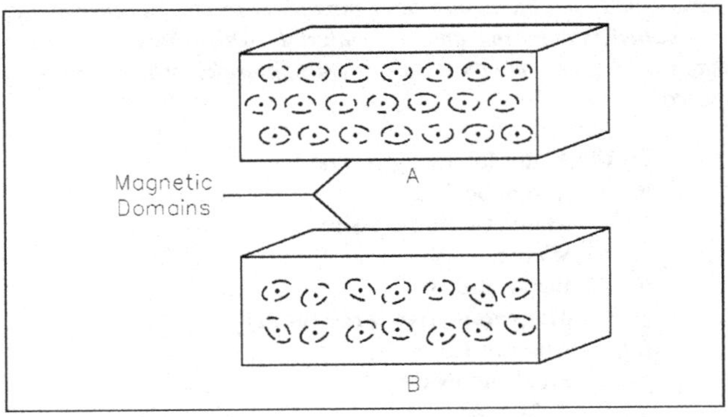

Figure 19 Magnetic Domains

The alignment of these domains results in the formation of magnetic poles at each end of the magnet. These poles are called the north pole and the south pole. The law of magnetism states that like magnetic poles repel and unlike magnetic poles attract one another (Figure 20).

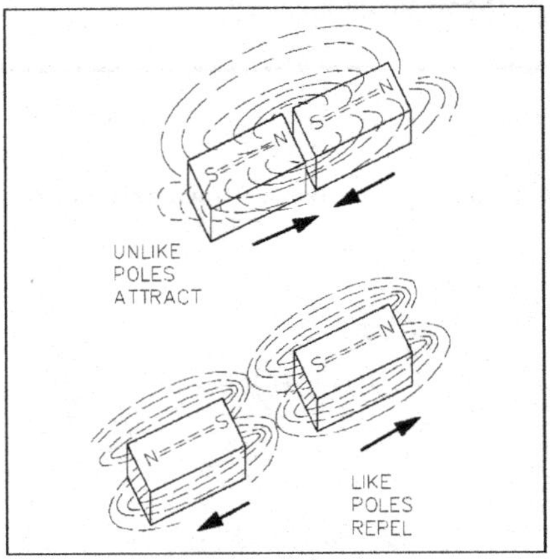

Figure 20 The Law of Magnetic Attraction and Repulsion

Magnetic Flux

The group of magnetic field lines emitted outward from the north pole of a magnet is called *magnetic flux*. The symbol for magnetic flux is Φ (phi).

The SI unit of magnetic flux is the weber (Wb). One *weber* is equal to 1×10^8 magnetic field lines.

Example: If a magnetic flux (Φ) has 5,000 lines, find the number of webers.

$$\Phi = \frac{5000 \text{ lines}}{1 \times 10^8 \text{ lines/Wb}} = \frac{5 \times 10^3}{10^8} = 50 \times 10^{-6} \text{ Wb} = 50 \ \mu\text{Wb}$$

Magnetic Flux Density

Magnetic flux density is the amount of magnetic flux per unit area of a section, perpendicular to the direction of flux. Equation (1-11) is the mathematical representation of magnetic flux density.

$$B = \frac{\Phi}{A} \tag{1-11}$$

where

B = magnetic flux density in teslas (T)

Φ = magnetic flux in webers (Wb)

A = area in square meters (m^2)

The result is that the SI unit for flux density is webers per square meter $\left(\frac{\text{Wb}}{\text{m}^2}\right)$. One weber per square meter equals one tesla.

Example: Find the flux density in teslas, when the flux is 800 μWb and the area is 0.004 m^2.

Given: $\Phi = 800\ \mu Wb = 8 \times 10^{-4}$ Wb

$A = 0.0004\ m^2 = 4 \times 10^{-4}\ m^2$

$$B = \frac{\Phi}{A} = \frac{8 \times 10^{-4}\ Wb}{4 \times 10^{-4}\ m^2} = 2\ Wb/m^2$$

Magnetic Materials

Magnetic materials are those materials that can be either attracted or repelled by a magnet and can be magnetized themselves. The most commonly used magnetic materials are iron and steel. A permanent magnet is made of a very hard magnetic material, such as cobalt steel, that retains its magnetism for long periods of time when the magnetizing field is removed. A temporary magnet is a material that will not retain its magnetism when the field is removed.

Permeability (μ) refers to the ability of a material to concentrate magnetic lines of flux. Those materials that can be easily magnetized are considered to have a high permeability. Relative permeability is the ratio of the permeability of a material to the permeability of a vacuum (μ_o). The symbol for relative permeability is μ_R (mu).

$$\mu_R = \frac{\mu_o}{\mu_o} \text{ where } \mu_o = 4\pi 10^{-7} H/m \quad (1\text{-}12)$$

Magnetic materials are classified as either magnetic or nonmagnetic based on the highly magnetic properties of iron. Because even weak magnetic materials may serve a useful purpose in some applications, classification includes the three groups described below.

Ferromagnetic Materials: Some of the ferromagnetic materials used are iron, steel, nickel, cobalt, and the commercial alloys, alnico and peralloy. Ferrites are nonmagnetic, but have the ferromagnetic properties of iron. Ferrites are made of ceramic material and have relative permeabilities that range from 50 to 200. They are commonly used in the coils for RF (radio frequency) transformers.

Paramagnetic Materials: These are materials such as aluminum, platinum, manganese, and chromium. These materials have a relative permeability of slightly more than one.

Diamagnetic Materials: These are materials such as bismuth, antimony, copper, zinc, mercury, gold, and silver. These materials have a relative permeability of less than one.

Electromagnetism

The relationship between magnetism and electrical current was discovered by a Danish scientist named Oersted in 1819. He found that if an electric current was caused to flow through a conductor, the conductor produced a magnetic field around that conductor (Figure 21).

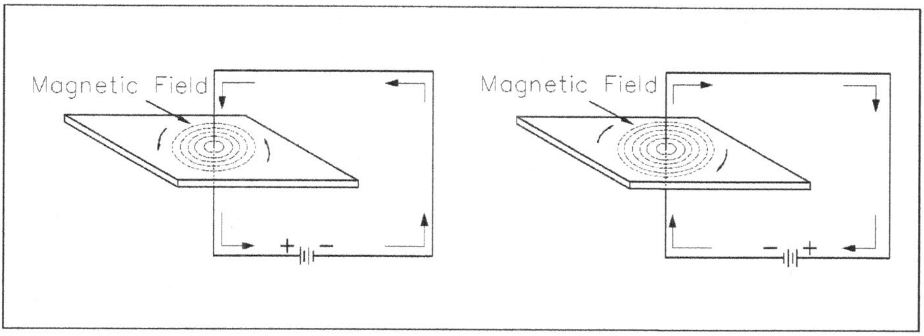

Figure 21 The Magnetic Field Produced by Current in a Conductor

Polarity of a Single Conductor

A convenient way to determine the relationship between the current flow through a conductor and the direction of the magnetic lines of force around the conductor is the left-hand rule for current carrying conductors, as illustrated in Figure 22. The student should verify that the left-hand rule holds true for the examples shown in Figure 21.

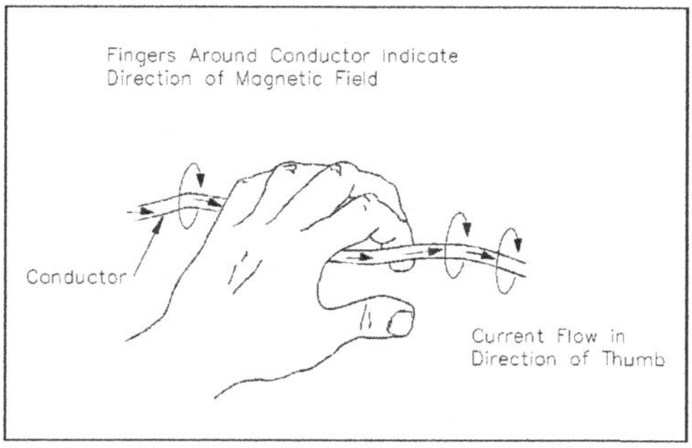

Figure 22 Left-hand Rule for Current Carrying Conductors

Magnetic Field and Polarity of a Coil

Bending a straight conductor into a loop has two results: (1) magnetic field lines become more dense inside the loop, and (2) all lines inside the loop are aiding in the same direction.

When a conductor is shaped into several loops, it is considered to be a coil. To determine the polarity of a coil, use the left-hand rule for coils (Figure 23).

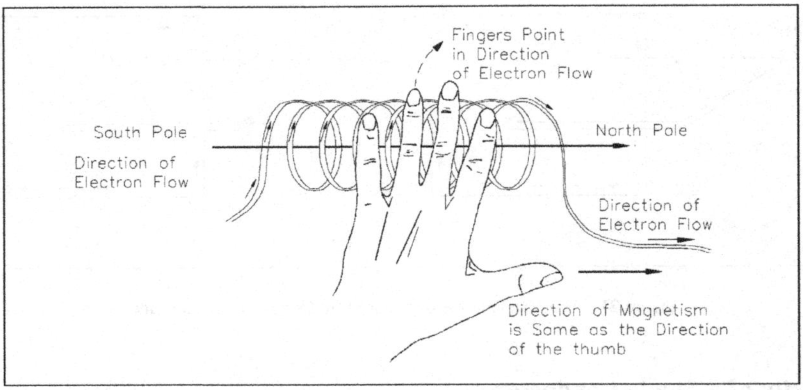

Figure 23 Left-hand Rule for Coils

Adding an iron core inside of a coil will increase the flux density. The polarity of the iron core will be the same as that of the coil. Current flow is from the negative side of the voltage source, through the coil, and back to the positive side of the source (Figure 24).

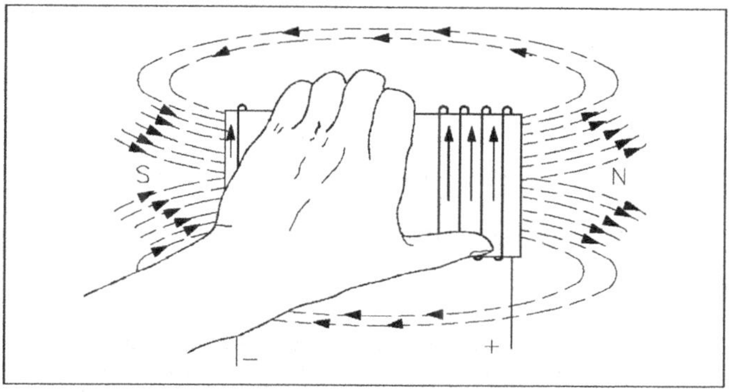

Figure 24 Left-hand Rule to Find North Pole of an Electromagnet

Magnetomotive Force

Magnetomotive force (mmf) is the strength of a magnetic field in a coil of wire. This is dependent on how much current flows in the turns of coil: the more current, the stronger the magnetic field; the more turns of wire, the more concentrated the lines of force. The current times the number of turns of the coil is expressed in units called "ampere-turns" (At), also known as mmf. Equation (1-13) is the mathematical representation for ampere-turns (At).

$$F_m = \text{ampere-turns} = NI \qquad (1\text{-}13)$$

where

F_m = magnetomotive force (mmf)

N = number of turns

I = current

Example: Calculate the ampere-turns for a coil with 1000 turns and a 5 mA current.

$N = 1000$ turns and $I = 5$ mA

substitute

$N = 1000$ turns and $I = 5 \times 10^{-3}$

$NI = 1000\ (5 \times 10^{-3}) = 5$ At

Field Intensity

When a coil with a certain number of ampere-turns is stretched to twice its length, the magnetic *field intensity*, or the concentration of its magnetic lines of force, will be half as great. Therefore, field intensity depends on the length of the coil. Equation (1-14) is the mathematical representation for field intensity, which is related to magnetomotive force as shown.

$$H = \frac{F_M}{L} = \frac{NI}{L} \qquad (1\text{-}14)$$

where

H = field intensity, $\frac{At}{m}$

NI = ampere-turns (At)

L = length between poles of coil (m)

F_M = Magnetomotive force (mmf)

Example 1: Find field intensity of an 80 turn, 20 cm coil, with 6A of current.

Solution:

$N = 80$, $I = 6A$, and $NI = 480$ At

$$H = \frac{480\ At}{0.2\ m} = \frac{2400\ At}{m}$$

Example 2: If the same coil in Example 1 were to be stretched to 40 cm with wire length and current remaining the same, find the new value of field intensity.

Solution:

$N = 80$, $I = 6A$, and $NI = 480$ At

$$H = \frac{480\ At}{0.4\ m} = \frac{1200\ At}{m}$$

Example 3: The 20 cm coil used in Example 1 with the same current is now wound around an iron core 40 cm in length. Find the field intensity.

Solution:

N = 80, I = 6A, and NI = 480 At

$$H = \frac{480\ \text{At}}{0.4\ \text{m}} = \frac{1200\ \text{At}}{\text{m}}$$

Note that field intensity for Examples 2 and 3 is the same.

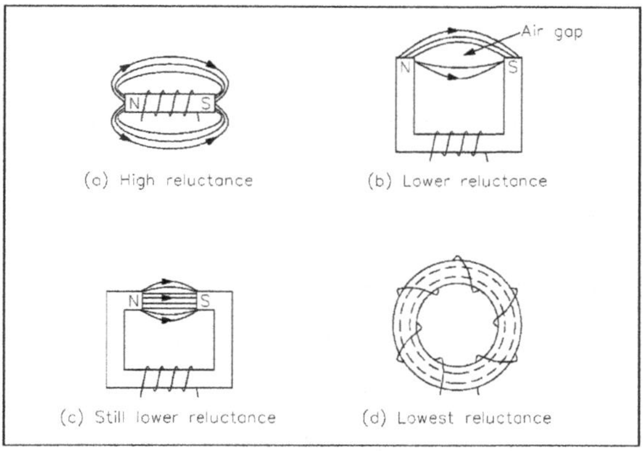

Figure 25 Different Physical Forms of Electromagnets

Reluctance

Opposition to the production of flux in a material is called *reluctance*, which corresponds to resistance. The symbol for reluctance is R, and it has the units of ampere-turns per weber (At/wb).

Reluctance is related to magnetomotive force, mmf, and flux, Φ, by the relationship shown in equation (1-15).

$$R = \frac{\text{mmf}}{\Phi} \qquad (1\text{-}15)$$

Reluctance is inversely proportional to permeability (μ). Iron cores have high permeability and, therefore, low reluctance. Air has a low permeability and, therefore, a high reluctance.

Generally, different types of materials have different values of reluctance (Figure 25). Air gap is the air space between two poles of a magnet. Since air has a very high reluctance, the size of the air gap affects the value of reluctance: the shorter the air gap, the stronger the field in the gap. Air is nonmagnetic and will not concentrate magnetic lines. The larger air gap only provides space for the magnetic lines to spread out.

Summary

The important information contained in this chapter is summarized below.

Magnetism Summary

- Magnetic flux - group of magnetic field lines that are emitted outward from the north pole of a magnet
- Magnetic flux density - amount of magnetic flux per unit area of a section, perpendicular to the direction of the flux
- Weber - measure of magnetic flux
- Permeability - ability of a material to concentrate magnetic lines of flux
- Ferromagnetic materials - iron, steel, nickel, cobalt, and commercial alloys with relative permeability ranging from 50-200
- Paramagnetic materials - aluminum, platinum, manganese, and chromium with relative permeability of slightly more than one
- Diamagnetic materials - bismuth, antimony, copper, zinc, mercury, gold, and silver with relative permeability of less than one
- Magnetomotive force (mmf) - strength of a magnetic field in a coil of wire dependent on current flowing through coil
- Ampere turns - current flowing through a coil times the number of turns in the coil
- Field intensity - identifies the magnetic flux density per unit length of a coil
- Reluctance - opposition to the production of flux in a material

MAGNETIC CIRCUITS

What is a magnetic circuit? To better understand magnetic circuits, a basic understanding of the physical qualities of magnetic circuits will be necessary.

EO 1.8 **EXPLAIN the physical qualities of a simple magnetic circuit, including relationships of qualities and units of measurements.**

EO 1.9 **Given the physical qualities of a simple magnetic circuit, CALCULATE the unknown values.**

EO 1.10 **DESCRIBE the shape and components of a BH magnetization curve.**

EO 1.11 **EXPLAIN the cause of hysteresis losses.**

EO 1.12 **Given Faraday's Law of induced voltage:**
- **a. DESCRIBE how varying parameters affect induced voltage.**
- **b. CALCULATE voltage induced in a conductor moving through a magnetic field.**

EO 1.13 **STATE Lenz's Law of induction.**

Magnetic Circuits

A magnetic circuit can be compared with an electric current in which EMF, or voltage, produces a current flow. The ampere-turns (NI), or the magnetomotive force (F_m or mmf), will produce a magnetic flux Φ (Figure 26). The mmf can be compared with EMF, and the flux (Φ) can be compared to current. Equation (1-16) is the mathematical representation of magnetomotive force derived using Ohm's Law, $I = \frac{E}{R}$.

$$\Phi = \frac{F_m}{R} = \frac{mmf}{R} \tag{1-16}$$

where

Φ = magnetic flux, Wb

F_m = magnetomotive force (mmf), At

R = reluctance, $\frac{At}{Wb}$

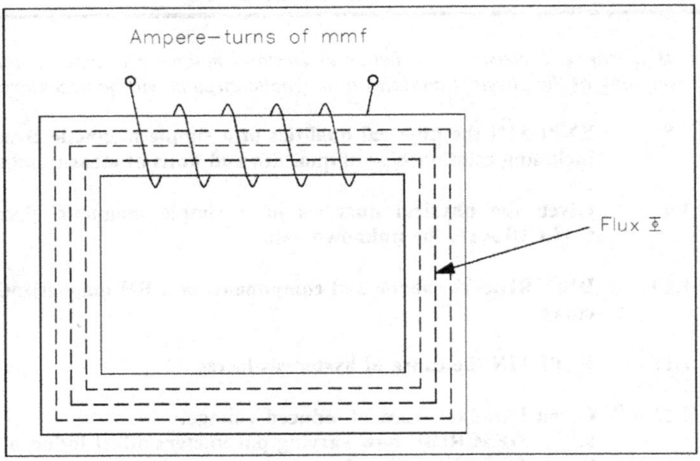

Figure 26 Magnetic Current with Closed Iron Path

Equation (1-17) is the mathematical representation for reluctance.

$$R = \frac{L}{\mu A} \tag{1-17}$$

where

R = reluctance, $\frac{At}{Wb}$

L = length of coil, m

μ = permeability of magnetic material, $\frac{(T \cdot m)}{At}$

A = cross-sectional area of coil, m^2

Example: A coil has an mmf of 600 At, and a reluctance of 3×10^6 At/Wb. Find the total flux Φ.

Solution:

$$\Phi = \frac{mmf}{R}$$

$$\Phi = \frac{600At}{3 \times 10^6 \ At/Wb} = 200 \times 10^{-6} \ Wb = 200 \mu Wb$$

BH Magnetization Curve

The BH Magnetization Curve (Figure 27) shows how much flux density (B) results from increasing the flux intensity (H). The curves in Figure 27 are for two types of soft iron cores plotted for typical values. The curve for soft iron 1 shows that flux density B increases rapidly with an increase in flux intensity H, before the core saturates, or develops a "knee." Thereafter, an increase in flux intensity H has little or no effect on flux density B. Soft iron 2 needs a much larger increase in flux intensity H before it reaches its saturation level at H = 5000 At/m, B = 0.3 T.

Air, which is nonmagnetic, has a very low BH profile, as shown in Figure 27.

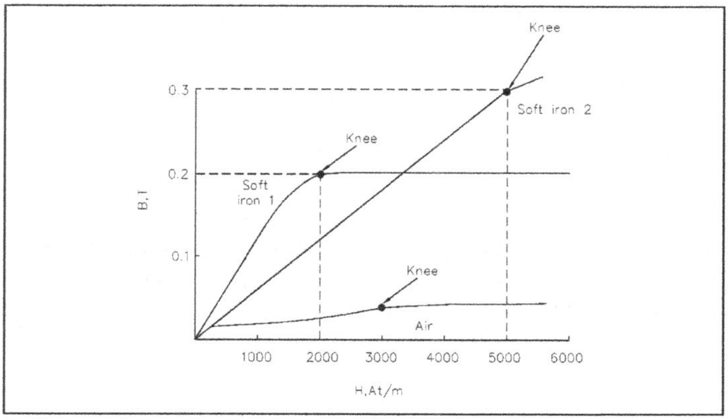

Figure 27 Typical BH Curve for Two Types of Soft Iron

The permeability (μ) of a magnetic material is the ratio of B to H. Equation (1-18) is the mathematical representation for magnetic material permeability.

$$\mu = \frac{B}{H} \qquad (1\text{-}18)$$

The average value of permeability is measured where the saturation point, or knee, is first established. Figure 27 shows that the normal or average permeability for the two irons as follows.

$$\mu \text{ soft iron } 1 = \frac{B}{H} = \frac{0.2}{2000} = 1 \times 10^{-4} \frac{(\text{T m})}{\text{At}}$$

$$\mu \text{ soft iron } 2 = \frac{B}{H} = \frac{0.3}{5000} = 6 \times 10^{-5} \frac{(\text{T m })}{\text{At}}$$

In SI units, the permeability of a vacuum is $\mu_o = 4 \pi \times 10^{-7}$ H/m or 1.26×10^{-6} or T-m/At. In order to calculate permeability, the value of relative permeability μ_r must be multiplied by μ_o. Equation (1-18) is the mathematical representation for permeability.

$$\mu = \mu_r \times \mu_o \qquad (1\text{-}18)$$

Example: Find the permeability of a material that has a relative permeability of 100.

$$\mu = \mu_r \times \mu_o = 100\ (1.26 \times 10^{-6})$$

$$= 126 \times 10^{-6} \frac{(\text{T-m})}{\text{At}}$$

Hysteresis

When current in a coil reverses direction thousands of times per second, hysteresis can cause considerable loss of energy. *Hysteresis* is defined as "a lagging behind." The magnetic flux in an iron core lags behind the magnetizing force.

The hysteresis loop is a series of curves that shows the characteristics of a magnetic material (Figure 28). Opposite directions of current will result in opposite directions of flux intensity shown as +H and -H. Opposite polarities are also shown for flux density as +B or -B. Current starts at the center (zero) when unmagnetized. Positive H values increase B to the saturation point, or $+B_{max}$, as shown by the dashed line. Then H decreases to zero, but B drops to the value of B_r due to hysteresis. By reversing the original current, H now becomes negative. B drops to zero and continues on to $-B_{max}$. As the -H values decrease (less negative), B is reduced to $-B_r$ when H is zero. With a positive swing of current, H once again becomes positive, producing saturation at $+B_{max}$. The hysteresis loop is completed. The loop does not return to zero because of hysteresis.

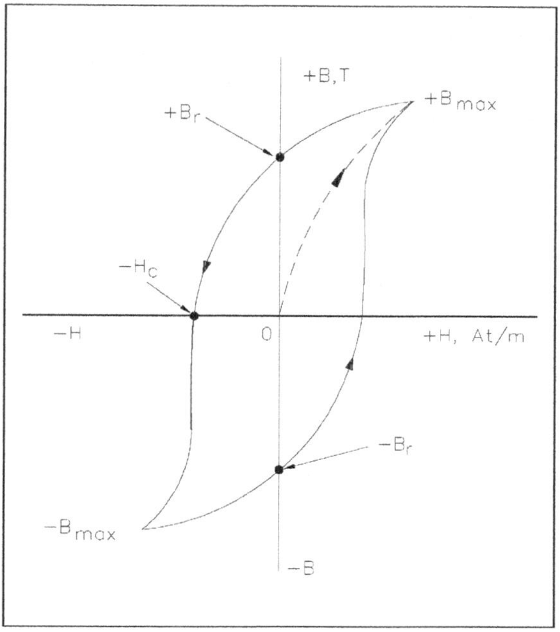

Figure 28 Hysteresis Loop for Magnetic Materials

The value of $+B_r$ or $-B_r$, which is the flux density remaining after the magnetizing force is zero, is called the *retentivity* of that magnetic material. The value of $-H_c$, which is the force that must be applied in the reverse direction to reduce flux density to zero, is called the *coercive force* of the material.

The greater the area inside the hysteresis loop, the larger the hysteresis losses.

Magnetic Induction

Electromagnetic induction was discovered by Michael Faraday in 1831. Faraday found that if a conductor "cuts across" lines of magnetic force, or if magnetic lines of force cut across a conductor, a voltage, or EMF, is induced into the conductor. Consider a magnet with its lines of force from the North Pole to the South Pole (Figure 29). A conductor C, which can be moved between the poles of the magnet, is connected to a galvanometer G, which can detect the presence of voltage, or EMF. When the conductor is not moving, zero EMF is indicated by the galvanometer.

If the conductor is moving outside the magnetic field at position 1, zero EMF is still indicated by the galvanometer. When the conductor is moved to position 2, the lines of magnetic force will be cut by the conductor, and the galvanometer will deflect to point A. Moving the conductor to position 3 will cause the galvanometer to return to zero. By reversing the direction in which the conductor is moved (3 to 1), the same results are noticed, but of opposite polarity. If we hold the conductor stationary in the magnetic lines of force, at position 2, the galvanometer indicates zero. This fact shows that there must be relative motion between the conductor and the magnetic lines of force in order to induce an EMF.

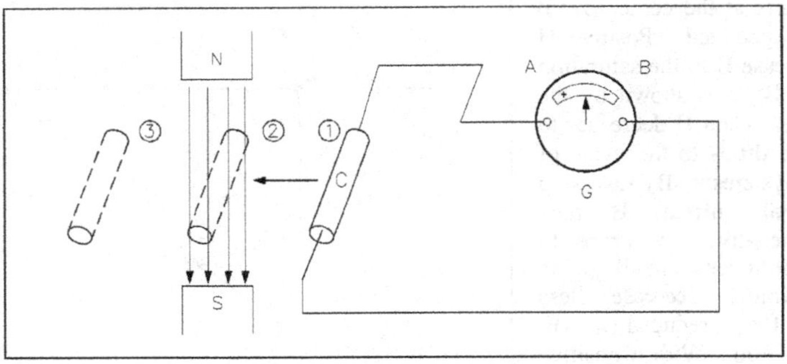

Figure 29 Induced EMF

The most important application of relative motion is seen in electric generators. In a DC generator, electromagnets are arranged in a cylindrical housing. Conductors, in the form of coils, are rotated on a core such that the coils continually cut the magnetic lines of force. The result is a voltage induced in each of the conductors. These conductors are connected in series, and the induced voltages are added together to produce the generator's output voltage.

Faraday's Law of Induced Voltage

The magnitude of the induced voltage depends on two factors: (1) the number of turns of a coil, and (2) how fast the conductor cuts across the magnetic lines of force, or flux. Equation (1-20) is the mathematical representation for Faraday's Law of Induced Voltage.

$$V_{ind} = N\left(\frac{\Delta\Phi}{\Delta t}\right) \qquad (1\text{-}20)$$

where

V_{ind} = induced voltage, V

N = number of turns in a coil

$\frac{\Delta\Phi}{\Delta t}$ = rate at which the flux cuts across the conductor, $\frac{Wb}{s}$

Example 1: Given: Flux = 4 Wb. The flux increases uniformly to 8 Wb in a period of 2 seconds. Find induced voltage in a coil that has 12 turns, if the coil is stationary in the magnetic field.

Solution:

$$V_{ind} = -N\left(\frac{\Delta\Phi}{\Delta t}\right)$$

$$\Delta\Phi = 8Wb - 4Wb = 4Wb$$

$$\Delta t = 2s$$

then

$$\frac{\Delta\Phi}{\Delta t} = \frac{4Wb}{2s} = \frac{2Wb}{s}$$

$$V_{ind} = -12\ (2) = -24 \text{ volts}$$

Example 2: In Example 1, what is the induced voltage, if the flux remains 4 Wb after 2 s?

Solution:

$$V_{ind} = -12\left(\frac{0}{2}\right) = 0 \text{ Volts}$$

No voltage is induced in Example 2. This confirms the principle that relative motion must exist between the conductor and the flux in order to induce a voltage.

Lenz's Law

Lenz's Law determines the polarity of the induced voltage. Induced voltage has a polarity that will oppose the change causing the induction. When current flows due to the induced voltage, a magnetic field is set up around that conductor so that the conductor's magnetic field reacts with the external magnetic field. This produces the induced voltage to oppose the change in the external magnetic field. The negative sign in equation (1-20) is an indication that the emf is in such a direction as to produce a current whose flux, if added to the original flux, would reduce the magnitude of the emf.

Summary

The important information contained in this chapter is summarized below.

Magnetic Circuits Summary

Simple magnetic circuit magnetic flux (Φ) is proportional to the magnetomotive force (F_m) and indirectly proportional to the reluctance (R) in a circuit.

$$\Phi\ (Wb) = \frac{F_m\ (At)}{R\left(\frac{At}{Wb}\right)}$$

A BH magnetization curve shows how much magnetic flux density (B) results from increasing magnetic flux intensity. The "knee" identifies the point where increasing flux intensity (H) results in a minimal increase in flux density (B).

Hysteresis losses are caused by reversing current direction thousands of times in a coil.

Faraday's Law of Induced Voltage depends on the number of turns of a coil and how fast the conductor cuts across the magnetic lines of force or flux.

$$V_{ind} = -N\frac{\Delta\Phi}{\Delta t}$$

Lenz's Law states that an induced voltage will have a polarity that will oppose the magnetic field that caused the induced voltage.

ELECTRICAL SYMBOLS

To read and interpret electrical system diagrams and schematics, one must be thoroughly familiar with the many symbols used. Once these symbols are mastered, most electrical diagrams and schematics will be understood with relative ease.

EO 1.14 Given a standard electrical symbol, IDENTIFY the component that the symbol represents. The symbols will be for the following components:

a. Resistor
b. Capacitor
c. Inductor
d. Relay
e. Contacts
f. Breaker
g. Switch
h. Transistor
i. Rheostat
j. Diode
k. Ground connections
l. Vacuum tube
m. Fuse
n. Junction
o. AC voltage source
p. Voltmeter
q. Ammeter
r. Wattmeter
s. Relay operated contacts
t. Potential transformer
u. Current transformer
v. Wye (Y) connection
w. Delta (Δ) connection
x. Light bulb
y. Battery

Symbols

The symbols for the various electrical components that will appear on electrical diagrams and schematics are shown in Figure 30.

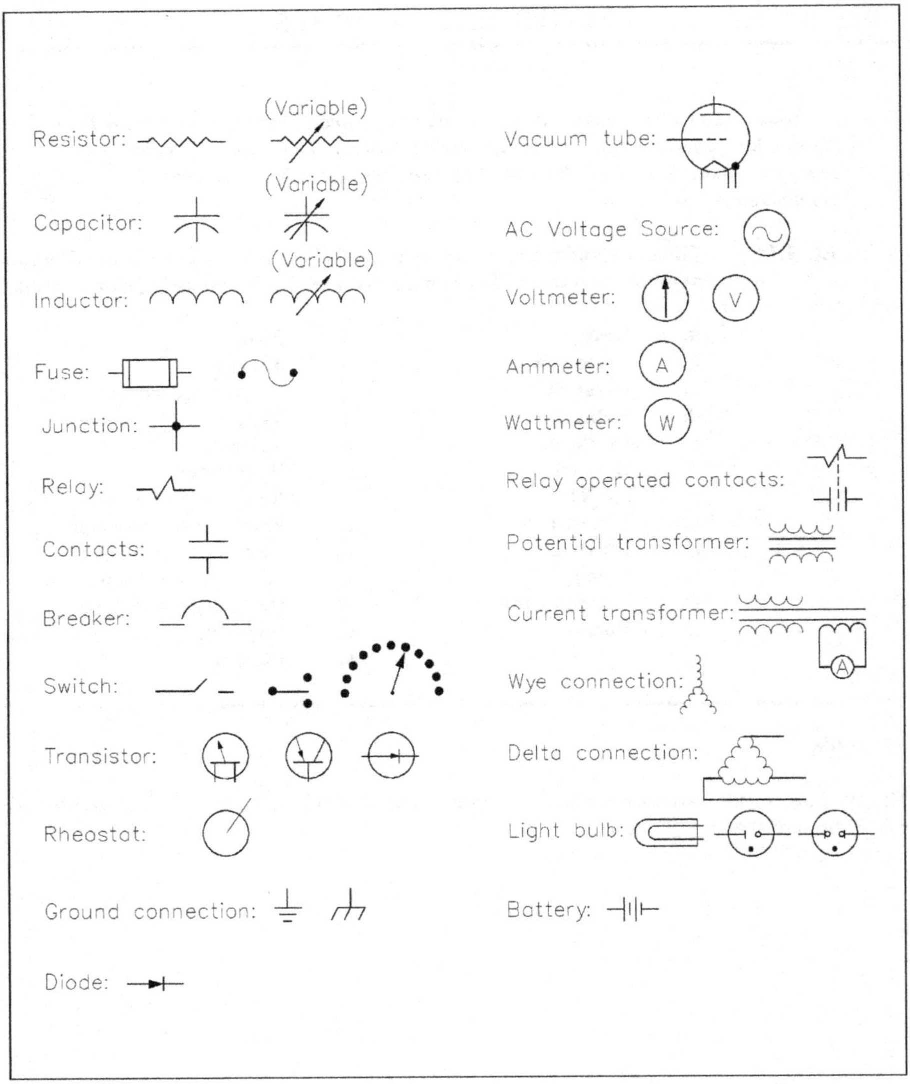

Figure 30 Electrical Symbols

Summary

The important information contained in this chapter is summarized below.

Electrical Symbols Summary

- To read and interpret electrical system diagrams and schematics, one must be thoroughly familiar with the many symbols used. Once these symbols are mastered, most electrical diagrams and schematics will be understood with relative ease.

Intentionally Left Blank

Appendix A
Metric System and Powers of Ten

BASIC ELECTRICAL THEORY

APPENDIX A

METRIC SYSTEM AND POWERS OF TEN

Metric System

Units of the international metric system, commonly called SI (system internationale), are used in electricity. The seven basic units are shown on Table A-1. Two supplementary units of SI are shown on Table A-2.

TABLE A-1
BASE UNITS of the
INTERNATIONAL METRIC SYSTEM

Quantity	Base Unit	Symbol
Length	meter	m
Mass	kilogram	Kg
Time	second	s
Electric current	ampere	A
Thermodynamic temperature	Kelvin	K
Light intensity	Candela	cd
Amount of substance	mole	mol

TABLE A-2
SUPPLEMENTARY SI UNITS

Quantity	Unit	Symbol
Plane angle	radian	rad
Socid angle	steradian	Sr

These base and supplemental units are used to derive other units. Most of the units of measure used in electricity are derived units. These units are shown on Table A-3. Some of these are derived from the base units, and some are derived from other derived units, or a combination of both. For example, the unit of current is the ampere, which is derived from the base units of second and coulomb. The derivation of these units is beyond the scope of this text. However, units commonly used in the study and use of electricity and their relationship to other units of measure are discussed in Chapter 3, *Units of Electrical Measurement.*

TABLE A-3
DERIVED SI UNITS

Quantity	Derived Unit	Symbol
Energy	joule	J
Force	newton	N
Power	watt	W
Electric charge	coulomb	C
Electric potential	volt	V
Electric resistance	ohm	Ω
Electric conductance	mho	℧
Electric capacitance	farad	F
Electric inductance	henry	H
Frequency	hertz	Hz
Magnetic flux	weber	Wb
Magnetic flux density	tesla	T

Metric Prefixes

When working with basic electrical measurement units, sometimes the values are too large or too small to express conveniently. For convenience, add metric prefixes (Table A-4) to the basic unit. For example, if we have a resistor that is 1,000,000 ohms (Ω), we can add the prefix kilo, or its designated symbol K, and express the value as 1,000 KΩ, or the metric prefix mega (symbol M) and express the value as 1 MΩ. In the case of capacitance or current, we may have a very small value such as 0.002 amperes (symbol A) or 0.000005 farads (symbol F). These values can be expressed by adding the prefix milli (symbol m) and expressing the current as 2 milli amperes or 2 mA, or adding the prefix micro (symbol μ) and expressing the capacitance as 5 micro farads, or 5 μF.

To use the prefixes listed in Table A-4, divide the value of the unit by the value of the prefix, and then add the prefix to the unit.

TABLE A-4
METRIC PREFIXES USED in ELECTRICITY

Prefix (letter symbol)	Value	
mega (M)	million	1,000,000
kilo (k)	thousand	1,000
milli (m)	thousandth	0.001
micro (μ)	millionth	0.000,001
nano (n)	thousand-millionth	0.000,000,001
pico (p)	million-millionth	0.000,000,000,001

Powers of Ten

Another way to express large and small values is to convert the value to powers of 10. This is a convenient way to express values and keep the units in the basic unit notation, rather than converting from one unit of measure to another unit that may be larger or smaller, as we did in metric prefixes above.

Examples of expressing numbers as power of 10 are shown in Table A-5.

TABLE A-5
POWERS of 10

Number = Power of 10	Commonly Read As
$0.000,001 = 10^{-6}$	10 to the minus sixth
$0.000,01 = 10^{-5}$	10 to the minus fifth
$0.000,1 = 10^{-4}$	10 to the minus fourth
$0.001 = 10^{-3}$	10 to the minus third
$0.01 = 10^{-2}$	10 to the minus two
$0.1 = 10^{-1}$	10 to the minus one
$1 = 10^{0}$	10 to the zero
$10 = 10^{1}$	10 to the first
$100 = 10^{2}$	10 to the second
$1,000 = 10^{3}$	10 to the third
$10,000 = 10^{4}$	10 to the fourth
$100,000 = 10^{5}$	10 to the fifth
$1,000,000 = 10^{6}$	10 to the sixth

Rules associated with powers of ten are as follows:

Rule 1: To express numbers larger than 1 as a small number times a power of 10, move the decimal point to the left as many places as desired. Then multiply the number obtained by 10 to a power that is equal to the number of places moved.

Example: To convert 6,000,000, move the decimal point 6 places to the left (6.000,000), then multiply 6 times 10 raised to a power equal to the number of decimal places moved, obtaining 6×10^6.

$$6{,}000{,}000 = 6 \times 10^6$$

Rule 2: To express numbers less than 1 as a whole number times a power of 10, move the decimal point to the right as many places as desired. Then multiply the number obtained by 10 to a negative power that is equal to the number of places moved.

Example: To convert 0.004A, move the decimal point 3 places to the right (004.), then multiply 4 by 10 raised to a negative power equal to the number of decimal places moved, obtaining 4×10^{-3}A.

$$0.004\text{A} = 4 \times 10^{-3}\text{A}$$

Rule 3: To convert a number expressed as a positive power of 10 to a decimal number, move the decimal point to the right as many places as the value of the exponent.

Example: To convert $1 \times 10^3\Omega$, move the decimal point 3 places to the right (1000.0), then drop the multiple of power of 10, obtaining 1,000.

$$1 \times 10^3 = 1{,}000\Omega$$

Rule 4: To convert a number expressed as a negative power of 10 to a decimal number, move the decimal point to the left as many places as the value of the exponent.

Example: To convert 5×10^{-3}A, move the decimal point 3 places to the left (0.005), then drop the multiple of the power of 10, obtaining 0.005A.

$$5 \times 10^{-3}A = 0.005A$$

Rule 5: To multiply 2 or more numbers expressed as powers of 10, multiply the coefficients to obtain the new coefficient, and add the exponents to obtain the new exponent of 10.

Example: To multiply 2×10^5 by 3×10^{-3}, multiply 2 x 3 to get 6, then add the exponents of the powers of 10, 5 + (-3), to get an exponent of 2, obtaining a product of 6×10^2.

$$(2 \times 10^5)(3 \times 10^{-3}) = 6 \times 10^2$$

Rule 6: To divide powers of 10, use the formula:

$$\frac{1}{10^n} = 1 \times 10^{-n}$$

We can transfer any power of 10 from numerator to denominator, or vice-versa, simply by changing the sign of the exponent.

Example: To divide 8×10^{-5} by 2×10^{-4}, divide the numerical number 8 by 2 = 4, then bring the power of 10, in the denominator, up to the numerator and change the sign of its exponent, then add the exponents together:

$$\frac{8x10^{-5}}{2x10^{-4}} = 4x10^{-5+4} = 4x10^{-1}$$

Table A-6 shows the metric prefixes expressed as powers of 10.

TABLE A-6
METRIC PREFIXES EXPRESSED as POWERS of 10

Metric Prefix	Power of 10
mega (M)	10^6
kilo (k)	10^3
milli (m)	10^{-3}
micro (μ)	10^{-6}
nano (n)	10^{-9}
pico (p)	10^{-12}

Electrical units can be expressed in different, but equivalent, units by using prefixes expressed in powers of 10.

Examples:

Express 780,000 Ω in Megohms (MΩ).

$1\Omega = 10^{-6}\ M\Omega$

$780{,}000\ \Omega = 780{,}000\ \Omega \times 10^{-6} = 0.78\ M\Omega$

Express 4.5 V in millivolts (mV).

$1\ V = 10^3\ mV$

$4.5\ V = 4.5 \times 10^3\ mV = 4500\ mV$

Express 0.015 A in milliamperes (mA).

$1 \text{ A} = 10^3 \text{ mA}$

$0.015 \text{ A} = 0.015 \times 10^3 = 15 \text{ mA}$

Intentionally Left Blank

Department of Energy
Fundamentals Handbook

ELECTRICAL SCIENCE
Module 2
Basic DC Theory

TABLE OF CONTENTS

LIST OF FIGURES iii

LIST OF TABLES vi

REFERENCES vii

OBJECTIVES viii

DC SOURCES 1

- Batteries 1
- DC Generator 2
- Thermocouples 3
- Rectifiers 5
- Forward Bias 5
- Reverse Bias 7
- Half-Wave Rectifier Circuit 8
- Full-Wave Rectifier Circuit 9
- Summary 11

DC CIRCUIT TERMINOLOGY 12

- Schematic Diagram 12
- One-Line Diagram 13
- Block Diagram 14
- Wiring Diagram 15
- Resistivity 16
- Temperature Coefficient of Resistance 17
- Electric Circuit 18
- Series Circuit 20
- Parallel Circuit 21
- Equivalent Resistance 21
- Summary 22

BASIC DC CIRCUIT CALCULATIONS 23

- Series Resistance 23
- Parallel Currents 29
- Resistance in Parallel 32
- Simplified Formulas 34

TABLE OF CONTENTS (Cont.)

Voltage Divider 36
Current Division 38
Summary 39

VOLTAGE POLARITY AND CURRENT DIRECTION 40

Conventional and Electron Flow 40
Polarities 40
Summary 41

KIRCHHOFF'S LAWS 42

Kirchhoff's Laws 42
Kirchhoff's Voltage Law 43
Applying Kirchhoff's Voltage Law 44
Kirchhoff's Current Law 45
Summary 48

DC CIRCUIT ANALYSIS 49

Loop Equations 49
Node Equations 54
Series-Parallel Circuit Analysis 58
Y and Delta Network Calculation 60
Summary 65

DC CIRCUIT FAULTS 66

Open Circuit (Series) 66
Open Circuit (Parallel) 67
Short Circuit (Series) 68
Short Circuit (Parallel) 68
Summary 69

LIST OF FIGURES

Figure 1 Basic Chemical Battery 2

Figure 2 Basic DC Generator 3

Figure 3 Production of a DC Voltage Using a Thermocouple 4

Figure 4 Forward-Biased Diode 6

Figure 5 Reverse-Biased Diode 7

Figure 6 Half-Wave Rectifier 8

Figure 7 Full-Wave Rectifier 9

Figure 8 Bridge Rectifier Circuit 10

Figure 9 Schematic Diagram 13

Figure 10 One-Line Diagram 14

Figure 11 Block Diagram 15

Figure 12 Wiring Diagram 15

Figure 13 Closed Circuit 19

Figure 14 Open Circuit 19

Figure 15 Short Circuit 20

Figure 16 Series Circuit 20

Figure 17 Parallel Circuit 21

Figure 18 Resistance in a Series Circuit 24

Figure 19 Voltage Drops in a Series Circuit 25

Figure 20 Voltage Total in a Series Circuit 26

LIST OF FIGURES (Cont.)

Figure 21 Example 1 Series Circuit 27

Figure 22 Example 2 Series Circuit 28

Figure 23 Example 1 Parallel Circuit 30

Figure 24 Example 2 Parallel Circuit 30

Figure 25 Example 3 Parallel Circuit 31

Figure 26 Equivalent Resistance in a Parallel Circuit 32

Figure 27 Total Resistance in a Parallel Circuit 33

Figure 28 Example Parallel Circuit 36

Figure 29 Voltage Divider 37

Figure 30 Current Division Example Circuit 38

Figure 31 Voltage Polarities 41

Figure 32 Closed Loop 43

Figure 33 Using Kirchhoff's Voltage Law to find Current with one Source 44

Figure 34 Using Kirchhoff's Voltage Law to find Current with Multiple Battery Sources 45

Figure 35 Illustration of Kirchhoff's Current Law 46

Figure 36 Using the Current Law 47

Figure 37 Example Circuit for Loop Equations 49

Figure 38 Assumed Direction of Current Flow 50

Figure 39 Marking Polarity 50

Figure 40 Applying Voltage Law to Loop 1 51

LIST OF FIGURES (Cont.)

Figure 41 Applying Voltage Laws to Outer Loop 53

Figure 42 Applying Current Law to Junction 54

Figure 43 Node Point 55

Figure 44 Circuit for Node Analysis 56

Figure 45 Node - Voltage Analysis 57

Figure 46 Redrawn Circuit Example 59

Figure 47 T or Y Network 60

Figure 48 π (pi) or Δ (delta) Network 61

Figure 49 Y - Δ Equivalent 62

Figure 50 Bridge Circuit 63

Figure 51 Y - Δ Redrawn Circuit 64

Figure 52 Steps to Simplify Redrawn Circuit 65

Figure 53 Open Series Circuit 66

Figure 54 Open Parallel Circuit - Total 67

Figure 55 Open Parallel Circuit - Branch 67

Figure 56 Shorted DC Circuit 68

Figure 57 Shorted Parallel Circuit 69

LIST OF TABLES

Table 1 Properties of Conducting Materials 17

Table 2 Temperature Coefficients for Various Materials 18

REFERENCES

- Gussow, Milton, Schaum's Outline Series, Basic Electricity, McGraw-Hill.
- Academic Program for Nuclear Power Plant Personnel, Volume IV, Columbia, MD: General Physics Corporation, Library of Congress Card #A 326517, 1982.
- Sienko and Plane, Chemical Principles and Properties, 2nd Edition, McGraw-Hill.
- Academic Program for Nuclear Power Plant Personnel, Volume II, Columbia, MD: General Physics Corporation Library of Congress Card #A 326517, 1982.
- Nasar and Unnewehr, Electromechanics and Electric Machines, John Wiley and Sons.
- Van Valkenburgh, Nooger, and Neville, Basic Electricity, Vol. 5, Hayden Book Company.
- Exide Industrial Marketing Division, The Storage Battery, Lead-Acid Type, The Electric Storage Battery Company.
- Lister, Eugene C., Electric Circuits and Machines, 5th Edition, McGraw-Hill.
- Croft, Carr, Watt, and Summers, American Electricians Handbook, 10th Edition, McGraw-Hill.
- Mason, C. Russel, The Art and Science of Protective Relaying, John Wiley and Sons.
- Mileaf, Harry, Electricity One - Seven, Revised 2nd Edition, Hayden Book Company.
- Buban and Schmitt, Understanding Electricity and Electronics, 3rd Edition, McGraw-Hill.
- Kidwell, Walter, Electrical Instruments and Measurements, McGraw-Hill.

TERMINAL OBJECTIVE

1.0 Using the rules associated with basic DC circuit characteristics, **ANALYZE** various DC circuits to find resistances, currents, and voltages at any given point within the circuit.

ENABLING OBJECTIVES

1.1 **LIST** the four ways to produce a DC voltage.

1.2 **STATE** the purpose of a rectifier.

1.3 **DESCRIBE** the outputs of the following circuits:
- a. Half-wave bridge rectifier
- b. Full-wave bridge rectifier

1.4 Given a diagram, **IDENTIFY** it as one of the following types:
- a. Schematic diagram
- b. One-line diagram
- c. Block diagram
- d. Wiring diagram

1.5 **DEFINE** the following terms:
- a. Resistivity
- b. Temperature coefficient of resistance
- c. Closed circuit
- d. Open circuit
- e. Short circuit
- f. Series circuit
- g. Parallel circuit
- h. Equivalent resistance

1.6 Given a circuit, **DETERMINE** whether the circuit is an open circuit or a closed circuit.

1.7 Given a circuit, **CALCULATE** total resistance for a series or parallel circuit.

1.8 **DESCRIBE** what is meant by the term "voltage divider."

1.9 **DESCRIBE** what is meant by the term "current division."

ENABLING OBJECTIVES (Cont.)

1.10 **DESCRIBE** the difference between electron flow and conventional current flow.

1.11 Given a circuit showing current flows, **IDENTIFY** the polarity of the voltage drops in the circuit.

1.12 **STATE** Kirchhoff's voltage law.

1.13 **STATE** Kirchhoff's current law.

1.14 Given a circuit, **SOLVE** problems for voltage and current using Kirchhoff's laws.

1.15 Given a simple DC circuit, **DETERMINE** the equivalent resistance of series and parallel combinations of elements.

1.16 **DESCRIBE** the voltage and current effects of an open in a DC circuit.

1.17 **DESCRIBE** the voltage and current effects in a shorted DC circuit.

Intentionally Left Blank

DC SOURCES

When most people think of DC, they usually think of batteries. In addition to batteries, however, there are other devices that produce DC which are frequently used in modern technology.

1.1 LIST the four ways to produce a DC voltage.

1.2 STATE the purpose of a rectifier.

1.3 DESCRIBE the outputs of the following circuits:
- **a. Half-wave bridge rectifier**
- **b. Full-wave bridge rectifier**

Batteries

A battery consists of two or more chemical cells connected in series. The combination of materials within a battery is used for the purpose of converting chemical energy into electrical energy. To understand how a battery works, we must first discuss the chemical cell.

The chemical cell is composed of two electrodes made of different types of metal or metallic compounds which are immersed in an electrolyte solution. The chemical actions which result are complicated, and they vary with the type of material used in cell construction. Some knowledge of the basic action of a simple cell will be helpful in understanding the operation of a chemical cell in general.

In the cell, electrolyte ionizes to produce positive and negative ions (Figure 1, Part A). Simultaneously, chemical action causes the atoms within one of the electrodes to ionize.

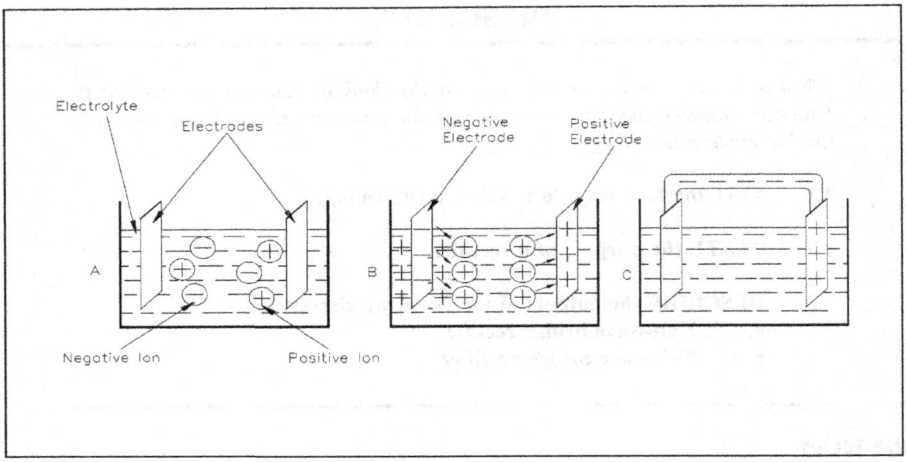

Figure 1 Basic Chemical Battery

Due to this action, electrons are deposited on the electrode, and positive ions from the electrode pass into the electrolyte solution (Part B). This causes a negative charge on the electrode and leaves a positive charge in the area near the electrode (Part C).

The positive ions, which were produced by ionization of the electrolyte, are repelled to the other electrode. At this electrode, these ions will combine with the electrons. Because this action causes removal of electrons from the electrode, it becomes positively charged.

DC Generator

A simple DC generator consists of an armature coil with a single turn of wire. The armature coil cuts across the magnetic field to produce a voltage output. As long as a complete path is present, current will flow through the circuit in the direction shown by the arrows in Figure 2. In this coil position, commutator segment 1 contacts with brush 1, while commutator segment 2 is in contact with brush 2.

Rotating the armature one-half turn in the clockwise direction causes the contacts between the commutator segments to be reversed. Now segment 1 is contacted by brush 2, and segment 2 is in contact with brush 1.

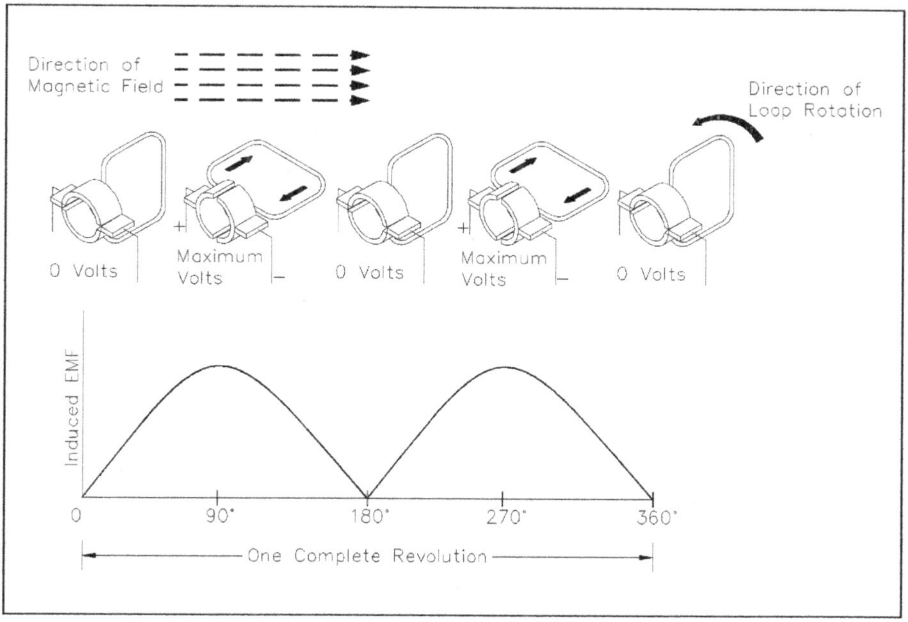

Figure 2 Basic DC Generator

Due to this commutator action, that side of the armature coil which is in contact with either of the brushes is always cutting the magnetic field in the same direction. Brushes 1 and 2 have a constant polarity, and pulsating DC is delivered to the load circuit.

Thermocouples

A thermocouple is a device used to convert heat energy into a voltage output. The thermocouple consists of two different types of metal joined at a junction (Figure 3).

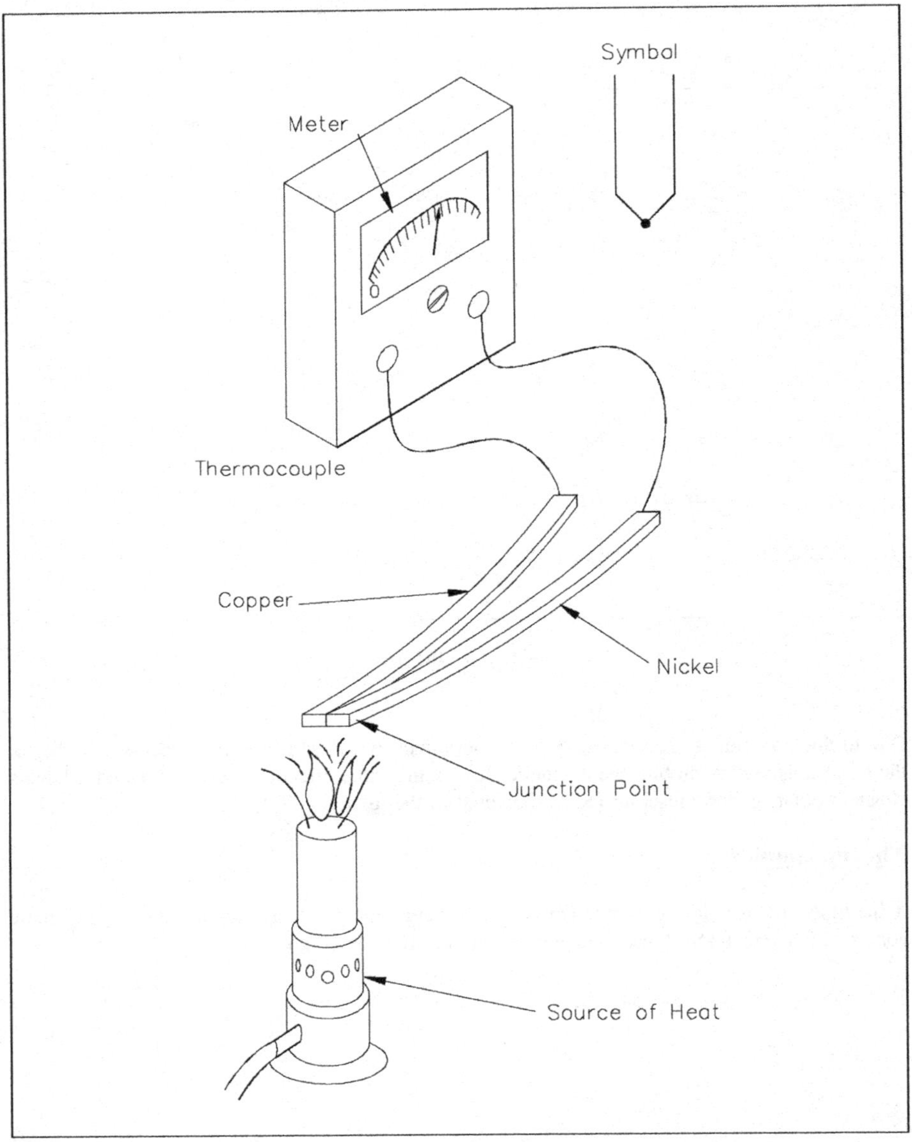

Figure 3 Production of a DC Voltage Using a Thermocouple

As the junction is heated, the electrons in one of the metals gain enough energy to become free electrons. The free electrons will then migrate across the junction and into the other metal. This displacement of electrons produces a voltage across the terminals of the thermocouple. The combinations used in the makeup of a thermocouple include: iron and constantan; copper and constantan; antimony and bismuth; and chromel and alumel.

Thermocouples are normally used to measure temperature. The voltage produced causes a current to flow through a meter, which is calibrated to indicate temperature.

Rectifiers

Most electrical power generating stations produce alternating current. The major reason for generating AC is that it can be transferred over long distances with fewer losses than DC; however, many of the devices which are used today operate only, or more efficiently, with DC. For example, transistors, electron tubes, and certain electronic control devices require DC for operation. If we are to operate these devices from ordinary AC outlet receptacles, they must be equipped with rectifier units to convert AC to DC. In order to accomplish this conversion, we use diodes in rectifier circuits. The purpose of a rectifier circuit is to convert AC power to DC.

The most common type of solid state diode rectifier is made of silicon. The diode acts as a gate, which allows current to pass in one direction and blocks current in the other direction. The polarity of the applied voltage determines if the diode will conduct. The two polarities are known as forward bias and reverse bias.

Forward Bias

A diode is forward biased when the positive terminal of a voltage source is connected to its anode, and the negative terminal is connected to the cathode (Figure 4A). The power source's positive side will tend to repel the holes in the p-type material toward the p-n junction by the negative side. A hole is a vacancy in the electron structure of a material. Holes behave as positive charges. As the holes and the electrons reach the p-n junction, some of them break through it (Figure 4B). Holes combine with electrons in the n-type material, and electrons combine with holes in the p-type material.

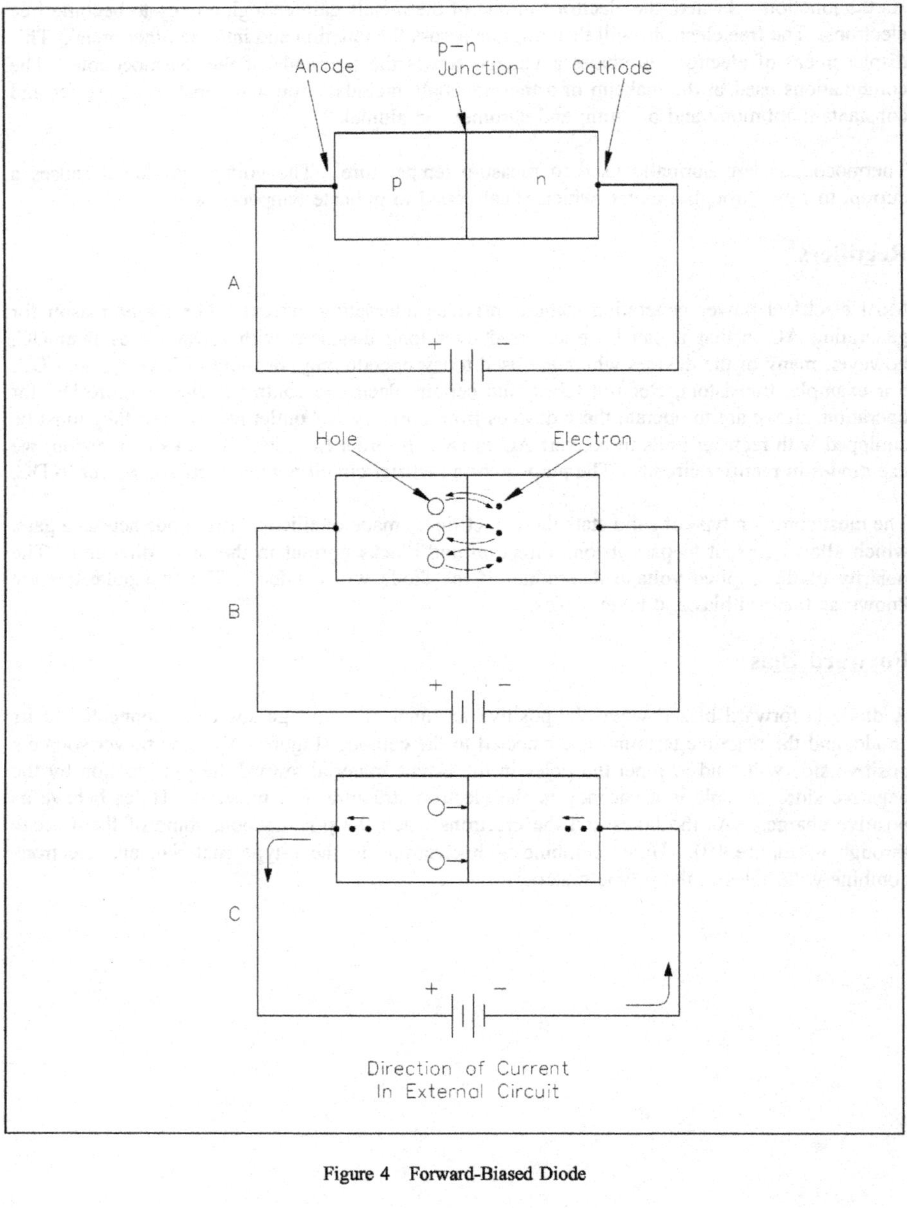

Figure 4 Forward-Biased Diode

When a hole combines with an electron, or an electron combines with a hole near the p-n junction, an electron from an electron-pair bond in the p-type material breaks its bond and enters the positive side of the source. Simultaneously, an electron from the negative side of the source enters the n-type material (Figure 4C). This produces a flow of electrons in the circuit.

Reverse Bias

Reverse biasing occurs when the diode's anode is connected to the negative side of the source, and the cathode is connected to the positive side of the source (Figure 5A). Holes within the p-type material are attracted toward the negative terminal, and the electrons in the n-type material are attracted to the positive terminal (Figure 5B). This prevents the combination of electrons and holes near the p-n junction, and therefore causes a high resistance to current flow. This resistance prevents current flow through the circuit.

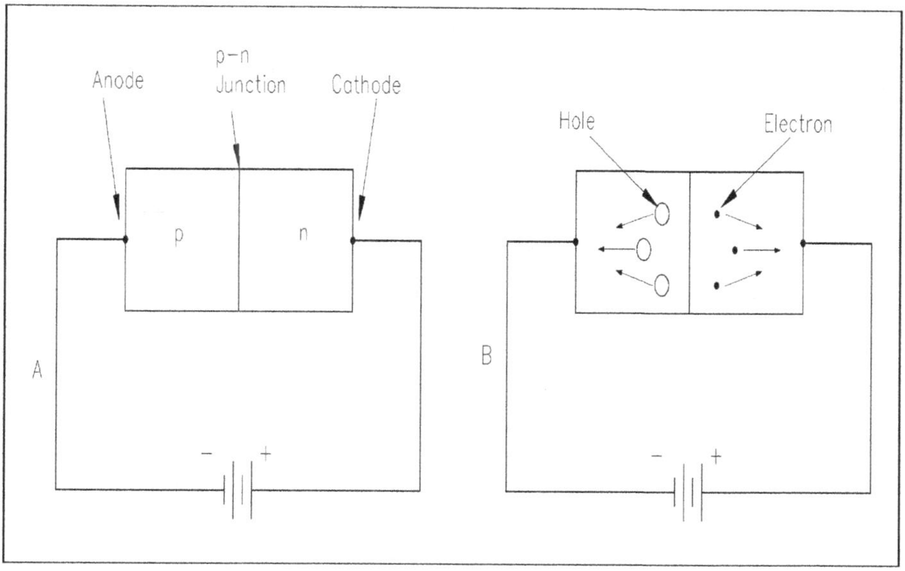

Figure 5 Reverse-Biased Diode

Half-Wave Rectifier Circuit

When a diode is connected to a source of alternating voltage, it will be alternately forward-biased, and then reverse-biased, during each cycle of the AC sine-wave. When a single diode is used in a rectifier circuit, current will flow through the circuit only during one-half of the input voltage cycle (Figure 6). For this reason, this rectifier circuit is called a half-wave rectifier. The output of a half-wave rectifier circuit is pulsating DC.

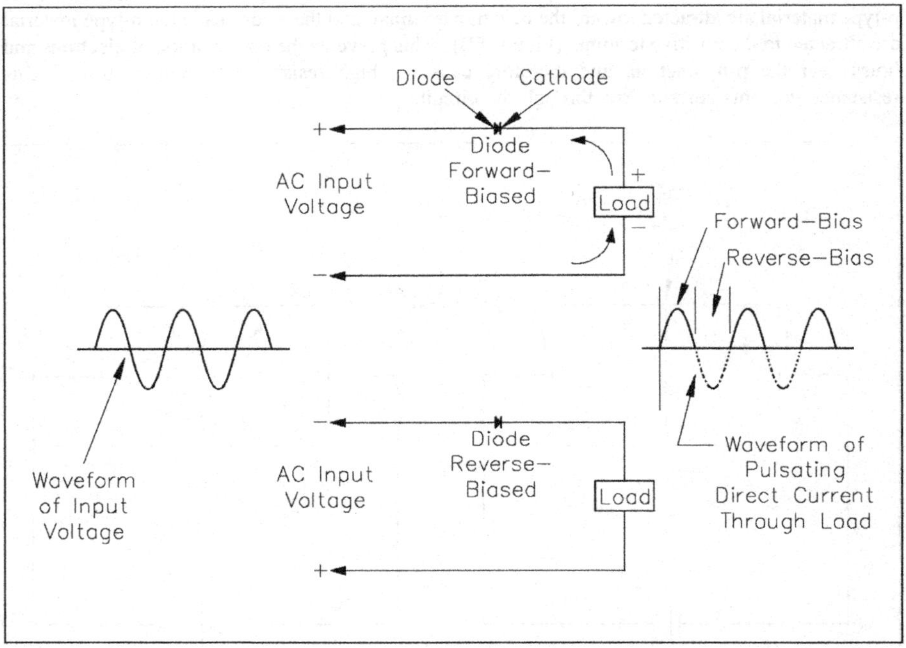

Figure 6 Half-Wave Rectifier

Full-Wave Rectifier Circuit

A full-wave rectifier circuit is a circuit that rectifies the entire cycle of the AC sine-wave. A basic full-wave rectifier uses two diodes. The action of these diodes during each half cycle is shown in Figure 7.

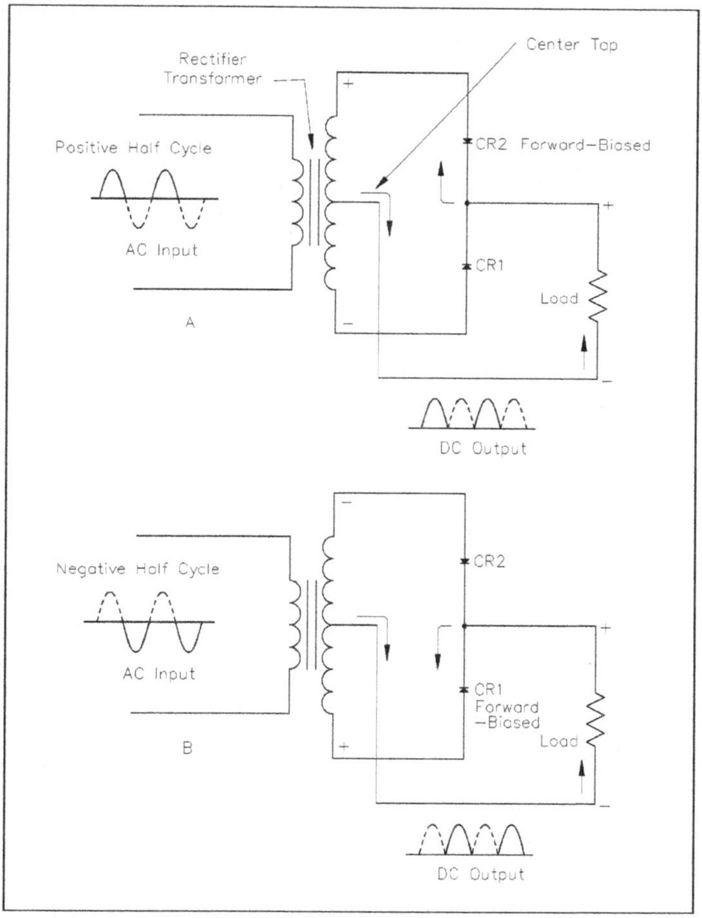

Figure 7 Full-Wave Rectifier

Another type of full-wave rectifier circuit is the full-wave bridge rectifier. This circuit utilizes four diodes. These diodes' actions during each half cycle of the applied AC input voltage are shown in Figure 8. The output of this circuit then becomes a pulsating DC, with all of the waves of the input AC being transferred. The output looks identical to that obtained from a full-wave rectifier (Figure 7).

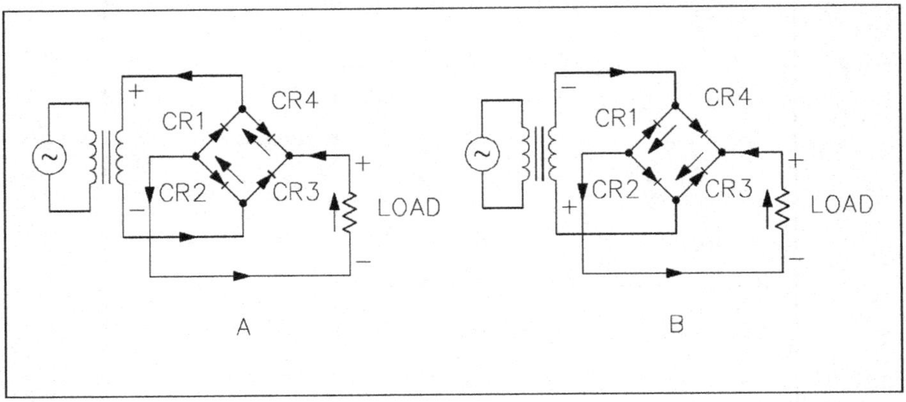

Figure 8 Bridge Rectifier Circuit

Summary

The important information concerning DC sources is summarized below.

DC Sources Summary

- There are four common ways that DC voltages are produced.

  - Batteries
  - DC Generators
  - Thermocouples
  - Rectifiers

- Thermocouples convert energy from temperature into a DC voltage. This voltage can be used to measure temperature.

- A rectifier converts AC to DC.

- There are two types of rectifiers.

  - Half-Wave rectifiers
  - Full-Wave rectifiers

- Half-wave rectifiers convert the AC to a pulsating DC and convert only one-half of the sine wave.

- Full-wave rectifiers convert the AC to a pulsating DC and convert all of the sine wave.

DC CIRCUIT TERMINOLOGY

Before operations with DC circuits can be studied, an understanding of the types of circuits and common circuit terminology associated with circuits is essential.

EO 1.4 Given a diagram, IDENTIFY it as one of the following types:
a. Schematic diagram
b. One-line diagram
c. Block diagram
d. Wiring diagram

EO 1.5 DEFINE the following terms:
a. Resistivity
b. Temperature coefficient of resistance
c. Closed circuit
d. Open circuit
e. Short circuit
f. Series circuit
g. Parallel circuit
h. Equivalent resistance

EO 1.6 Given a circuit, DETERMINE whether the circuit is an open circuit or a closed circuit.

Schematic Diagram

Schematic diagrams are the standard means by which we communicate information in electrical and electronics circuits. On schematic diagrams, the component parts are represented by graphic symbols, some of which were presented earlier in Module 1. Because graphic symbols are small, it is possible to have diagrams in a compact form. The symbols and associated lines show how circuit components are connected and the relationship of those components with one another.

As an example, let us look at a schematic diagram of a two-transistor radio circuit (Figure 9). This diagram, from left to right, shows the components in the order they are used to convert radio waves into sound energy. By using this diagram it is possible to trace the operation of the circuit from beginning to end. Due to this important feature of schematic diagrams, they are widely used in construction, maintenance, and servicing of all types of electronic circuits.

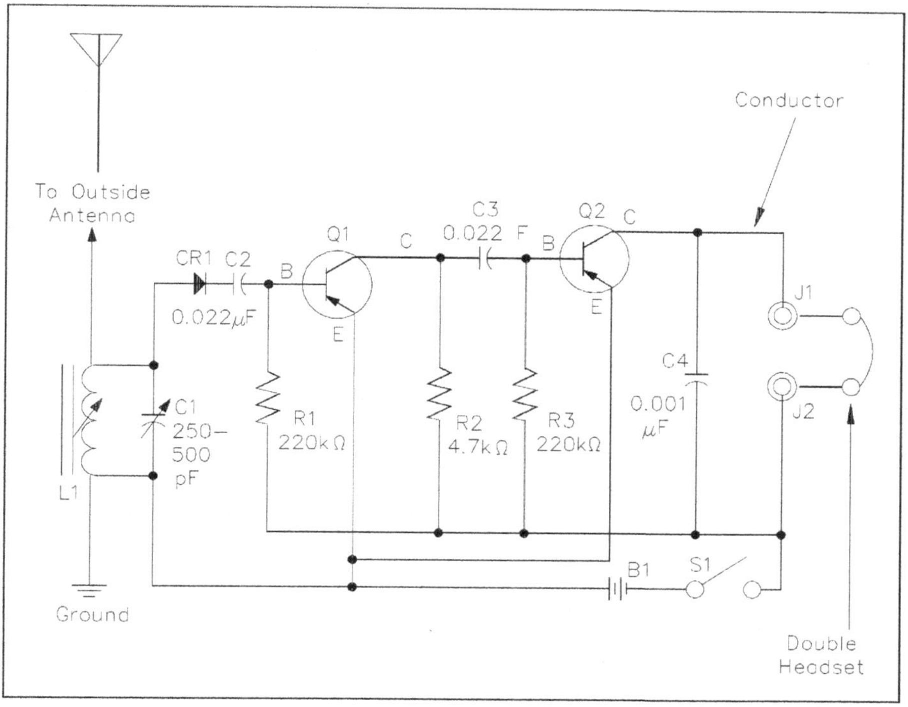

Figure 9 Schematic Diagram

One-Line Diagram

The one-line, or single-line, diagram shows the components of a circuit by means of single lines and the appropriate graphic symbols. One-line diagrams show two or more conductors that are connected between components in the actual circuit. The one-line diagram shows all pertinent information about the sequence of the circuit, but does not give as much detail as a schematic diagram. Normally, the one-line diagram is used to show highly complex systems without showing the actual physical connections between components and individual conductors.

As an example, Figure 10 shows a typical one-line diagram of an electrical substation.

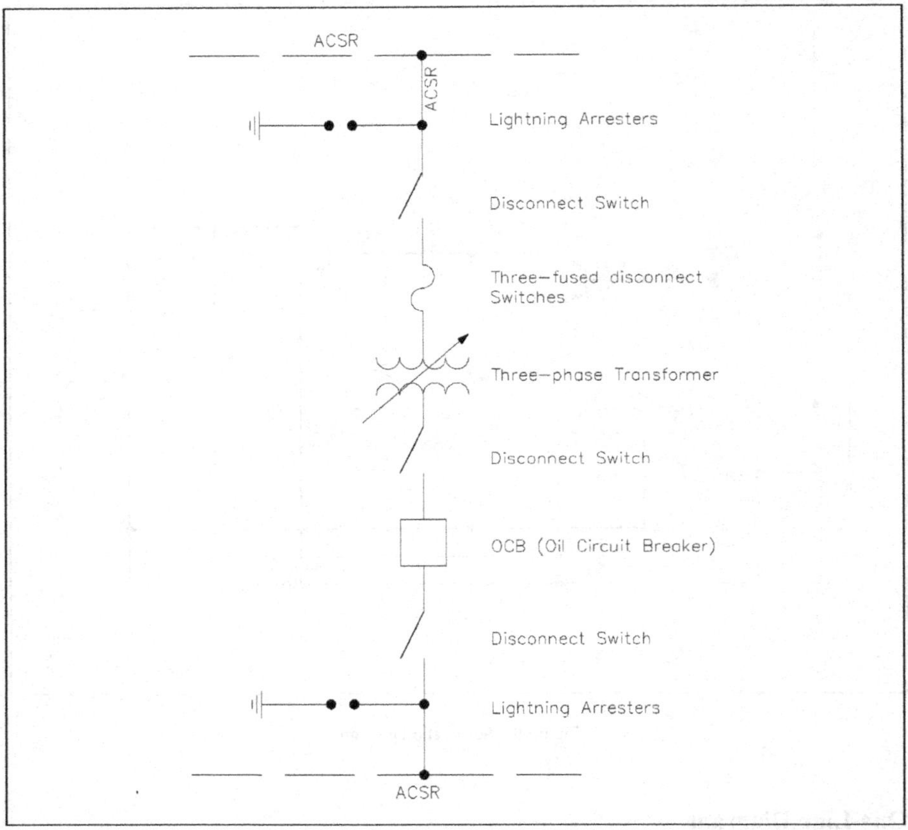

Figure 10 One-Line Diagram

Block Diagram

A block diagram is used to show the relationship between component groups, or stages in a circuit. In block form, it shows the path through a circuit from input to output (Figure 11). The blocks are drawn in the form of squares or rectangles connected by single lines with arrowheads at the terminal end, showing the direction of the signal path from input to output. Normally, the necessary information to describe the stages of components is contained in the blocks.

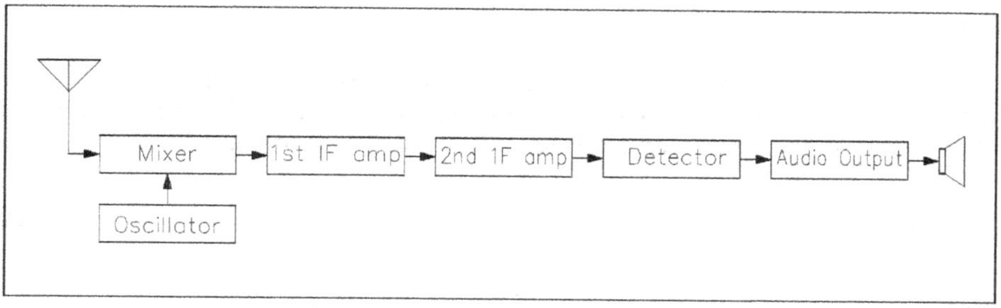

Figure 11 Block Diagram

Wiring Diagram

A wiring diagram is a very simple way to show wiring connections in an easy-to-follow manner. These types of diagrams are normally found with home appliances and automobile electrical systems (Figure 12). Wiring diagrams show the component parts in pictorial form, and the components are identified by name. Most wiring diagrams also show the relative location of component parts and color coding of conductors or leads.

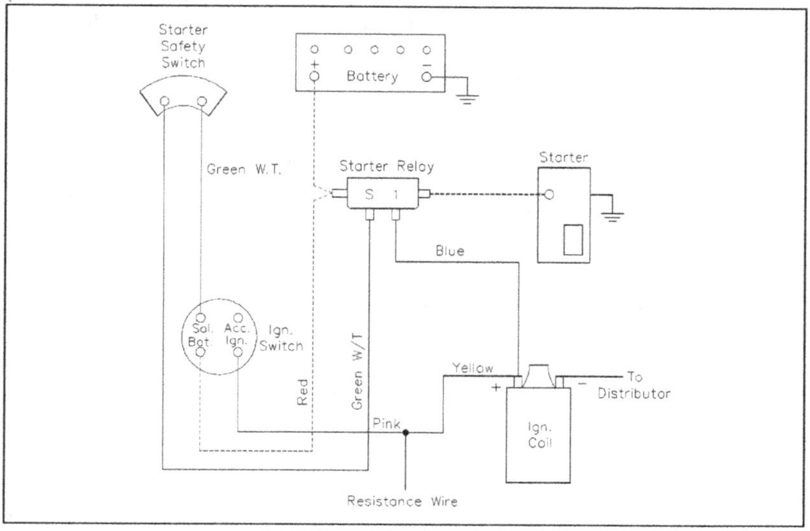

Figure 12 Wiring Diagram

Resistivity

Resistivity is defined as the measure of the resistance a material imposes on current flow. The resistance of a given length of conductor depends upon the resistivity of that material, the length of the conductor, and the cross-sectional area of the conductor, according to Equation (2-1).

$$R = \rho \frac{L}{A} \qquad (2\text{-}1)$$

where

R = resistance of conductor, Ω

ρ = specific resistance or resistivity cm-Ω/ft

L = length of conductor, ft

A = cross-sectional area of conductor, cm

The resistivity ρ (rho) allows different materials to be compared for resistance, according to their nature, without regard to length or area. The higher the value of ρ, the higher the resistance.

Table 1 gives resistivity values for metals having the standard wire size of one foot in length and a cross-sectional area of 1 cm.

TABLE 1
Properties of Conducting Materials

Material	ρ = Resistivity at 20°C-cm-Ω/ft (a)
Aluminum	17
Carbon	(b)
Constantan	295
Copper	10.4
Gold	14
Iron	58
Nichrome	676
Nickel	52
Silver	9.8
Tungsten	33.8

(a) Precise values depend on exact composition of material.
(b) Carbon has 2500-7500 times the resistance of copper.

Temperature Coefficient of Resistance

Temperature coefficient of resistance, α (alpha), is defined as the amount of change of the resistance of a material for a given change in temperature. A positive value of α indicates that R increases with temperature; a negative value of α indicates R decreases; and zero α indicates that R is constant. Typical values are listed in Table 2.

TABLE 2
Temperature Coefficients for Various Materials

Material	Temperature Coefficient, Ω per °C
Aluminum	0.004
Carbon	-0.0003
Constantan	0 (avg)
Copper	0.004
Gold	0.004
Iron	0.006
Nichrome	0.0002
Nickel	0.005

For a given material, α may vary with temperature; therefore, charts are often used to describe how resistance of a material varies with temperature.

An increase in resistance can be approximated from equation (2-2).

$$R_t = R_o + R_o(\alpha\Delta T) \qquad (2\text{-}2)$$

where

R_t = higher resistance at higher temperatures

R_o = resistance at 20°C

α = temperature coefficient

ΔT = temperature rise above 20°C

Electric Circuit

Each electrical circuit has at least four basic parts: (1) a source of electromotive force, (2) conductors, (3) load or loads, and (4) some means of control. In Figure 13, the source of EMF is the battery; the conductors are wires which connect the various component parts; the resistor is the load; and a switch is used as the circuit control device.

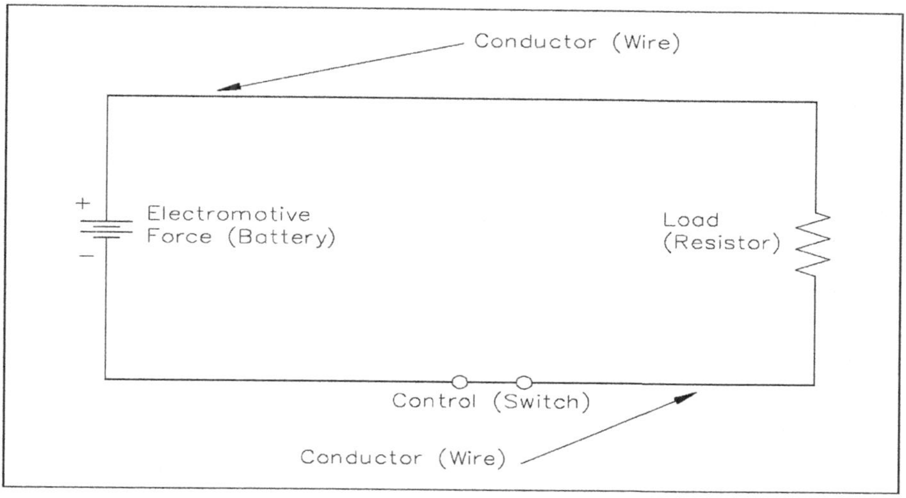

Figure 13 Closed Circuit

A *closed circuit* (Figure 13) is an uninterrupted, or unbroken, path for current from the source (EMF), through the load, and back to the source.

An *open circuit*, or incomplete circuit, (Figure 14) exists if a break in the circuit occurs; this prevents a complete path for current flow.

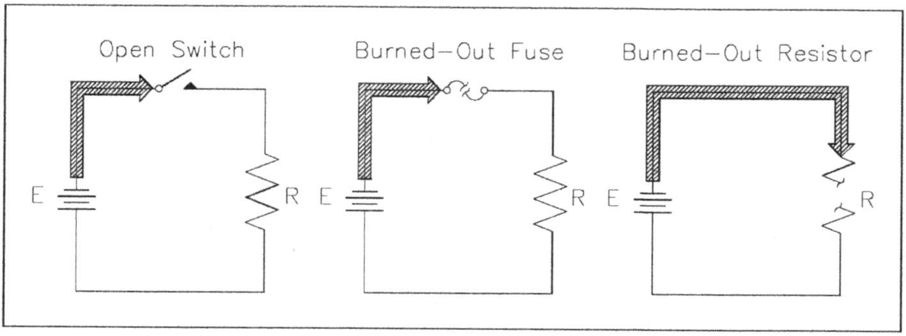

Figure 14 Open Circuit

A *short circuit* is a circuit which offers very little resistance to current flow and can cause dangerously high current flow through a circuit (Figure 15). Short circuits are usually caused by an inadvertent connection between two points in a circuit which offers little or no resistance to current flow. Shorting resistor R in Figure 15 will probably cause the fuse to blow.

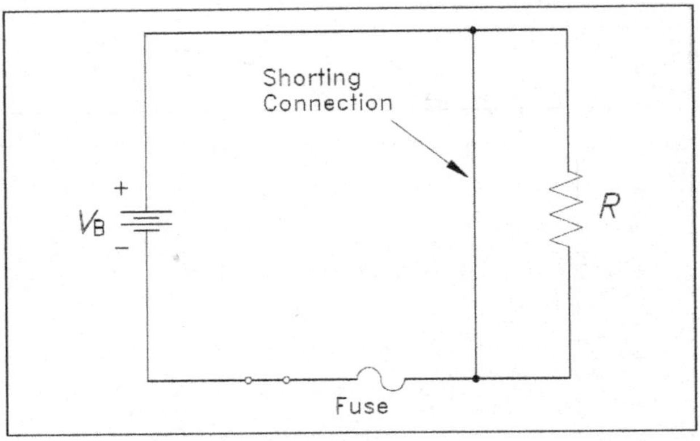

Figure 15 Short Circuit

Series Circuit

A *series circuit* is a circuit where there is only one path for current flow. In a series circuit (Figure 16), the current will be the same throughout the circuit. This means that the current flow through R_1 is the same as the current flow through R_2 and R_3.

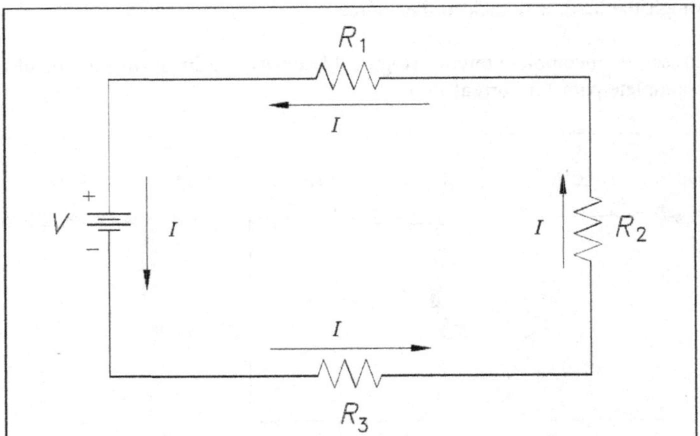

Figure 16 Series Circuit

Parallel Circuit

Parallel circuits are those circuits which have two or more components connected across the same voltage source (Figure 17). Resistors R_1, R_2, and R_3 are in parallel with each other and the source. Each parallel path is a branch with its own individual current. When the current leaves the source V, part I_1 of I_T will flow through R_1; part I_2 will flow through R_2; and part I_3 will flow through R_3. Current through each branch can be different; however, voltage throughout the circuit will be equal.

$$V = V_1 = V_2 = V_3.$$

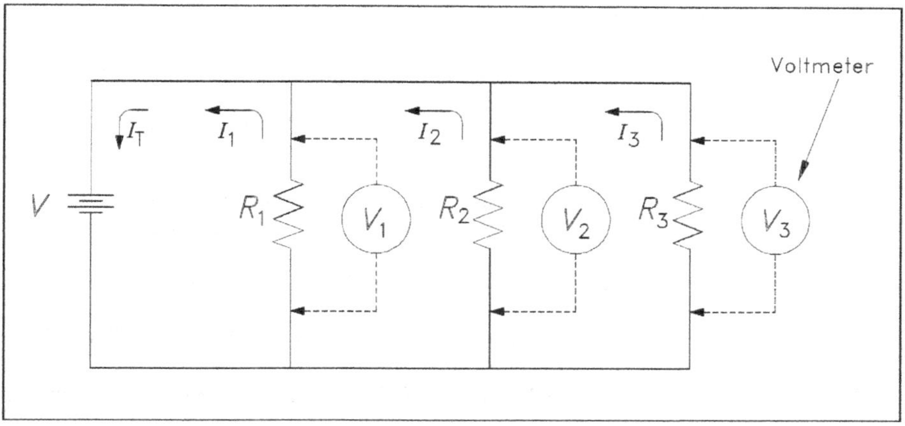

Figure 17 Parallel Circuit

Equivalent Resistance

In a parallel circuit, the total resistance of the resistors in parallel is referred to as *equivalent resistance*. This can be described as the total circuit resistance as seen by the voltage source. In all cases, the equivalent resistance will be less than any of the individual parallel circuit resistors. Using Ohm's Law, equivalent resistance (R_{EQ}) can be found by dividing the source voltage (V) by the total circuit current (I_T), as shown in Figure 17.

$$R_{EQ} = \frac{V}{I_t}$$

Summary

The important information concerning basic DC circuits is summarized below.

DC Circuit Terminology Summary

- There are four types of circuit diagrams.

  - Schematic diagram
  - One-line diagram
  - Block diagram
  - Wiring diagram

- Resistivity is defined as the measure of the resistance a material imposes on current flow.

- Temperature coefficient of resistance, α (alpha), is defined as the amount of change of the resistance of a material for a given change in temperature.

- A closed circuit is one that has a complete path for current flow.

- An open circuit is one that does not have a complete path for current flow.

- A short circuit is a circuit with a path that has little or no resistance to current flow.

- A series circuit is one where there is only one path for current flow.

- A parallel circuit is one which has two or more components connected across the same voltage source.

- Equivalent resistance is the total resistance of the resistors in parallel.

BASIC DC CIRCUIT CALCULATIONS

Each type of DC circuit contains certain characteristics that determine the way its voltage and current behave. To begin analysis of the voltages and currents at each part of a circuit, an understanding of these characteristics is necessary.

EO 1.7 **Given a circuit, CALCULATE total resistance for a series or parallel circuit.**

EO 1.8 **DESCRIBE what is meant by the term "voltage divider."**

EO 1.9 **DESCRIBE what is meant by the term "current division."**

Series Resistance

The total resistance in a series circuit is equal to the sum of all the parts of that circuit, as shown in equation (2-3).

$$R_T = R_1 + R_2 + R_3 \ldots \text{etc.} \tag{2-3}$$

where

R_T = resistance total

R_1, R_2, and R_3 = resistance in series

Example: A series circuit has a 60Ω, a 100Ω, and a 150Ω resistor in series (Figure 18). What is the total resistance of the circuit?

Solution:

$$\begin{aligned} R_T &= R_1 + R_2 + R_3 \\ &= 60 + 100 + 150 \\ &= 310\ \Omega \end{aligned}$$

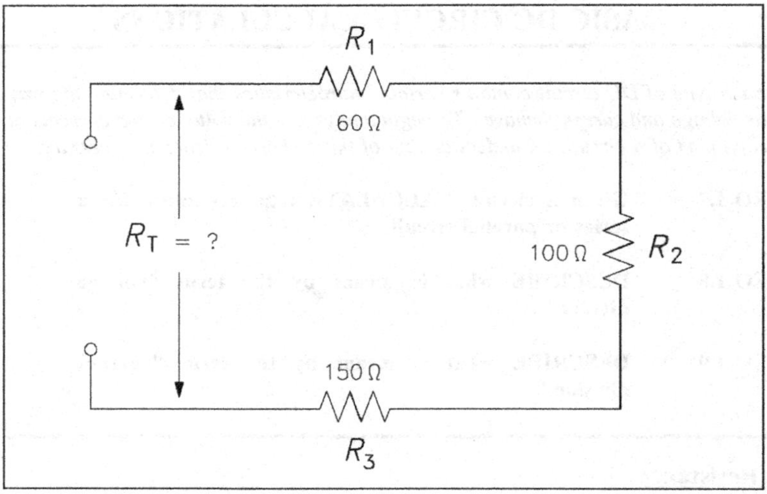

Figure 18 Resistance in a Series Circuit

The total voltage across a series circuit is equal to the sum of the voltages across each resistor in the circuit (Figure 19) as shown in equation (2-4).

$$V_T = V_1 + V_2 + V_3 \text{ ... etc.} \tag{2-4}$$

where

V_T = total voltage

V_1 = voltage across R_1

V_2 = voltage across R_2

V_3 = voltage across R_3

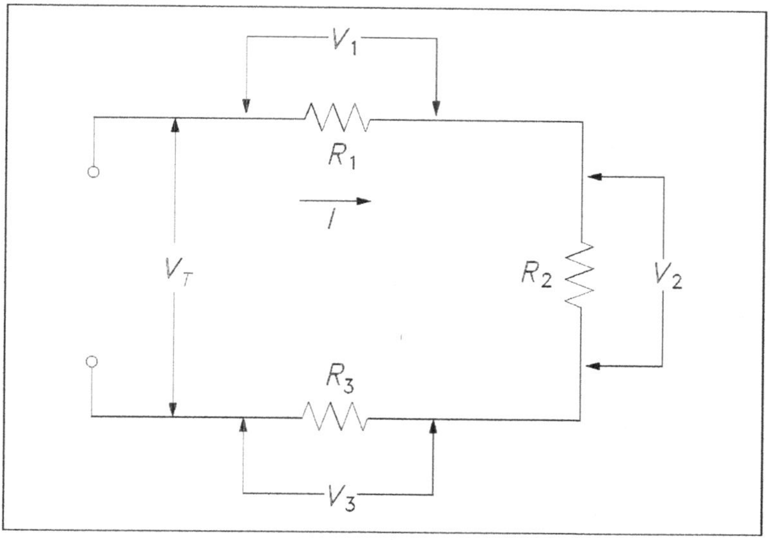

Figure 19 Voltage Drops in a Series Circuit

Ohm's law may now be applied to the entire series circuit or to individual component parts of the circuit. When used on individual component parts, the voltage across that part is equal to the current times the resistance of that part. For the circuit shown in Figure 20, the voltage can be determined as shown below.

$V_1 = IR_1$

$V_2 = IR_2$

$V_3 = IR_3$

$V_T = V_1 + V_2 + V_3$

$V_T = 10 \text{ volts} + 24 \text{ volts} + 36 \text{ volts}$

$V_T = 70 \text{ volts}$

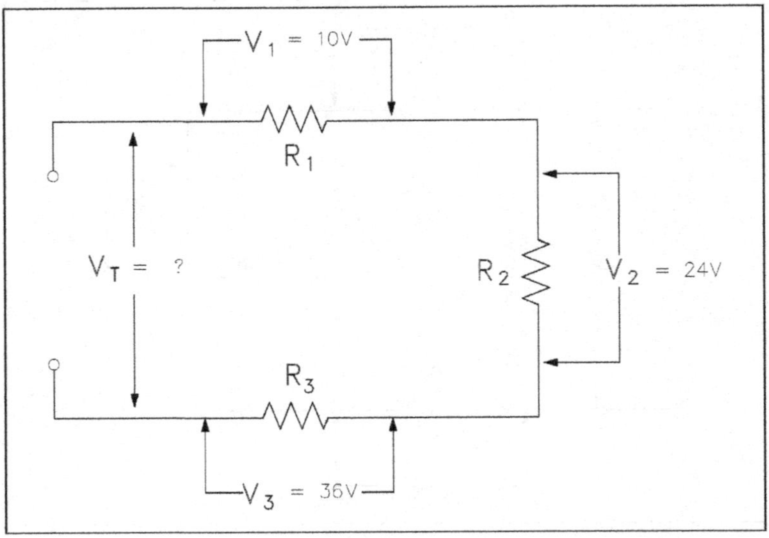

Figure 20 Voltage Total in a Series Circuit

To find the total voltage across a series circuit, multiply the current by the total resistance as shown in equation (2-5).

$$V_T = IR_T \tag{2-5}$$

where

V_T = total voltage

I = current

R_T = total resistance

Example 1: A series circuit has a 50Ω, a 75Ω, and a 100Ω resistor in series (Figure 21). Find the voltage necessary to produce a current of 0.5 amps.

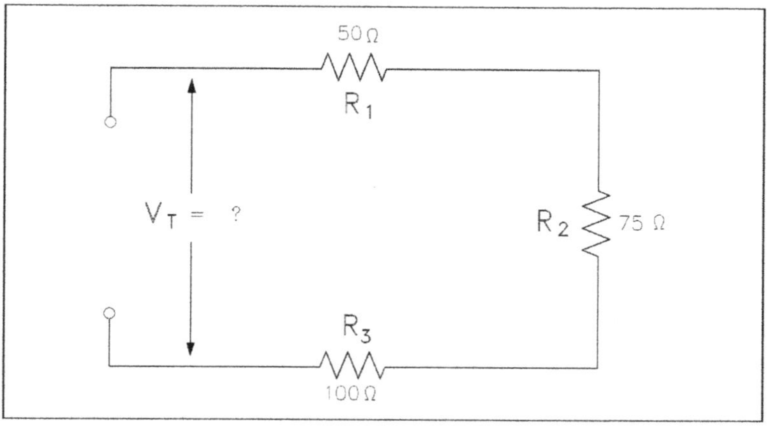

Figure 21 Example 1 Series Circuit

Solution:

Step 1: Find circuit current. As we already know, current is the same throughout a series circuit, which is already given as 0.5 amps.

Step 2: Find R_T.

$$R_T = R_1 + R_2 + R_3$$

$$R_T = 50\ \Omega + 75\ \Omega + 100\ \Omega$$

$$R_T = 225\ \Omega$$

Step 3: Find V_T. Use Ohm's law.

$$V_T = IR_T$$

$$V_T = (0.5 \text{ amps})(225\ \Omega)$$

$$V_T = 112.5 \text{ volts}$$

Example 2: A 120 V battery is connected in series with three resistors: 40Ω, 60Ω, and 100Ω (Figure 22). Find the voltage across each resistor.

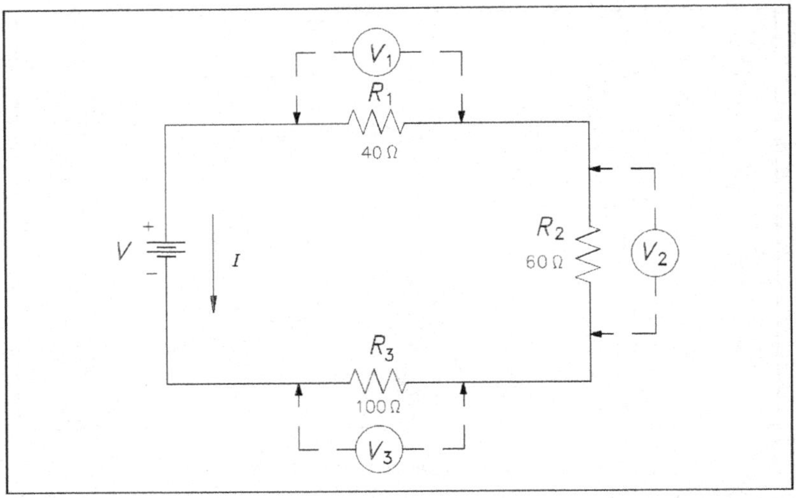

Figure 22 Example 2 Series Circuit

Solution:

Step 1: Find total resistance.

$$R_T = R_1 + R_2 + R_3$$

$$R_T = 40\ \Omega + 60\ \Omega + 100\ \Omega$$

$$R_T = 200 \text{ ohms}$$

Step 2: Find circuit current (I).

$$V_T = IR_T$$

Solving for I:

$$I = \frac{V_T}{R_T}$$

$$I = \frac{120 \text{ volts}}{200\ \Omega}$$

$$I = 0.6 \text{ amps}$$

Step 3: Find the voltage across each component.

$V_1 = IR_1$
$V_1 = (0.6 \text{ amps})(40\ \Omega)$
$V_1 = 24 \text{ volts}$

$V_2 = IR_2$
$V_2 = (0.6 \text{ amps})(60\ \Omega)$
$V_2 = 36 \text{ volts}$

$V_3 = IR_3$
$V_3 = (0.6 \text{ amps})(100\ \Omega)$
$V_3 = 60 \text{ volts}$

The voltages of V_1, V_2, and V_3 in Example 2 are known as "voltage drops" or "IR drops." Their effect is to reduce the available voltage to be applied across the other circuit components. The sum of the voltage drops in any series circuit is always equal to the applied voltage. We can verify our answer in Example 2 by using equation (2-4).

$$V_T = V_1 + V_2 + V_3$$
$$120 \text{ volts} = 24 \text{ volts} + 36 \text{ volts} + 60 \text{ volts}$$
$$120 \text{ volts} = 120 \text{ volts}$$

Parallel Currents

The sum of the currents flowing through each branch of a parallel circuit is equal to the total current flow in the circuit. Using Ohm's Law, the branch current for a three branch circuit equals the applied voltage divided by the resistance as shown in equations (2-6), (2-7), and (2-8).

Branch 1: $$I_1 = \frac{V_1}{R_1} = \frac{V}{R_1} \quad (2\text{-}6)$$

Branch 2: $$I_2 = \frac{V_2}{R_2} = \frac{V}{R_2} \quad (2\text{-}7)$$

Branch 3: $$I_3 = \frac{V_2}{R_2} = \frac{V}{R_2} \quad (2\text{-}8)$$

Example 1: Two resistors, each drawing 3A, and a third resistor, drawing 2A, are connected in parallel across a 115 volt source (Figure 23). What is total current?

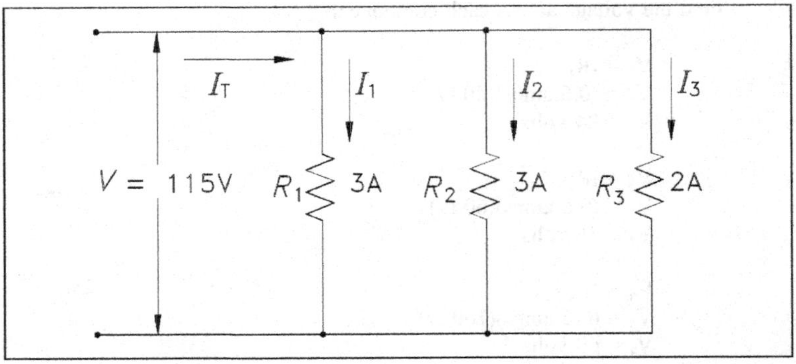

Figure 23 Example 1 Parallel Circuit

$I_T = I_1 + I_2 + I_3$

$I_T = 3A + 3A + 2A$

$I_T = 8A$

Example 2: Two branches, R_1 and R_2, are across a 120 V power source. The total current flow is 30 A (Figure 24). Branch R_1 takes 22 amps. What is the current flow in Branch R_2?

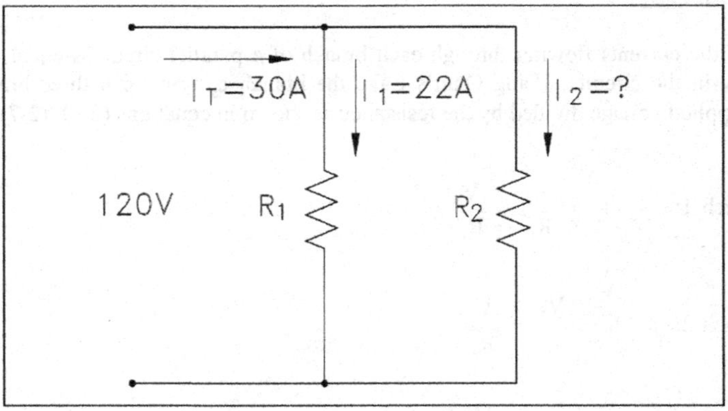

Figure 24 Example 2 Parallel Circuit

$I_T = I_1 + I_2$
$I_2 = I_T - I_1$
$I_2 = 30 - 22$
$I_2 = 8$ amps

Example 3: A parallel circuit consists of $R_1 = 15\Omega$, $R_2 = 20\Omega$ and $R_3 = 10\Omega$, with an applied voltage of 120 V (Figure 25). What current will flow through each branch?

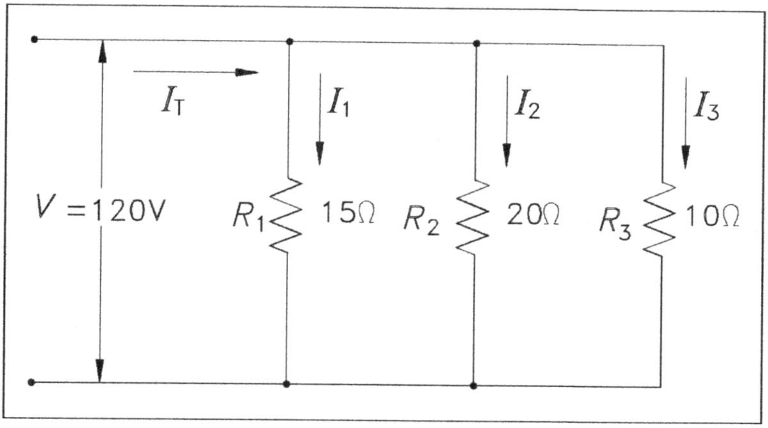

Figure 25 Example 3 Parallel Circuit

$$I_1 = \frac{V}{R_1} = \frac{120}{15} = 8A$$

$$I_2 = \frac{V}{R_2} = \frac{120}{20} = 6A$$

$$I_3 = \frac{V}{R_3} = \frac{120}{10} = 12A$$

$$I_T = I_1 + I_2 + I_3$$

$$I_T = 8A + 6A + 12A$$

$$I_T = 26A$$

Resistance in Parallel

Total resistance in a parallel circuit can be found by applying Ohm's Law. Divide the voltage across the parallel resistance by the total line current as shown in equation (2-9).

$$R_T = \frac{V}{I_T} \qquad (2\text{-}9)$$

Example: Find the total resistance of the circuit shown in Figure 25 if the line voltage is 120 V and total current is 26A.

$$R_T = \frac{V}{I_T} = \frac{120}{26} = 4.62\ \Omega$$

The total load connected to a 120 V source is the same as the single "equivalent resistance" of 4.62Ω connected across the source (Figure 26). Equivalent resistance is the total resistance a combination of loads present to a circuit.

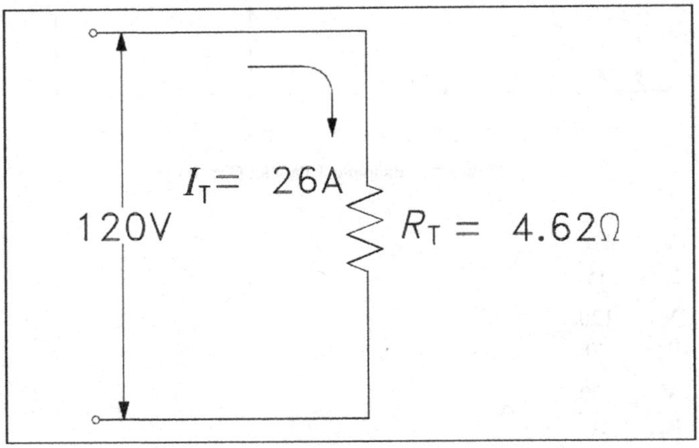

Figure 26 Equivalent Resistance in a Parallel Circuit

The total resistance in a parallel circuit can also be found by using the equation (2-10).

$$\frac{1}{R_T} = \frac{1}{R_1} + \frac{1}{R_2} + \frac{1}{R_3} + \dots \frac{1}{R_N} \tag{2-10}$$

Example 1: Find the total resistance of a 4Ω, an 8Ω, and a 16Ω resistor in parallel (Figure 27).

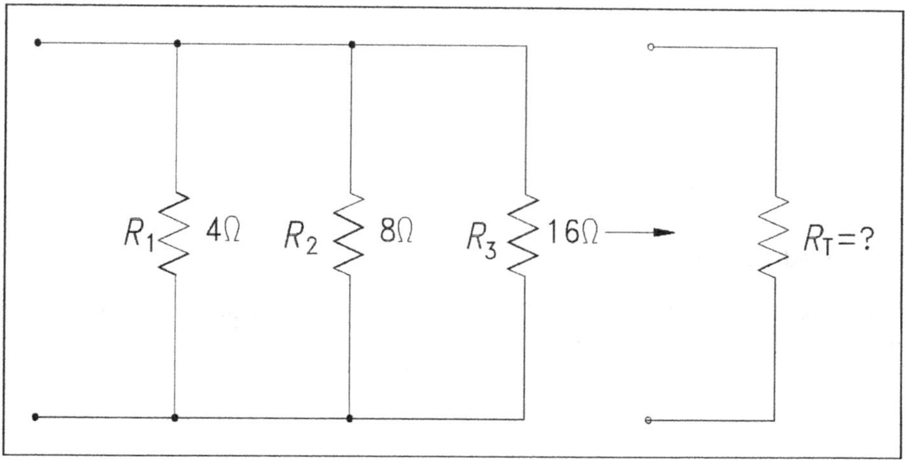

Figure 27 Total Resistance in a Parallel Circuit

Solution:

$$\frac{1}{R_T} = \frac{1}{R_1} + \frac{1}{R_2} + \frac{1}{R_3}$$

$$\frac{1}{R_T} = \frac{1}{4} + \frac{1}{8} + \frac{1}{16}$$

$$\frac{1}{R_T} = \frac{4}{16} + \frac{2}{16} + \frac{1}{16} = \frac{7}{16}$$

$$R_T = \frac{16}{7} = 2.29\Omega$$

Note: Whenever resistors are in parallel, the total resistance is always smaller than any single branch.

Example 2: Now add a fourth resistance of 4Ω in parallel to the circuit in Figure 27. What is the new total resistance of the circuit?

Solution:

$$\frac{1}{R_T} = \frac{1}{R_1} + \frac{1}{R_2} + \frac{1}{R_3} + \frac{1}{R_4}$$

$$\frac{1}{R_T} = \frac{1}{4} + \frac{1}{8} + \frac{1}{16} + \frac{1}{4}$$

$$\frac{1}{R_T} = \frac{4}{16} + \frac{2}{16} + \frac{1}{16} + \frac{4}{16} = \frac{11}{16}$$

$$R_T = \frac{16}{11} = 1.45\Omega$$

Simplified Formulas

Total resistance of equal resistors in a parallel circuit is equal to the resistance of one resistor divided by the number of resistors.

$$R_T = \frac{R}{N}$$

where

R_T = total resistance

R = resistance of one resistor

N = number of resistors

Example: Five lamps, each with a resistance of 40Ω, are connected in parallel. Find total resistance.

$$R = R_1 = R_2 = R_3 = R_4 = R_5 = 40\Omega$$

$$N = 5$$

$$R_T = \frac{R}{N} = \frac{40}{5} = 8\Omega$$

When any two resistors are unequal in a parallel circuit, it is easier to calculate R_T by multiplying the two resistances and then dividing the product by the sum, as shown in equation (2-11). As shown in equation (2-11), this is valid when there are only two resistors in parallel.

$$R_T = \frac{R_1 R_2}{R_1 + R_2} \qquad (2\text{-}11)$$

Example: Find the total resistance of a parallel circuit which has one 12Ω and one 4Ω resistor.

$$R_T = \frac{R_1 R_2}{R_1 + R_2} = \frac{(12)(4)}{12 + 4} = \frac{48}{16} = 3\Omega$$

In certain cases involving two resistors in parallel, it is useful to find an unknown resistor, R_x, to obtain a certain R_T. To find the appropriate formula, we start with equation (2-10) and let the known resistor be R and the unknown resistor be R_x.

$$R_T = \frac{R R_X}{R + R_X}$$

Cross multiply: $R_T R + R_T R_X = R R_X$

Transpose: $R R_X - R_T R_X = R_T R$

Factor: $R_X (R - R_T) = R_T R$

Solve for R_x: $R_X = \frac{R_T R}{R - R_T}$

Example: What value of resistance must be added, in parallel, with an 8Ω resistor to provide a total resistance of 6Ω (Figure 28)?

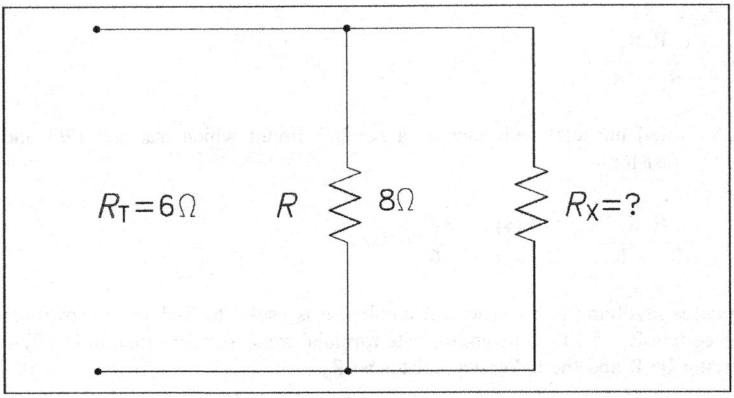

Figure 28 Example Parallel Circuit

Solution:

$$R_X = \frac{RR_T}{R - R_T} = \frac{(8)(6)}{8-6} = \frac{48}{2} = 24\Omega$$

Voltage Divider

A *voltage divider*, or network, is used when it is necessary to obtain different values of voltage from a single energy source. A simple voltage divider is shown in Figure 29. In this circuit, 24 volts is applied to three resistors in series. The total resistance limits the current through the circuit to one ampere. Individual voltages are found as follows using equation (2-12).

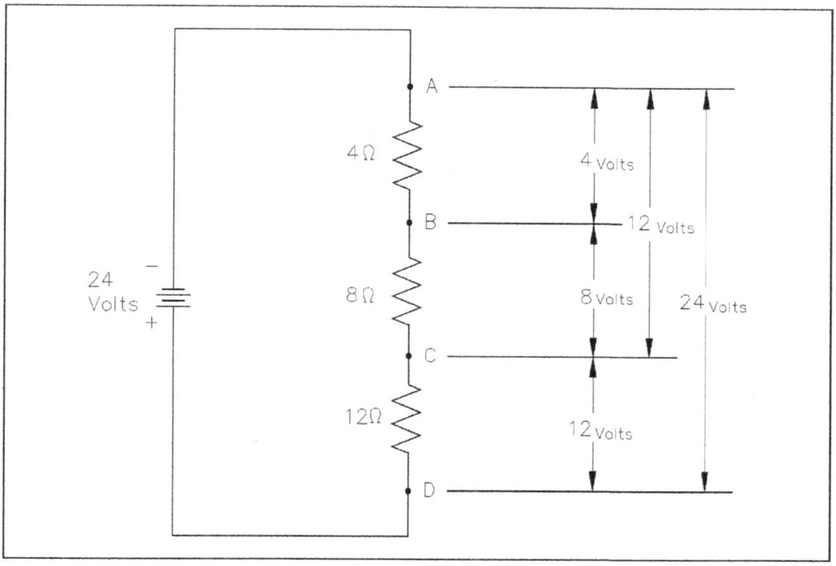

Figure 29 Voltage Divider

Total current: $I = \frac{V}{R} = \frac{24}{4 + 8 + 12} = \frac{24}{24} = 1 \text{ amp}$ (2-12)

Voltage drop across AB:

$$V = IR$$
$$= (1)(4)$$
$$V = 4 \text{ Volts}$$

Voltage drop across BC:

$$V = IR$$
$$= (1)(8)$$
$$V = 8 \text{ Volts}$$

Voltage drop across CD:

$$V = IR$$
$$= (1)(12)$$
$$V = 12 \text{ Volts}$$

Total voltage drop AC:

$$V = IR$$
$$= (1)(8 + 4) = (1)(12)$$
$$V = 12 \text{ Volts}$$

Current Division

Sometimes it is necessary to find the individual branch currents in a parallel circuit when only resistance and total current are known. When only two branches are involved, the current in one branch will be some fraction of I_T. The resistance in each circuit can be used to divide the total current into fractional currents in each branch. This process is known as *current division.*

$$I_1 = \frac{R_2}{R_1 + R_2} I_T$$

$$I_2 = \frac{R_1}{R_1 + R_2} I_T \qquad (2\text{-}13)$$

Note that the equation for each branch current has the opposite R in the numerator. This is because each branch current is inversely proportional to the branch resistance.

Example: Find branch current for I_1 and I_2 for the circuit shown in Figure 30.

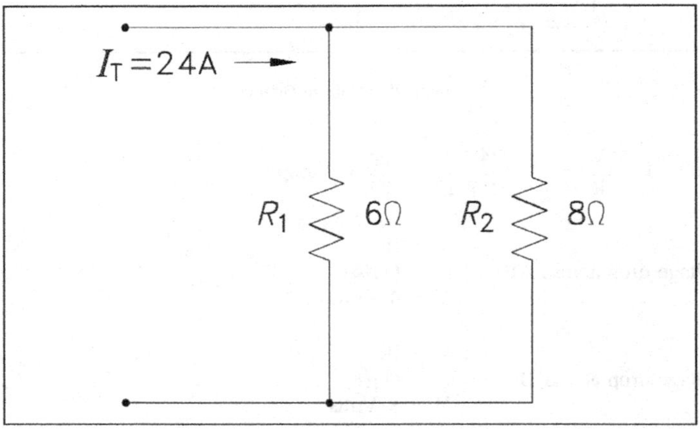

Figure 30 Current Division Example Circuit

Solution:

$$I_1 = \frac{R_2}{R_1 + R_2} I_T = \frac{8}{6 + 8}(24) = \frac{8}{14}(24) = 13.71 \text{ amps}$$

$$I_2 = \frac{R_1}{R_1 + R_2} I_T = \frac{6}{6 + 8}(24) = \frac{6}{14}(24) = 10.29 \text{ amps}$$

Since I_1 and I_T were known, we could have also simply subtracted I_1 from I_T to find I_2:

$$\begin{aligned} I_T &= I_1 + I_2 \\ I_2 &= I_T - I_1 \\ &= 24 - 13.71 \\ &= 10.29 \text{ amps} \end{aligned}$$

Summary

The important information in this chapter is summarized below.

Basic DC Circuit Calculations Summary

- Equivalent resistance is a term used to represent the total resistance a combination of loads presents to a circuit.
- A voltage divider is used to obtain different values of voltage from a single energy source.
- Current division is used to determine the current flowing through each leg of a parallel circuit.

VOLTAGE POLARITY AND CURRENT DIRECTION

Before introducing the laws associated with complex DC circuit analysis, the importance of voltage polarity and current direction must be understood. This chapter will introduce the polarities and current direction associated with DC circuits.

EO 1.10 **DESCRIBE the difference between electron flow and conventional current flow.**

EO 1.11 **Given a circuit showing current flows, IDENTIFY the polarity of the voltage drops in the circuit.**

Conventional and Electron Flow

The direction of electron flow is from a point of negative potential to a point of positive potential. The direction of positive charges, or holes, is in the opposite direction of electron flow. This flow of positive charges is known as *conventional flow*. All of the electrical effects of electron flow from negative to positive, or from a high potential to a lower potential, are the same as those that would be created by flow of positive charges in the opposite direction; therefore, it is important to realize that both conventions are in use, and they are essentially equivalent. In this manual, the electron flow convention is used.

Polarities

All voltages and currents have polarity as well as magnitude. In a series circuit, there is only one current, and its polarity is from the negative battery terminal through the rest of the circuit to the positive battery terminal. Voltage drops across loads also have polarities. The easiest way to find these polarities is to use the direction of the electron current as a basis. Then, where the electron current enters the load, the voltage is negative (Figure 31). This holds true regardless of the number or type of loads in the circuit. The drop across the load is opposite to that of the source. The voltage drops oppose the source voltage and reduce it for the other loads. This is because each load uses energy, leaving less energy for other loads.

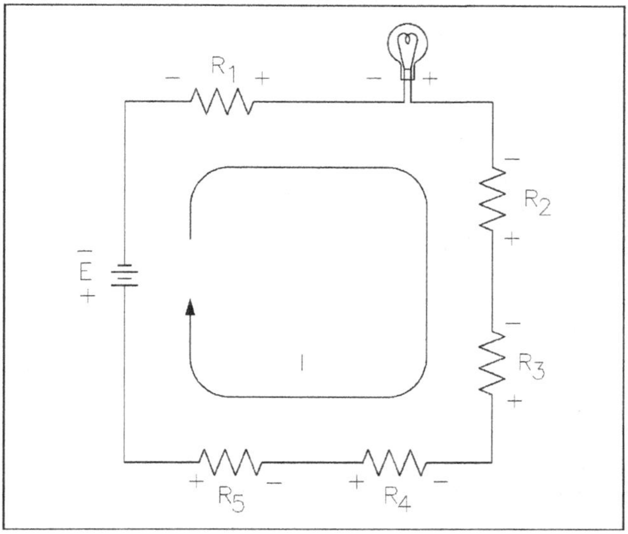

Figure 31 Voltage Polarities

Summary

The important information in this chapter is summarized below.

Voltage Polarity and Current Direction Summary

- The direction of electron flow is from a point of negative potential to a point of positive potential.
- The direction of positive charges, or holes, is in the opposite direction of electron flow. This flow of positive charges is known as "conventional flow."
- Where the electron current enters the load, the voltage is negative.

KIRCHHOFF'S LAWS

Kirchhoff's two laws reveal a unique relationship between current, voltage, and resistance in electrical circuits that is vital to performing and understanding electrical circuit analysis.

EO 1.12 STATE Kirchhoff's voltage law.

EO 1.13 STATE Kirchhoff's current law.

EO 1.14 Given a circuit, SOLVE problems for voltage and current using Kirchhoff's laws.

Kirchhoff's Laws

In all of the circuits examined so far, Ohm's Law described the relationship between current, voltage, and resistance. These circuits have been relatively simple in nature. Many circuits are extremely complex and cannot be solved with Ohm's Law. These circuits have many power sources and branches which would make the use of Ohm's Law impractical or impossible.

Through experimentation in 1857 the German physicist Gustav Kirchhoff developed methods to solve complex circuits. Kirchhoff developed two conclusions, known today as Kirchhoff's Laws.

Law 1: The sum of the voltage drops around a closed loop is equal to the sum of the voltage sources of that loop (Kirchhoff's Voltage Law).

Law 2: The current arriving at any junction point in a circuit is equal to the current leaving that junction (Kirchhoff's Current Law).

Kirchhoff's two laws may seem obvious based on what we already know about circuit theory. Even though they may seem very simple, they are powerful tools in solving complex and difficult circuits.

Kirchhoff's laws can be related to conservation of energy and charge if we look at a circuit with one load and source. Since all of the power provided from the source is consumed by the load, energy and charge are conserved. Since voltage and current can be related to energy and charge, then Kirchhoff's laws are only restating the laws governing energy and charge conservation.

The mathematics involved becomes more difficult as the circuits become more complex. Therefore, the discussion here will be limited to solving only relatively simple circuits.

Kirchhoff's Voltage Law

Kirchhoff's first law is also known as his "voltage law." The voltage law gives the relationship between the "voltage drops" around any closed loop in a circuit, and the voltage sources in that loop. The total of these two quantities is always equal. In equation form:

$$E_{source} = E_1 + E_2 + E_3 + \text{etc.} = I_1R_1 + I_2R_2 + I_3R_3 + \text{etc.}$$

$$\Sigma E_{source} = \Sigma IR \quad (2\text{-}14)$$

where the symbol Σ (the Greek letter sigma) means "the sum of."

Kirchhoff's voltage law can be applied only to closed loops (Figure 32). A closed loop must meet two conditions:

1. It must have one or more voltage sources.
2. It must have a complete path for current flow from any point, around the loop, and back to that point.

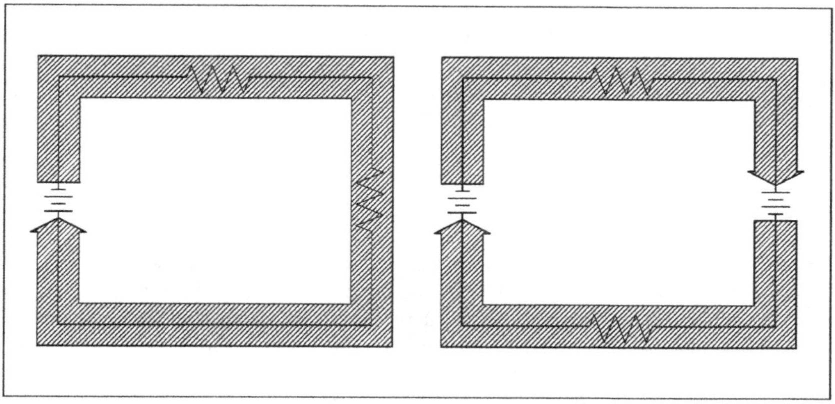

Figure 32 Closed Loop

You will remember that in a simple series circuit, the sum of the voltage drops around the circuit is equal to the applied voltage. Actually, this is Kirchhoff's voltage law applied to the simplest case, that is, where there is only one loop and one voltage source.

Applying Kirchhoff's Voltage Law

For a simple series circuit, Kirchhoff's voltage law corresponds to Ohm's Law. To find the current in a circuit (Figure 33) by using Kirchhoff's voltage law, use equation (2-15).

$$\Sigma E_{source} = \Sigma IR \tag{2-15}$$

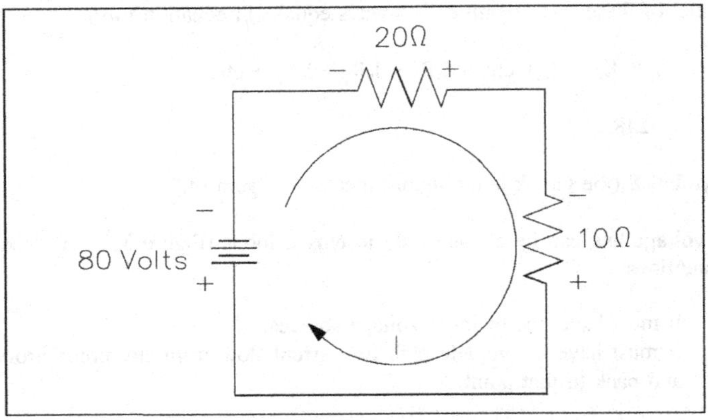

Figure 33 Using Kirchhoff's Voltage Law to find Current with one Source

$80 = 20(I) + 10(I)$

$80 = 30(I)$

$I = 80/30 = 2.66$ amperes

In the problem above, the direction of current flow was known before solving the problem. When there is more than one voltage source, the direction of current flow may or may not be known. In such a case, a direction of current flow must be assumed in the beginning of the problem. All the sources that would aid the current in the assumed direction of current flow are then positive, and all that would oppose current flow are negative. If the assumed direction is correct, the answer will be positive. The answer would be negative if the direction assumed was wrong. In any case, the correct magnitude will be attained.

For example, what is the current flow in Figure 34? Assume that the current is flowing in the direction shown.

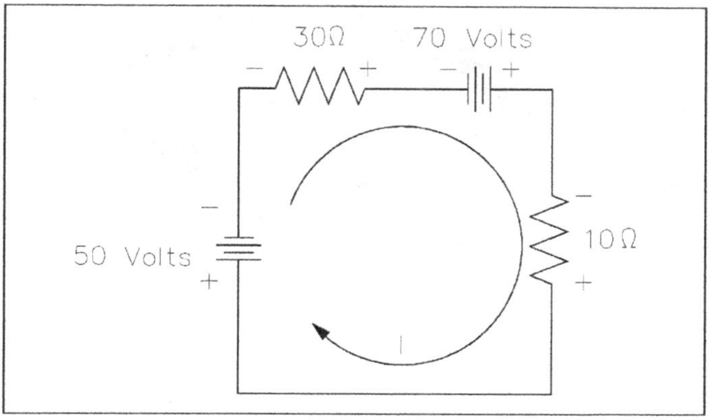

Figure 34 Using Kirchhoff's Voltage Law to find Current with Multiple Battery Sources

Using Kirchhoff's Voltage Law:

$$\Sigma E_{source} = \Sigma IR$$

$$50 - 70 = 30I + 10I$$

$$-20 = 40I$$

$$I = \frac{-20}{40}$$

$$I = -0.5$$

The result is negative. The current is actually 0.5 ampere in the opposite direction to that of the assumed direction.

Kirchhoff's Current Law

Kirchhoff's second law is called his current law and states: "At any junction point in a circuit, the current arriving is equal to the current leaving." Thus, if 15 amperes of current arrives at a junction that has two paths leading away from it, 15 amperes will divide among the two branches, but a total of 15 amperes must leave the junction. We are already familiar with Kirchhoff's current law from parallel circuits, that is, the sum of the branch currents is equal to the total current entering the branches, as well as the total current leaving the branches (Figure 35).

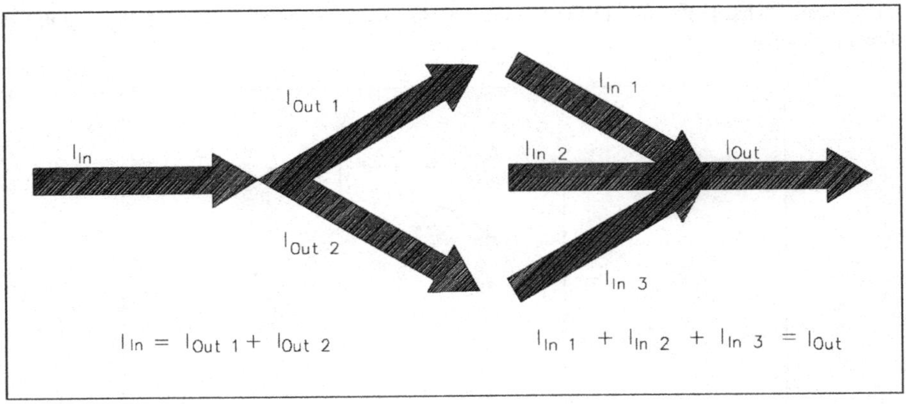

Figure 35 Illustration of Kirchhoff's Current Law

In equation form, Kirchhoff's current law may be expressed:

$$I_{IN} - I_{OUT} = 0 \qquad (2\text{-}16)$$

or

$$I_{IN} = I_{OUT}$$

Normally Kirchhoff's current law is not used by itself, but with the voltage law, in solving a problem.

Example: Find I_2 in the circuit shown in Figure 36 using Kirchhoff's voltage and current laws.

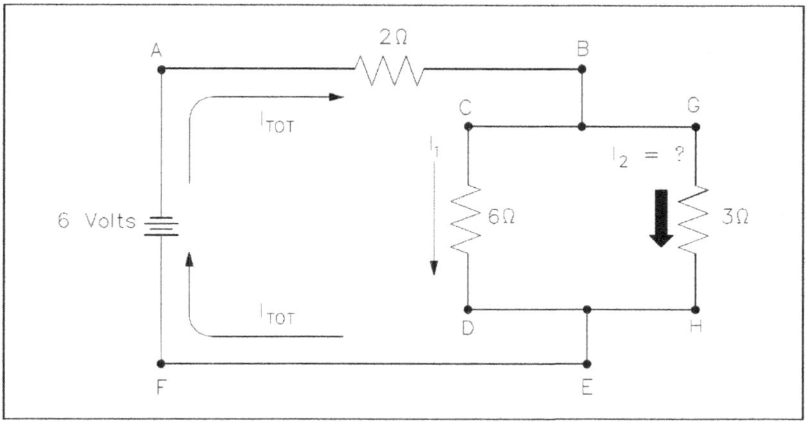

Figure 36 Using the Current Law

Solution:

First, apply Kirchhoff's voltage law to both loops.

Loop ABCDEF	Loop ABGHEF
$\Sigma IR = \Sigma E_{source}$	$\Sigma IR = \Sigma E_{source}$
$2\ I_{total} + 6\ I_1 = 6$	$2\ I_{total} + 3\ I_2 = 6$

Since Kirchhoff'a current law states $I_{total} = I_1 + I_2$, substitute $(I_1 + I_2)$ in the place of I_{total} in both loop equations and simplify.

Loop ABCDEF	Loop ABGHEF
$2\ (I_1 + I_2) + 6\ I_1 = 6$	$2\ (I_1 + I_2) + 3\ I_2 = 6$
$2\ I_1 + 2\ I_2 + 6\ I_1 = 6$	$2\ I_1 + 2\ I_2 + 3\ I_2 = 6$
$8\ I_1 + 2\ I_2 = 6$	$2\ I_1 + 5\ I_2 = 6$

We now have two equations and two unknowns and must eliminate I_1 to find I_2. One way is to multiply Loop ABGHEF equation by four, and subtract Loop ABCDEF equation from the result.

Multiply by 4:

$4\ (2\ I_1 + 5\ I_2 = 6)$
$8\ I_1 + 20\ I_2 + 24$

Subtract:

$$\begin{array}{r} 8I_1 + 20I_2 = 24 \\ -(8I_1 + 2I_2 = 6) \\ \hline 18I_2 = 18 \end{array}$$

Now we have an equation with only I_2, which is the current we are looking for.

$$18I_2 = 18$$

$$I_2 = \frac{18}{18} = 1 \text{ ampere}$$

This circuit could have been solved simply by using Ohm's Law, but we used Kirchhoff's Laws to show the techniques used in solving complex circuits when Ohm's Law cannot be used.

Summary

The important information in this chapter is summarized below.

Kirchhoff's Laws Summary

- Kirchhoff's voltage law states that the sum of the voltage drops around a closed loop is equal to the sum of the voltage sources of that loop.
- Kirchhoff's current law states that the current arriving at any junction point in a circuit is equal to the current leaving that junction.
- Since voltage and current can be related to energy and charge, then Kirchhoff's laws are only restating the laws governing energy and charge conservation.

DC CIRCUIT ANALYSIS

All of the rules governing DC circuits that have been discussed so far can now be applied to analyze complex DC circuits. To apply these rules effectively, loop equations, node equations, and equivalent resistances must be used.

EO 1.15 Given a simple DC circuit, DETERMINE the equivalent resistance of series and parallel combinations of elements.

Loop Equations

As we have already learned, Kirchhoff's Laws provide a practical means to solve for unknowns in a circuit. Kirchhoff's current law states that at any junction point in a circuit, the current arriving is equal to the current leaving. In a series circuit the current is the same at all points in that circuit. In parallel circuits, the total current is equal to the sum of the currents in each branch. Kirchhoff's voltage law states that the sum of all potential differences in a closed loop equals zero.

Using Kirchhoff's laws, it is possible to take a circuit with two loops and several power sources (Figure 37) and determine loop equations, solve loop currents, and solve individual element currents.

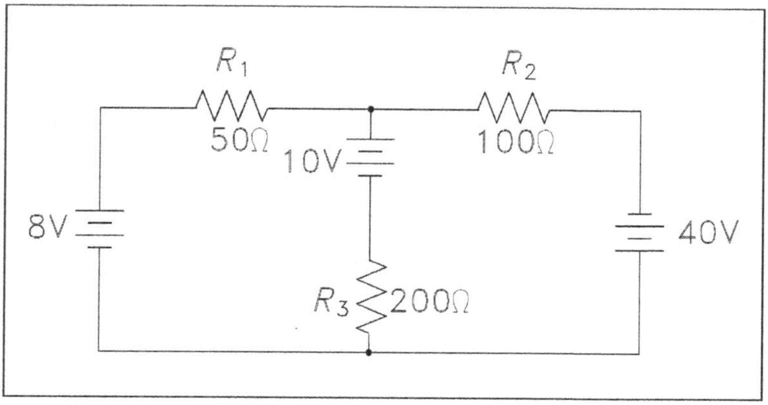

Figure 37 Example Circuit for Loop Equations

The first step is to draw an assumed direction of current flow (Figure 38). It does not matter whether the direction is correct. If it is wrong, the resulting value for current will be negative.

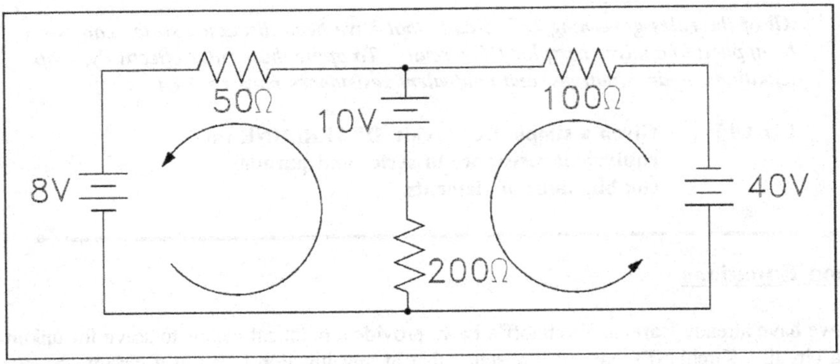

Figure 38 Assumed Direction of Current Flow

Second, mark the polarity of voltage across each component (Figure 39). It is necessary to choose a direction for current through the center leg, but it is not necessary to put in a new variable. It is simply I_2 - I_1.

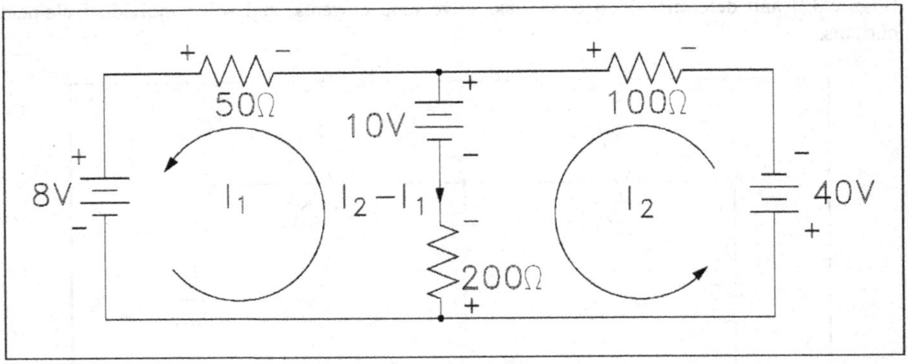

Figure 39 Marking Polarity

Third, apply Kirchhoff's voltage law to loops one and two by picking a point in each loop and writing a loop equation of the voltage drops around the loop; then set the equation equal to zero.

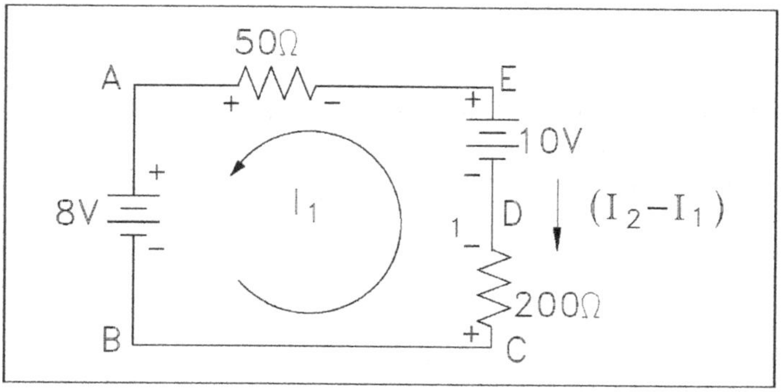

Figure 40 Applying Voltage Law to Loop 1

Figure 40 shows Loop one.

From Point A to Point B, there is an increase in voltage of 8 volts. From Point C to Point D, there is an increase in voltage of 200 $(I_2 - I_1)$. From Point D to Point E, there is a decrease in voltage of 10 volts. From Point E to Point A, there is a voltage decrease of $50I_1$ volts. The result in equation form is illustrated in equation (2-16).

$$8 + 200\,(I_2 - I_1) - 50\,I_1 - 10 = 0 \qquad (2\text{-}17)$$

Using the same procedure for Loop 2 of Figure 39, the resulting equation is shown in equation (2-18).

$$10 - 200\,(I_2 - I_1) + 40 - 100\,I_2 = 0 \qquad (2\text{-}18)$$

Fourth, solve equations (2-17) and (2-18) simultaneously. First, rearrange and combine like terms in the equation for Loop 1.

$$-50\,I_1 + 200\,I_2 - 200\,I_1 = 10 - 8$$
$$-250\,I_1 + 200\,I_2 = 2$$

Divide both sides by two.

$$-125\,I_1 + 100\,I_2 = 1$$

Rearrange and combine like terms in the Loop 2 equation.

$$-200\ I_2 + 200\ I_1 - 100\ I_2 = -10 - 40$$
$$200\ I_1 - 300\ I_2 = -50$$

Multiplying the Loop 1 equation by 3, and add it to the Loop 2 equation.

$$3(-125\ I_1 + 100\ I_2 = 1) = -375\ I_1 + 300 I_2 = 3$$
$$+\ 200\ I_2 - 300 I_2 = -50$$
$$\overline{-175\ I_1 = -47}$$

Solving for I_1:

$$-175\ I_1 = -47$$
$$I_1 = \frac{-47}{-175} = 0.2686\ \text{amp} = 268.6\ \text{mA}$$

Solving for I_2 using the Loop 1 equation:

$$-125\ (0.2686) + 100\ I_2 = 1$$
$$100\ I_2 = 1 + 33.58$$
$$I_2 = \frac{34.58}{100}$$
$$I_2 = 0.3458\ \text{amp} = 345.8\ \text{mA}$$

The current flow through R_1 (50Ω) is I_1. The current flow through R_2(100Ω) is I_2, and through R_3(200Ω) is I_2 - I_1:

$$I_3 = I_2 - I_1 = 345.8\ \text{mA} - 268.6\ \text{mA}$$
$$I_3 = I_2 - I_1 = 77.2\ \text{mA}$$

Fifth, apply Ohm's Law to obtain the voltage drops across Resistors R_1, R_2, and R_3:

$$V_1 = I_1R_1 = (0.2686 \text{ amps})(50\Omega) = 13.43 \text{ Volts}$$

$$V_2 = I_2R_2 = (0.3458 \text{ amps})(100\Omega) = 34.58 \text{ Volts}$$

$$V_3 = (I_2 - I_1)\,R_3 = (0.0772 \text{ amps})(200\Omega) = 15.44 \text{ Volts}$$

Sixth, check the calculations by applying Kirchhoff's Laws:

Check 1: Apply Kirchhoff's voltage law to the larger outer loop (Figure 41).

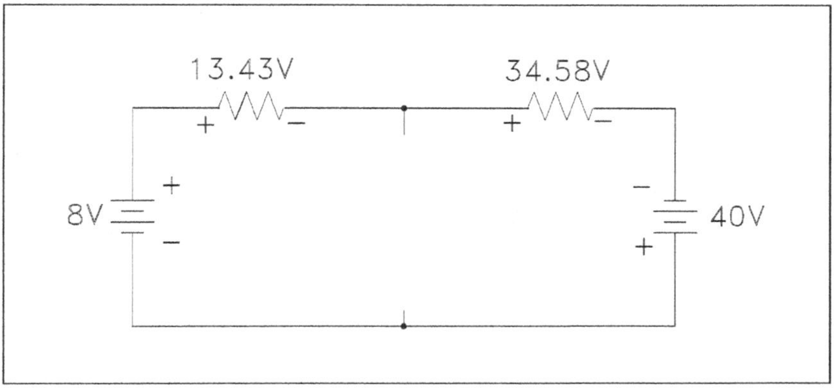

Figure 41 Applying Voltage Laws to Outer Loop

The sum of the voltage drops around the loop is essentially zero. (Not exactly zero due to rounding off.)

$$8 - 13.43 - 34.58 + 40 = 0$$
$$-0.01 \cong 0$$

Therefore, the solution checks.

Check 2: Use Kirchhoff's current law at one of the junctions (Figure 42).

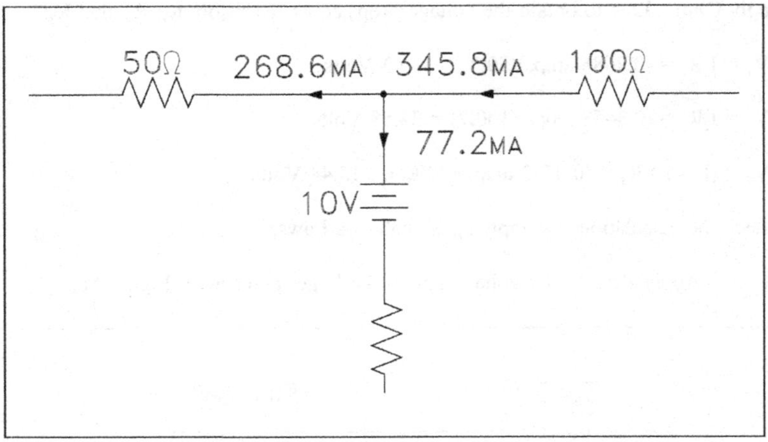

Figure 42 Applying Current Law to Junction

The sum of the currents out of the junction is:

0.2686 + 0.0772	=	0.3458 a
	=	345.8 ma

The current into the junction is 345.8 ma.

The current into the junction is equal to the current out of the junction. Therefore, the solution checks.

Node Equations

Kirchhoff's current law, as previously stated, says that at any junction point in a circuit the current arriving is equal to the current leaving. Let us consider five currents entering and leaving a junction shown as P (Figure 43). This junction is also considered a node.

Assume that all currents entering the node are positive, and all currents that leave the node are negative. Therefore, I_1, I_3, and I_4 are positive, and I_2 and I_5 are negative. Kirchhoff's Law also states that the sum of all the currents meeting at the node is zero. For Figure 43, Equation (2-19) represents this law mathematically.

$$I_1 + I_2 + I_3 + I_4 + I_5 = 0 \qquad (2\text{-}19)$$

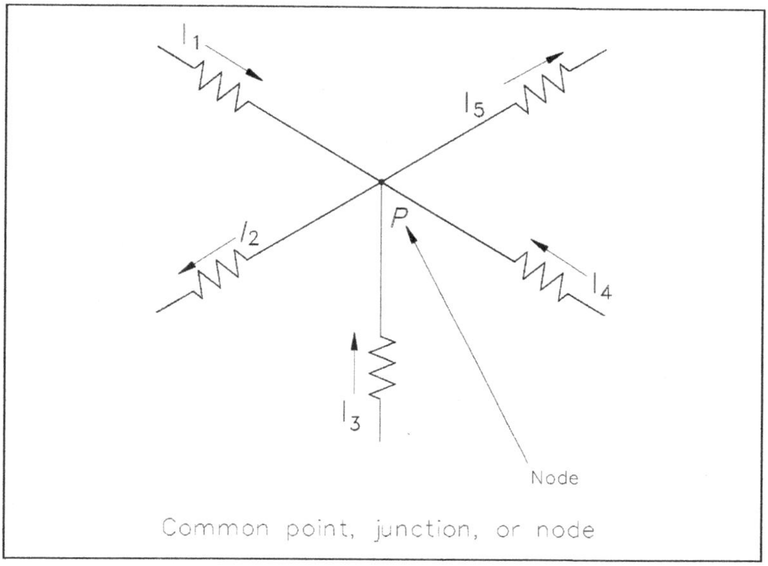

Figure 43 Node Point

By solving node equations, we can calculate the unknown node voltages. To each node in a circuit we will assign a letter or number. In Figure 44, A, B, C, and N are nodes, and N and C are principal nodes. Principal nodes are those nodes with three or more connections. Node C will be our selected reference node. V_{AC} is the voltage between Nodes A and C; V_{BC} is the voltage between Nodes B and C; and V_{NC} is the voltage between Nodes N and C. We have already determined that all node voltages have a reference node; therefore, we can substitute V_A for V_{AC}, V_B for V_{BC}, and V_N for V_{NC}.

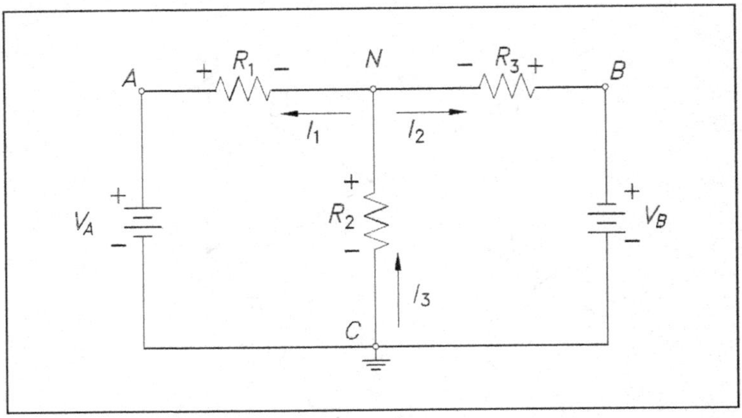

Figure 44 Circuit for Node Analysis

Assume that loop currents I_1 and I_2 leave Node N, and that I_3 enters Node N (Figure 44).

From Kirchhoff's current law:

$$\Sigma I = 0$$

$$I_1 + I_2 + I_3 = 0 \quad (2\text{-}20)$$

$$I_3 = I_1 + I_2$$

Using Ohm's Law and solving for the current through each resistor we obtain the following.

$I = \frac{V_R}{R}$ where V_R is the voltage across resistor, R.

$$I_3 = \frac{V_N}{R_2}$$

$$I_1 = \frac{V_A - V_N}{R_1}$$

$$I_2 = \frac{V_B - V_N}{R_3}$$

Substitute these equations for I_1, I_2, and I_3 into Kirchhoff's current equation (2-20) yields the following.

$$\frac{V_N}{R_2} = \frac{V_A - V_N}{R_1} + \frac{V_B - V_N}{R_3}$$

The circuit shown in Figure 45 can be solved for voltages and currents by using the node-voltage analysis.

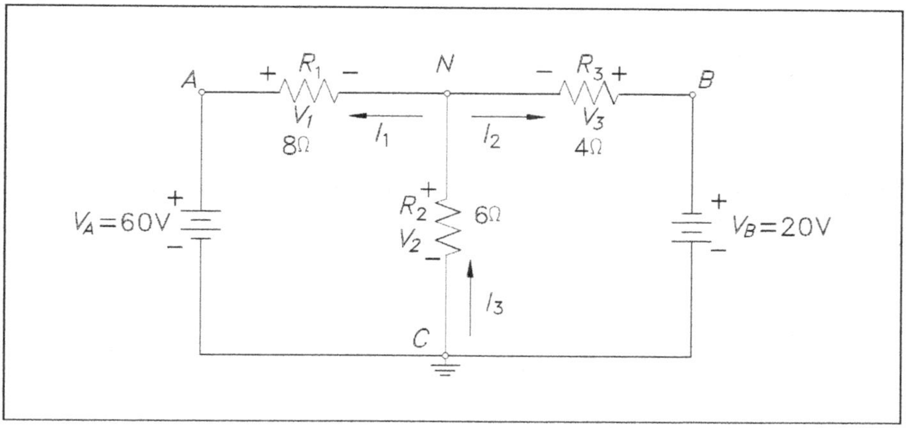

Figure 45 Node - Voltage Analysis

First, assume direction of current flow shown. Mark nodes A, B, C, and N, and mark the polarity across each resistor.

Second, using Kirchhoff's current law at Node N, solve for V_N.

$$I_3 = I_1 + I_2$$

$$\frac{V_N}{R_2} = \frac{V_A - V_N}{R_1} + \frac{V_B - V_N}{R_3}$$

$$\frac{V_N}{6} = \frac{60 - V_N}{8} + \frac{20 - V_N}{4}$$

Clear the fraction so that we have a common denominator:

$$4V_N = 3\,(60 - V_N) + 6\,(20 - V_N)$$
$$4V_N = 180 - 3V_N + 120 - 6V_N$$
$$13V_N = 300$$
$$V_N = 23.077$$

Third, find all voltage drops and currents.

$$V_1 = V_A - V_N = 60 - 23.077 = 36.923 \text{ Volts}$$
$$V_2 = V_N = 23.077 \text{ Volts}$$
$$V_3 = V_B - V_N = 20 - 23.077 = -3.077 \text{ Volts}$$

The negative value for V_3 shows that the current flow through R_3 is opposite that which was assumed and that the polarity across R_3 is reversed.

$$I_1 = \frac{V_1}{R_1} = \frac{36.923 \text{ V}}{8\Omega} = 4.65 \text{ amps}$$
$$I_2 = \frac{V_3}{R_3} = \frac{-3.077 \text{ V}}{4\Omega} = -0.769 \text{ amps}$$
$$I_3 = \frac{V_2}{R_2} = \frac{23.077 \text{ V}}{6\Omega} = 3.846 \text{ amps}$$

The negative value for I_3 shows that the current flow through R_3 is opposite that which was assumed.

Series-Parallel Circuit Analysis

When solving for voltage, current, and resistance in a series-parallel circuit, follow the rules which apply to the series part of the circuit, and follow the rules which apply to the parallel part of the circuit. Solving these circuits can be simplified by reducing the circuit to a single equivalent resistance circuit, and redrawing the circuit in simplified form. The circuit is then called an equivalent circuit (Figure 46).

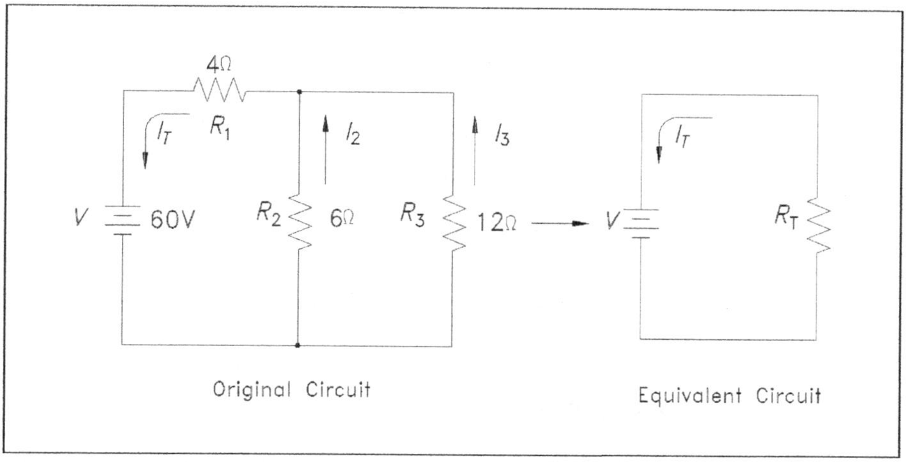

Figure 46 Redrawn Circuit Example

The easiest way to solve these types of circuits is to do it in steps.

Step 1: Find the equivalent resistance of the parallel branch:

$$R_p = \frac{R_2 R_3}{R_2 + R_3} = \frac{(6)(12)}{6 + 12} = \frac{72}{18} = 4\Omega$$

Step 2: Find the resistance of the equivalent series circuit:

$$R_T = R_1 + R_P = 4\Omega + 4\Omega = 8\Omega$$

Step 3: Find total current (I_T):

$$I_T = \frac{V}{R_T} = \frac{60\ V}{8\Omega} = 7.5 \text{ amps}$$

Step 4: Find I_2 and I_3. The voltage across R_1 and R_2 is equal to the applied voltage (V), minus the voltage drop across R_1.

$$V_2 = V_3 = V - I_T R_1 = 60 - (7.5 \text{ X } 4) = 30 \text{ V}$$

Then, I_2 and I_3 are calculated.

$$I_2 = \frac{V_2}{R_2} = \frac{30}{6} = 5 \text{ amps}$$

$$I_3 = \frac{V_3}{R_3} = \frac{30}{12} = 2.5 \text{ amps}$$

Y and Delta Network Calculation

Because of its shape, the network shown in Figure 47 is called a T (tee) or Y (wye) network. These are different names for the same network.

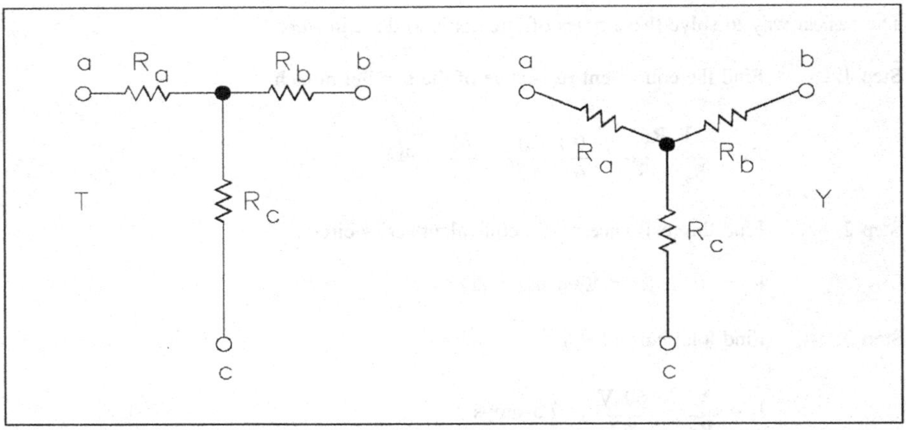

Figure 47 T or Y Network

The network shown in Figure 48 is called π (pi) or Δ (delta) because the shapes resemble Greek letters π and Ω. These are different names for the same network.

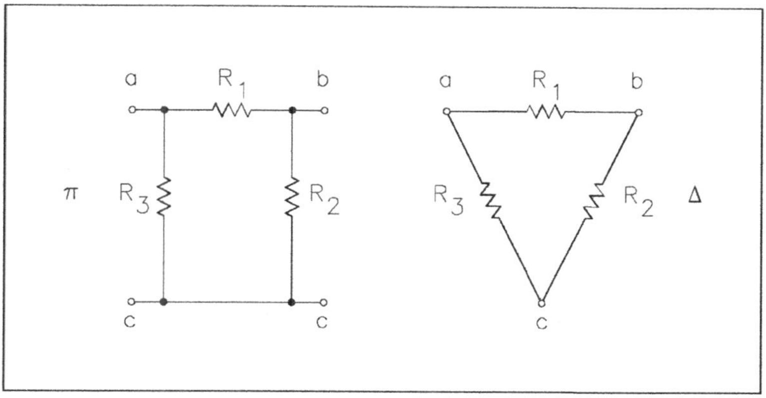

Figure 48 π (pi) or Δ (delta) Network

In order to analyze the circuits, it may be helpful to convert Y to Δ, or Δ to Y, to simplify the solution. The formulas that will be used for these conversions are derived from Kirchhoff's laws. The resistances in these networks are shown in a three-terminal network. After we use the conversion formulas, one network is equivalent to the other because they have equivalent resistances across any one pair of terminals (Figure 49).

Δ to Y conversion:

$$R_a = \frac{R_1 R_3}{R_1 + R_2 + R_3}$$

$$R_b = \frac{R_1 R_2}{R_1 + R_2 + R_3}$$

$$R_c = \frac{R_2 R_3}{R_1 + R_2 + R_3}$$

Rule 1: The resistance of any branch of a Y network is equal to the product of the two adjacent sides of a Δ network, divided by the sum of the three Δ resistances.

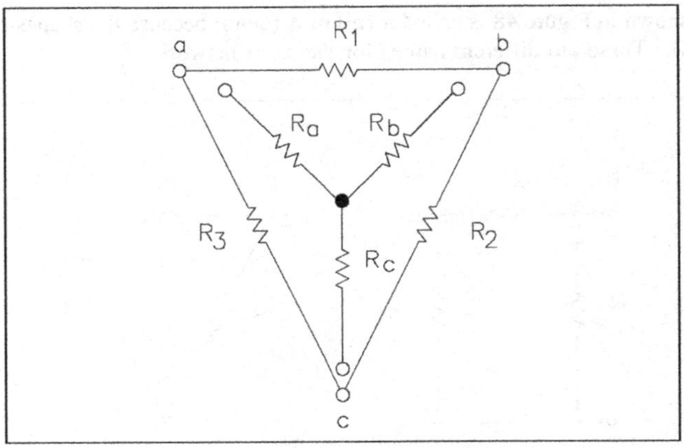

Figure 49 Y - Δ Equivalent

Y to Δ conversion:

$$R_1 = \frac{R_a R_b + R_b R_c + R_c R_a}{R_c}$$

$$R_2 = \frac{R_a R_b + R_b R_c + R_c R_a}{R_a}$$

$$R_3 = \frac{R_a R_b + R_b R_c + R_c R_a}{R_b}$$

Rule 2: **The resistance of any side of a Δ network is equal to the sum of the Y network resistance, multiplied in pairs, divided by the opposite branch of the Y network.**

Let us consider a bridge circuit (Figure 50).

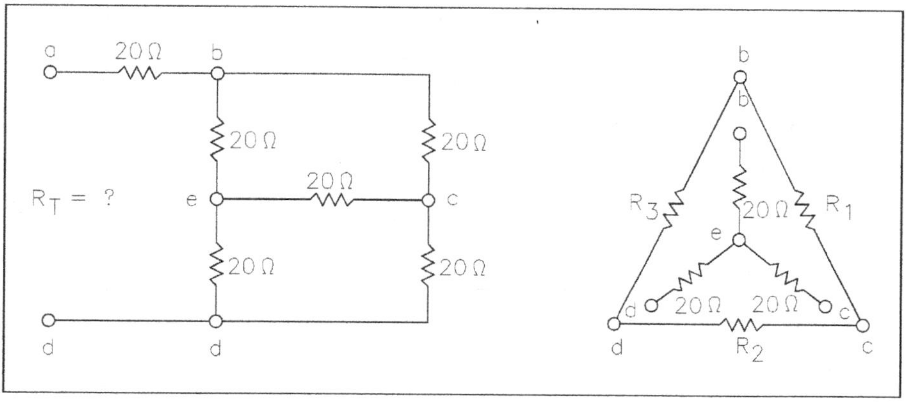

Figure 50 Bridge Circuit

Find R_t at terminals a and d.

Step 1: Convert the Y network (b-e, e-c, e-d) to the equivalent Δ network.

Using Rule 2:

$$R_1 = \frac{(20)(20) + (20)(20) + (20)(20)}{20} = \frac{1200}{20} = 60\Omega$$

$$R_2 = \frac{1200}{20} = 60\Omega$$

$$R_3 = \frac{1200}{20} = 60\Omega$$

Step 2: Now, we can redraw the Y circuit as a Δ circuit and reconnect it to the original circuit (Figure 51):

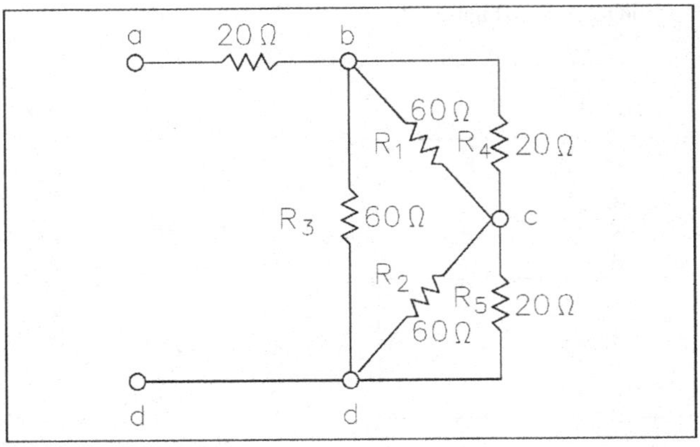

Figure 51 Y - Δ Redrawn Circuit

Step 3: Reduce and simplify the circuit. Note that the 20Ω and 60Ω branches are in parallel in Figure 51. Refer to Figures 51 and 52 for redrawing the circuit in each step below.

$$R_P = \frac{R_1 R_4}{R_1 + R_4} = \frac{(20)(60)}{20 + 60} = \frac{1200}{80} = 15\Omega$$

$$R_q = \frac{R_1 R_5}{R_1 + R_5} = \frac{(20)(60)}{20 + 60} = \frac{1200}{80} = 15\Omega$$

$$R_r = \frac{R_3(R_p + R_Q)}{R_3 + (R_p + R_Q)} = \frac{(60)(15+15)}{60 + 30} = \frac{1800}{90} = 20\Omega$$

$$R_T = 20 + 20 = 40\Omega$$

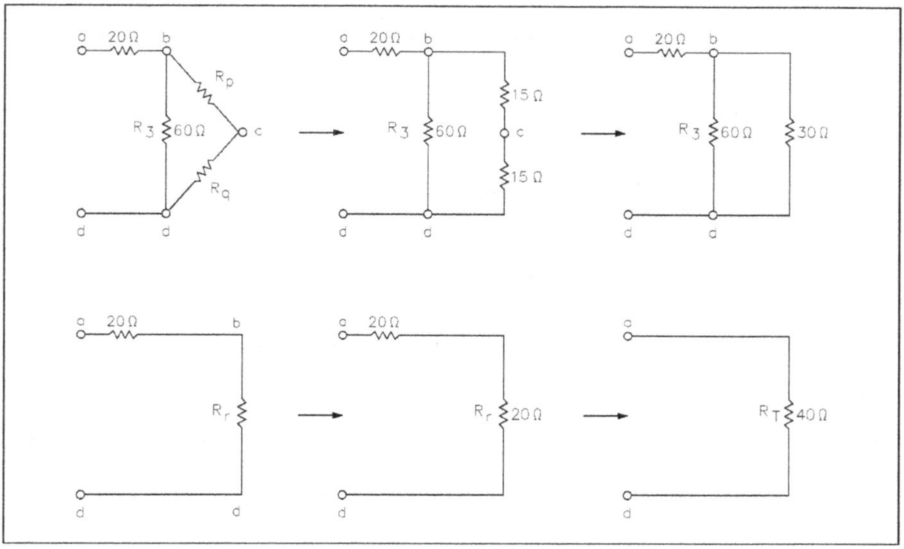

Figure 52 Steps to Simplify Redrawn Circuit

Summary

The important information in this chapter is summarized below.

DC Circuit Analysis Summary

- The current flow at any element in a DC circuit can be determined using loop equations.
- The voltage at any point in a DC circuit can be determined using node equations.
- The equivalent resistance of series and parallel combinations of elements can be used to simplify DC circuit analysis.

DC CIRCUIT FAULTS

Faults within a DC circuit will cause various effects, depending upon the nature of the fault. An understanding of the effects of these faults is necessary to fully understand DC circuit operation.

EO 1.16 DESCRIBE the voltage and current effects of an open in a DC circuit.

EO 1.17 DESCRIBE the voltage and current effects in a shorted DC circuit.

Open Circuit (Series)

A circuit must have a "complete" path for current flow, that is, from the negative side to the positive side of a power source. A series circuit has only one path for current to flow. If this path is broken, no current flows, and the circuit becomes an open circuit (Figure 53).

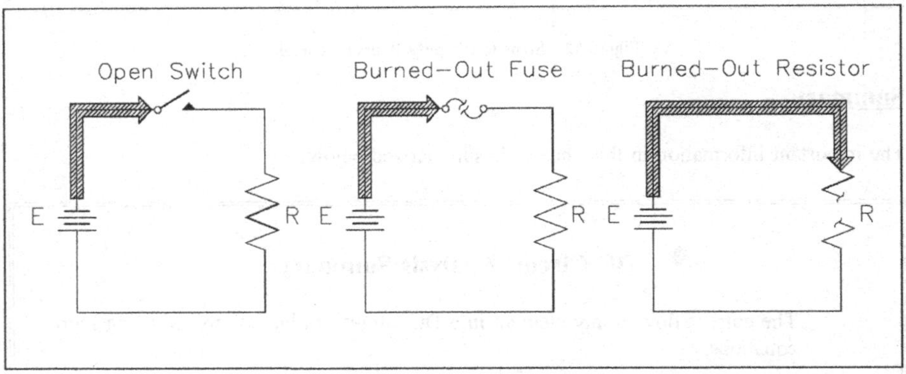

Figure 53 Open Series Circuit

Circuits can be opened deliberately, such as by the use of a switch, or they may be opened by a defect, such as a broken wire or a burned-out resistor.

Since no current flows in an open series circuit, there are no voltage drops across the loads. No power is consumed by the loads, and total power consumed by the circuit is zero.

Open Circuit (Parallel)

A parallel circuit has more than one path for current to flow. If one of the paths is opened, current will continue to flow as long as a complete path is provided by one or more of the remaining paths. It does not mean that you cannot stop current flow through a parallel circuit by opening it at one point; it means that the behavior of a parallel circuit depends on where the opening occurs (Figure 54).

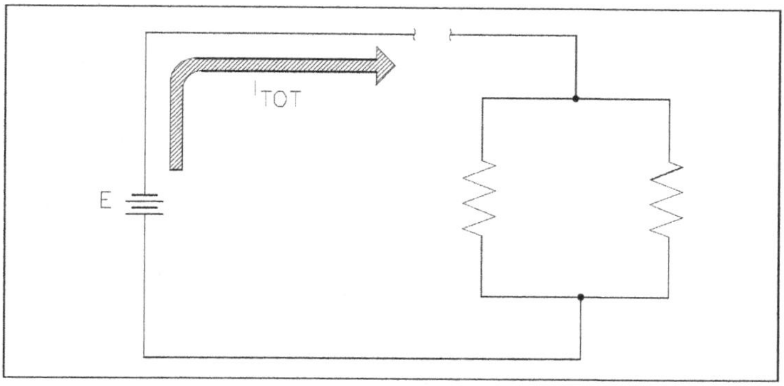

Figure 54 Open Parallel Circuit - Total

If a parallel circuit is opened at a point where only a branch current flows, then only that branch is open, and current continues to flow in the rest of the circuit (Figure 55).

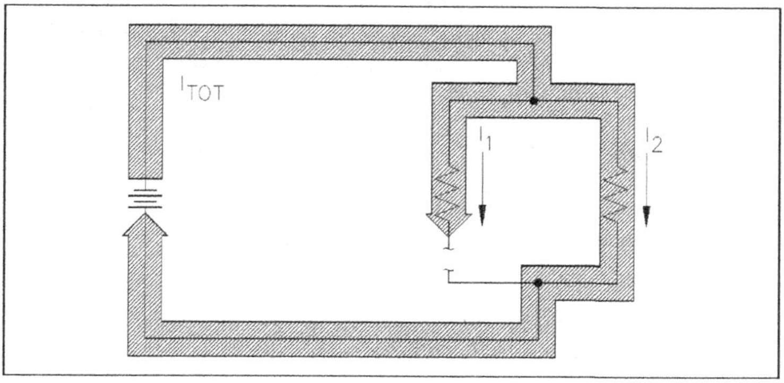

Figure 55 Open Parallel Circuit - Branch

Short Circuit (Series)

In a DC circuit, the only current limit is the circuit resistance. If there is no resistance in a circuit, or if the resistance suddenly becomes zero, a very large current will flow. This condition of very low resistance and high current flow is known as a "short circuit" (Figure 56).

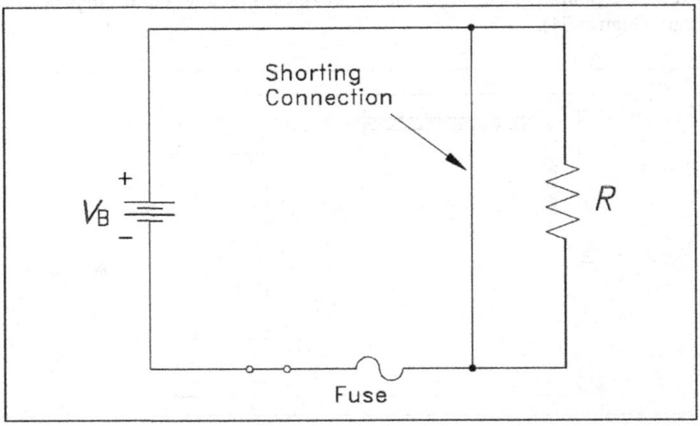

Figure 56 Shorted DC Circuit

A short circuit is said to exist if the circuit resistance is so low that current increases to a point where damage can occur to circuit components. With an increase in circuit current flow, the terminal voltage of the energy source will decrease. This occurs due to the internal resistance of the energy source causing an increased voltage drop within the energy source. The increased current flow resulting from a short circuit can damage power sources, burn insulation, and start fires. Fuses are provided in circuits to protect against short circuits.

Short Circuit (Parallel)

When a parallel circuit becomes short circuited, the same effect occurs as in a series circuit: there is a sudden and very large increase in circuit current (Figure 57).

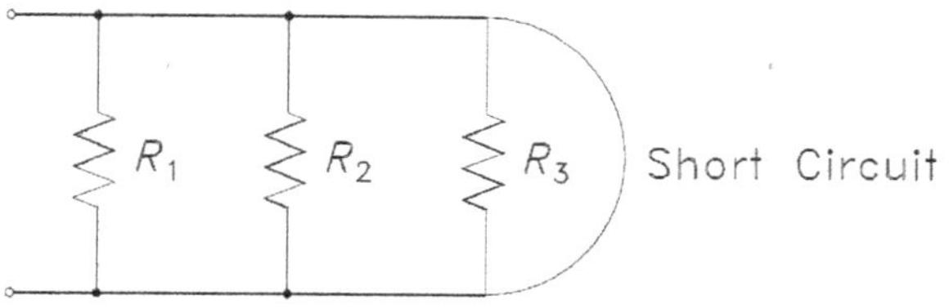

Figure 57 Shorted Parallel Circuit

Parallel circuits are more likely than series circuits to develop damaging short circuits. This is because each load is connected directly across the power source. If any of the load becomes shorted, the resistance between the power source terminals is practically zero. If a series load becomes shorted, the resistance of the other loads keeps the circuit resistance from dropping to zero.

Summary

The important information in this chapter is summarized below.

DC Circuit Faults Summary

- An open series DC circuit will result in no power being consumed by any of the loads.
- The effect of an open in a parallel circuit is dependent upon the location of the open.
- A shorted DC circuit will result in a sudden and very large increase in circuit current.

Intentionally Left Blank

DOE-HDBK-1011/2-92
JUNE 1992

DOE FUNDAMENTALS HANDBOOK

ELECTRICAL SCIENCE

Volume 2 of 4

U.S. Department of Energy **FSC-6910**

Washington, D.C. 20585

This document has been reproduced directly from the best available copy.

Available to DOE and DOE contractors from the Office of Scientific and Technical Information. P. O. Box 62, Oak Ridge, TN 37831; prices available from (615) 576-8401. FTS 626-8401.

Available to the public from the National Technical Information Service, U.S. Department of Commerce, 5285 Port Royal Rd., Springfield, VA 22161.

Order No. DE92019786

ABSTRACT

The *Electrical Science Fundamentals Handbook* was developed to assist nuclear facility operating contractors provide operators, maintenance personnel, and the technical staff with the necessary fundamentals training to ensure a basic understanding of electrical theory, terminology, and application. The handbook includes information on alternating current (AC) and direct current (DC) theory, circuits, motors, and generators; AC power and reactive components; batteries; AC and DC voltage regulators; transformers; and electrical test instruments and measuring devices. This information will provide personnel with a foundation for understanding the basic operation of various types of DOE nuclear facility electrical equipment.

Key Words: Training Material, Magnetism, DC Theory, DC Circuits, Batteries, DC Generators, DC Motors, AC Theory, AC Power, AC Generators, Voltage Regulators, AC Motors, Transformers, Test Instruments, Electrical Distribution

FOREWORD

The *Department of Energy (DOE) Fundamentals Handbooks* consist of ten academic subjects, which include Mathematics; Classical Physics; Thermodynamics, Heat Transfer, and Fluid Flow; Instrumentation and Control; Electrical Science; Material Science; Mechanical Science; Chemistry; Engineering Symbology, Prints, and Drawings; and Nuclear Physics and Reactor Theory. The handbooks are provided as an aid to DOE nuclear facility contractors.

These handbooks were first published as Reactor Operator Fundamentals Manuals in 1985 for use by DOE category A reactors. The subject areas, subject matter content, and level of detail of the Reactor Operator Fundamentals Manuals were determined from several sources. DOE Category A reactor training managers determined which materials should be included, and served as a primary reference in the initial development phase. Training guidelines from the commercial nuclear power industry, results of job and task analyses, and independent input from contractors and operations-oriented personnel were all considered and included to some degree in developing the text material and learning objectives.

The *DOE Fundamentals Handbooks* represent the needs of various DOE nuclear facilities' fundamental training requirements. To increase their applicability to nonreactor nuclear facilities, the Reactor Operator Fundamentals Manual learning objectives were distributed to the Nuclear Facility Training Coordination Program Steering Committee for review and comment. To update their reactor-specific content, DOE Category A reactor training managers also reviewed and commented on the content. On the basis of feedback from these sources, information that applied to two or more DOE nuclear facilities was considered generic and was included. The final draft of each of the handbooks was then reviewed by these two groups. This approach has resulted in revised modular handbooks that contain sufficient detail such that each facility may adjust the content to fit their specific needs.

Each handbook contains an abstract, a foreword, an overview, learning objectives, and text material, and is divided into modules so that content and order may be modified by individual DOE contractors to suit their specific training needs. Each subject area is supported by a separate examination bank with an answer key.

The *DOE Fundamentals Handbooks* have been prepared for the Assistant Secretary for Nuclear Energy, Office of Nuclear Safety Policy and Standards, by the DOE Training Coordination Program. This program is managed by EG&G Idaho, Inc.

OVERVIEW

The *Department of Energy Fundamentals Handbook* entitled *Electrical Science* was prepared as an information resource for personnel who are responsible for the operation of the Department's nuclear facilities. A basic understanding of electricity and electrical systems is necessary for DOE nuclear facility operators, maintenance personnel, and the technical staff to safely operate and maintain the facility and facility support systems. The information in the handbook is presented to provide a foundation for applying engineering concepts to the job. This knowledge will help personnel more fully understand the impact that their actions may have on the safe and reliable operation of facility components and systems.

The *Electrical Science* handbook consists of fifteen modules that are contained in four volumes. The following is a brief description of the information presented in each module of the handbook.

Volume 1 of 4

Module 1 - Basic Electrical Theory

This module describes basic electrical concepts and introduces electrical terminology.

Module 2 - Basic DC Theory

This module describes the basic concepts of direct current (DC) electrical circuits and discusses the associated terminology.

Volume 2 of 4

Module 3 - DC Circuits

This module introduces the rules associated with the reactive components of inductance and capacitance and how they affect DC circuits.

Module 4 - Batteries

This module introduces batteries and describes the types of cells used, circuit arrangements, and associated hazards.

Module 5 - DC Generators

This module describes the types of DC generators and their application in terms of voltage production and load characteristics.

Module 6 - DC Motors

This module describes the types of DC motors and includes discussions of speed control, applications, and load characteristics.

Volume 3 of 4

Module 7 - Basic AC Theory

This module describes the basic concepts of alternating current (AC) electrical circuits and discusses the associated terminology.

Module 8 - AC Reactive Components

This module describes inductance and capacitance and their effects on AC circuits.

Module 9 - AC Power

This module presents power calculations for single-phase and three-phase AC circuits and includes the power triangle concept.

Module 10 - AC Generators

This module describes the operating characteristics of AC generators and includes terminology, methods of voltage production, and methods of paralleling AC generation sources.

Module 11 - Voltage Regulators

This module describes the basic operation and application of voltage regulators.

Volume 4 of 4

Module 12 - AC Motors

This module explains the theory of operation of AC motors and discusses the various types of AC motors and their application.

Module 13 - Transformers

This module introduces transformer theory and includes the types of transformers, voltage/current relationships, and application.

Module 14 - Test Instruments and Measuring Devices

This module describes electrical measuring and test equipment and includes the parameters measured and the principles of operation of common instruments.

Module 15 - Electrical Distribution Systems

This module describes basic electrical distribution systems and includes characteristics of system design to ensure personnel and equipment safety.

The information contained in this handbook is by no means all encompassing. An attempt to present the entire subject of electrical science would be impractical. However, the *Electrical Science* handbook does present enough information to provide the reader with a fundamental knowledge level sufficient to understand the advanced theoretical concepts presented in other subject areas, and to better understand basic system and equipment operations.

Department of Energy
Fundamentals Handbook

ELECTRICAL SCIENCE
Module 3
DC Circuits

TABLE OF CONTENTS

LIST OF FIGURES ii

LIST OF TABLES iii

REFERENCES iv

OBJECTIVES v

INDUCTANCE 1

Inductors 1
Summary 8

CAPACITANCE 9

Capacitor 9
Capacitance 11
Types of Capacitors 12
Capacitors in Series and Parallel 13
Capacitive Time Constant 16
Summary 18

LIST OF FIGURES

Figure 1 Induced EMF 1

Figure 2 Induced EMF in Coils 2

Figure 3 Self-Induced EMF 2

Figure 4 Inductors in Series 3

Figure 5 Inductors in Parallel 4

Figure 6 DC Current Through an Inductor 4

Figure 7 Time Constant 5

Figure 8 Voltage Applied to an Inductor 6

Figure 9 Inductor and Resistor in Parallel 7

Figure 10 Capacitor and Symbols 9

Figure 11 Charging a Capacitor 10

Figure 12 Discharging a Capacitor 10

Figure 13 Capacitors Connected in Series 13

Figure 14 Capacitors Connected in Parallel 14

Figure 15 Example 1 - Capacitors Connected in Series 15

Figure 16 Example 2 - Capacitors Connected in Series 15

Figure 17 Example 3 - Capacitors Connected in Parallel 16

Figure 18 Capacitive Time Constant for Charging Capacitor 17

Figure 19 Capacitive Time Constant for Discharging Capacitor 17

Figure 20 Example - Capacitive Time Constant 18

LIST OF TABLES

Table 1 Types of Capacitors 13

REFERENCES

- Gussow, Milton, Schaum's Outline Series, Basic Electricity, McGraw-Hill.
- Academic Program for Nuclear Power Plant Personnel, Volume IV, Columbia, MD: General Physics Corporation, Library of Congress Card #A 326517, 1982.
- Academic Program for Nuclear Power Plant Personnel, Volume II, Columbia, MD: General Physics Corporation, Library of Congress Card #A 326517, 1982.
- Nasar and Unnewehr, Electromechanics and Electric Machines, John Wiley and Sons.
- Van Valkenburgh, Nooger, and Neville, Basic Electricity, Vol. 5, Hayden Book Company.
- Lister, Eugene C., Electric Circuits and Machines, 5th Edition, McGraw-Hill.
- Croft, Carr, Watt, and Summers, American Electricians Handbook, 10th Edition, McGraw-Hill.
- Mileaf, Harry, Electricity One - Seven, Revised 2nd Edition, Hayden Book Company.
- Buban and Schmitt, Understanding Electricity and Electronics, 3rd Edition, McGraw-Hill.
- Kidwell, Walter, Electrical Instruments and Measurements, McGraw-Hill.

TERMINAL OBJECTIVE

1.0 Using the rules associated with inductors and capacitors, **DESCRIBE** the characteristics of these elements when they are placed in a DC circuit.

ENABLING OBJECTIVES

1.1 **DESCRIBE** how current flow, magnetic field, and stored energy in an inductor relate to one another.

1.2 **DESCRIBE** how an inductor opposes a change in current flow.

1.3 Given a circuit containing inductors, **CALCULATE** total inductance for series and parallel circuits.

1.4 Given an inductive resistive circuit, **CALCULATE** the time constant for the circuit.

1.5 **DESCRIBE** the construction of a capacitor.

1.6 **DESCRIBE** how a capacitor stores energy.

1.7 **DESCRIBE** how a capacitor opposes a change in voltage.

1.8 Given a circuit containing capacitors, **CALCULATE** total capacitance for series and parallel circuits.

1.9 Given a circuit containing capacitors and resistors, **CALCULATE** the time constant of the circuit.

Intentionally Left Blank

INDUCTANCE

Experiments investigating the unique behavioral characteristics of inductance led to the invention of the transformer.

EO 1.1 **DESCRIBE how current flow, magnetic field, and stored energy in an inductor relate to one another.**

EO 1.2 **DESCRIBE how an inductor opposes a change in current flow.**

EO 1.3 **Given a circuit containing inductors, CALCULATE total inductance for series and parallel circuits.**

EO 1.4 **Given an inductive resistive circuit, CALCULATE the time constant for the circuit.**

Inductors

An inductor is a circuit element that will store electrical energy in the form of a magnetic field. It is usually a coil of wire wrapped around a core of permeable material. The magnetic field is generated when current is flowing through the wire. If two circuits are arranged as in Figure 1, a magnetic field is generated around Wire A, but there is no electromotive force (EMF) induced into Wire B because there is no relative motion between the magnetic field and Wire B.

If we now open the switch, the current stops flowing in Wire A, and the magnetic field collapses. As the field collapses, it moves relative to Wire B. When this occurs, an EMF is induced in Wire B.

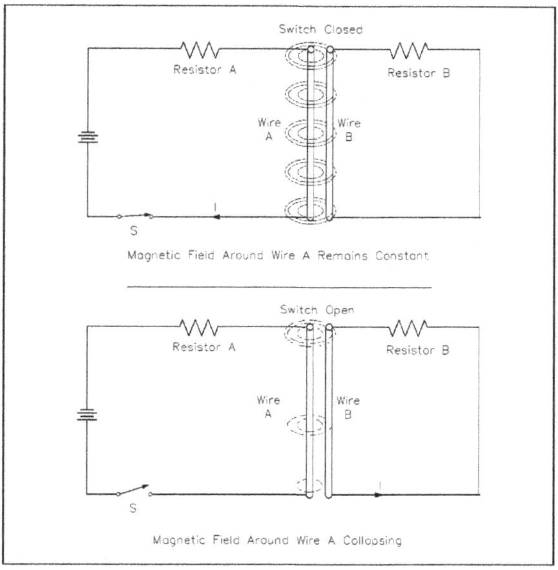

Figure 1 Induced EMF

This is an example of Faraday's Law, which states that a voltage is induced in a conductor when that conductor is moved through a magnetic field, or when the magnetic field moves past the conductor. When the EMF is induced in Wire B, a current will flow whose magnetic field opposes the change in the magnetic field that produced it.

For this reason, an induced EMF is sometimes called counter EMF or CEMF. This is an example of Lenz's Law, which states that the induced EMF opposes the EMF that caused it.

The three requirements for inducing an EMF are:

1. a conductor,
2. a magnetic field, and
3. relative motion between the two.

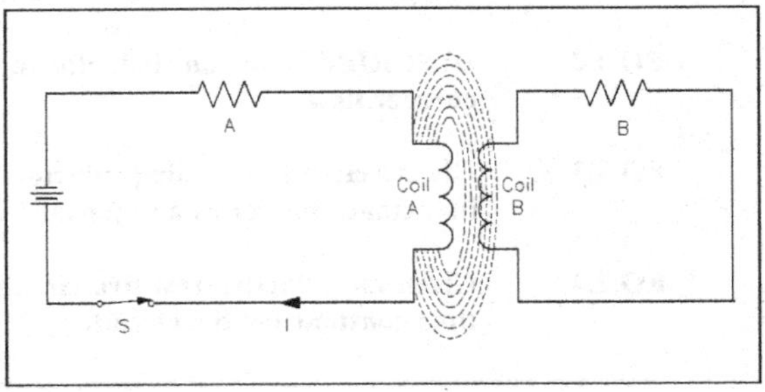

Figure 2 Induced EMF in Coils

The faster the conductor moves, or the faster the magnetic field collapses or expands, the greater the induced EMF. The induction can also be increased by coiling the wire in either Circuit A or Circuit B, or both, as shown in Figure 2.

Self-induced EMF is another phenomenon of induction. The circuit shown in Figure 3 contains a coil of wire called an inductor (L). As current flows through the circuit, a large magnetic field is set up around the coil. Since the current is not changing, there is no EMF produced. If we open the switch, the field around the inductor collapses. This collapsing magnetic field produces a voltage in the coil. This is called self-induced EMF.

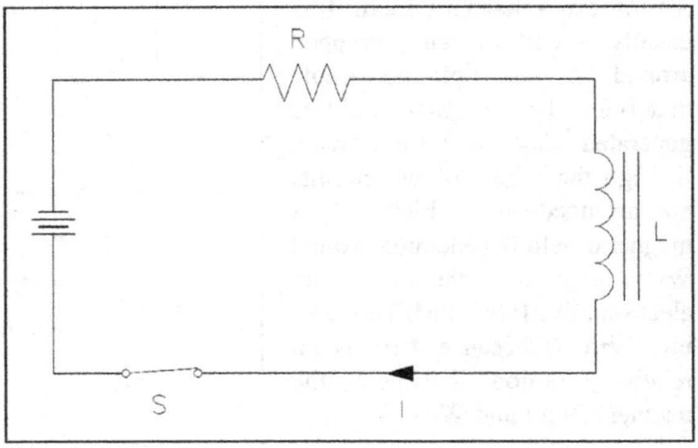

Figure 3 Self-Induced EMF

The polarity of self-induced EMF is given to us by Lenz's Law. The polarity is in the direction that opposes the change in the magnetic field that induced the EMF. The result is that the current caused by the induced EMF tends to maintain the same current that existed in the circuit before the switch was opened. It is commonly said that an inductor tends to oppose a change in current flow.

The induced EMF, or counter EMF, is proportional to the time rate of change of the current. The proportionality constant is called the "inductance" (L). Inductance is a measure of an inductor's ability to induce CEMF. It is measured in henries (H). An inductor has an inductance of one henry if one amp per second change in current produces one volt of CEMF, as shown in Equation (3-1).

$$\text{CEMF} = -L \frac{\Delta I}{\Delta t} \tag{3-1}$$

where

CEMF = induced voltage (volts)

L = inductance (henries)

$\frac{\Delta I}{\Delta t}$ = time rate of change of current (amp/sec)

The minus sign shows that the CEMF is opposite in polarity to the applied voltage.

Example: A 4-henry inductor is in series with a variable resistor. The resistance is increased so that the current drops from 6 amps to 2 amps in 2 seconds. What is the CEMF induced?

$$\text{CEMF} = -L \frac{\Delta I}{\Delta t}$$

$$= -4\left(\frac{2A - 6A}{2}\right)$$

$$= -4(-2)$$

$$\text{CEMF} = +8 \text{ volts}$$

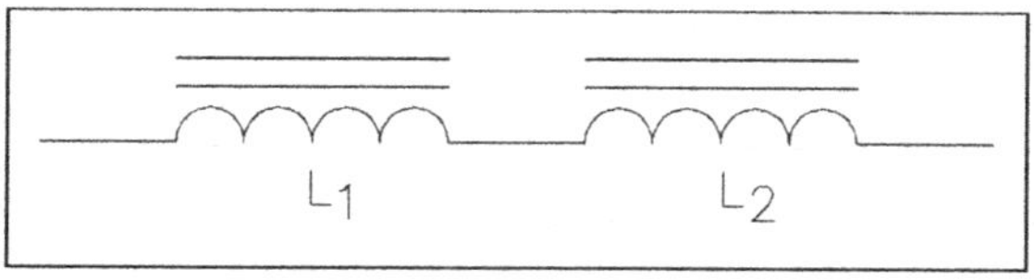

Figure 4 Inductors in Series

Inductors in series are combined like resistors in series. Equivalent inductance (L_{eq}) of two inductors in series (Figure 4) is given by Equation (3-2).

$$L_{eq} = L_1 + L_2 + \ldots L_n \tag{3-2}$$

Inductors in parallel are combined like resistors in parallel as given by Equation (3-3).

$$\frac{1}{L_{eq}} = \frac{1}{L_1} + \frac{1}{L_2} + \ . \ . \ . \ \frac{1}{L_N} \qquad (3\text{-}3)$$

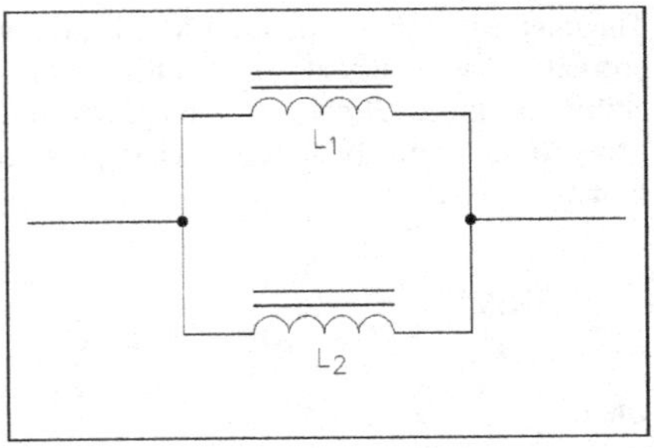

Figure 5 Inductors in Parallel

When only two inductors are in parallel, as shown in Figure 5, Equation (3-3) may be simplified as given in Equation (3-4). As shown in Equation (3-4), this is valid when there are only two inductors in parallel.

$$\frac{1}{L_{eq}} = \frac{L_1 \, L_2}{L_1 + L_2} \qquad (3\text{-}4)$$

Inductors will store energy in the form of a magnetic field. Circuits containing inductors will behave differently from a simple resistance circuit. In circuits with elements that store energy, it is common for current and voltage to exhibit exponential increase and decay (Figure 6).

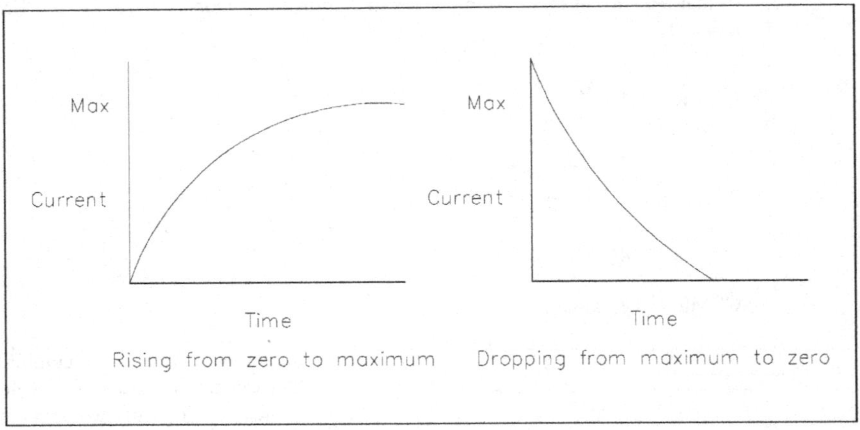

Figure 6 DC Current Through an Inductor

The relationship between values of current reached and the time it takes to reach them is called a time constant. The time constant for an inductor is defined as the time required for the current either to increase to 63.2 percent of its maximum value or to decrease by 63.2 percent of its maximum value (Figure 7).

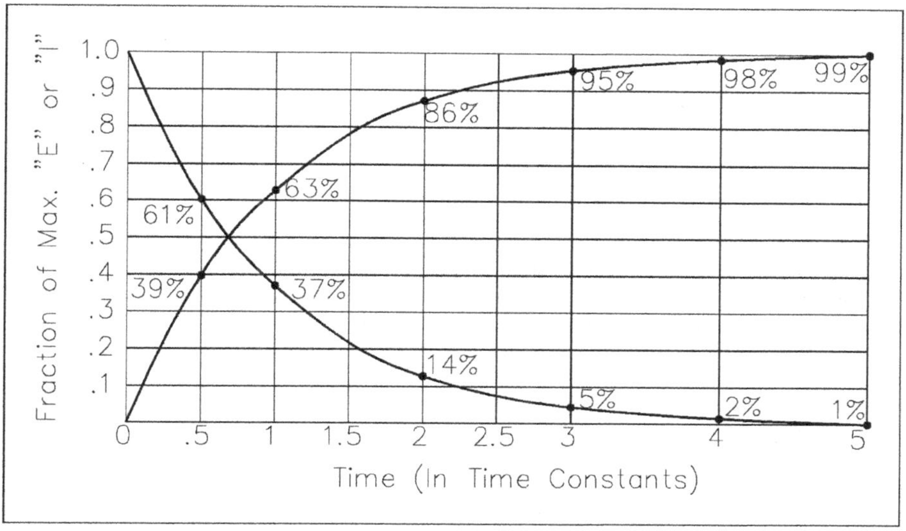

Figure 7 Time Constant

The value of the time constant is directly proportional to the inductance and inversely proportional to the resistance. If these two values are known, the time constant can be found using Equation (3-5).

$$T_L = \frac{L}{R} \tag{3-5}$$

where

T_L = time constant (seconds)
L = inductance (henries)
R = resistance (ohms)

The voltage drop across an inductor is directly proportional to the product of the inductance and the time rate of change of current through the inductor, as shown in Equation (3-6).

$$V_L = L \frac{\Delta I}{\Delta t} \tag{3-6}$$

where

V_L = voltage drop across the inductor (volts)

L = inductance (henries)

$\frac{\Delta I}{\Delta t}$ = time rate of change of current (amp/sec)

After five time constants, circuit parameters normally reach their final value. Circuits that contain both inductors and resistors are called RL circuits. The following example will illustrate how an RL circuit reacts to changes in the circuit (Figure 8).

1. Initially, the switch is in Position 1, and no current flows through the inductor.

2. When we move the switch to Position 2, the battery attempts to force a current of 10v/100Ω = 0.1A through the inductor. But as current begins to flow, the inductor generates a magnetic field. As the field increases, a counter EMF is induced that opposes the battery voltage. As a steady state is reached, the counter EMF goes to zero exponentially.

3. When the switch is returned to Position 1, the magnetic field collapses, inducing an EMF that tends to maintain current flow in the same direction through the inductor. Its polarity will be opposite to that induced when the switch was placed in Position 2.

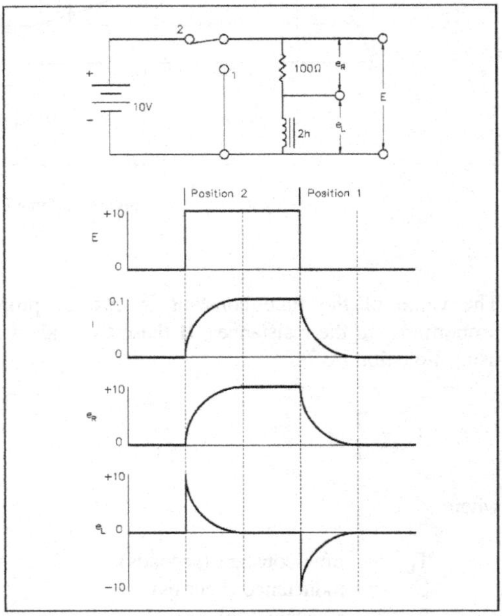

Figure 8 Voltage Applied to an Inductor

The example that follows shows how a circuit with an inductor in parallel with a resistor reacts to changes in the circuit. Inductors have some small resistance, and this is shown schematically as a 1Ω resistor (Figure 9).

1. While the switch is closed, a current of 20 v/1Ω = 20 amps flows through the inductor. This causes a very large magnetic field around the inductor.

2. When we open the switch, there is no longer a current through the inductor. As the magnetic field begins to collapse, a voltage is induced in the inductor. The change in applied voltage is instantaneous; the counter EMF is of exactly the right magnitude to prevent the current from changing initially. In order to maintain the current at 20 amps flowing through the inductor, the self-induced voltage in the inductor must be enough to push 20 amps through the 101Ω of resistance. The CEMF = (101)(20) = 2020 volts.

3. With the switch open, the circuit looks like a series RL circuit without a battery. The CEMF induced falls off, as does the current, with a time constant T_L of:

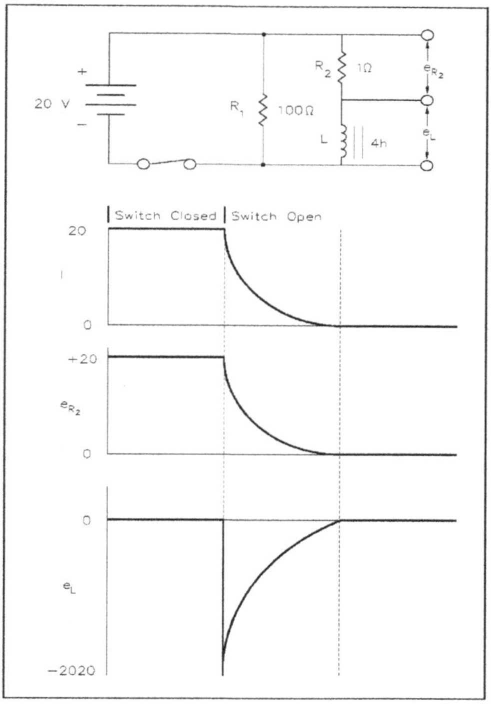

Figure 9 Inductor and Resistor in Parallel

$$T_L = \frac{L}{R}.$$

$$T_L = \frac{4H}{101\Omega} = 0.039 \text{ sec}$$

Summary

The important information on inductors is summarized below.

Inductance Summary

- When an inductor has a DC current flowing through it, the inductor will store energy in the form of a magnetic field.
- An inductor will oppose a change in current flow by the CEMF induced when the field collapses or expands.
- Inductors in series are combined like resistors in series.
- Inductors in parallel are combined like resistors in parallel.
- The time constant for an inductor is defined as the required time for the current either to increase to 63.2 percent of its maximum value or to decrease by 63.2 percent of its maximum value.

CAPACITANCE

Because of the effect of capacitance, an electrical circuit can store energy, even after being de-energized.

EO 1.5	**DESCRIBE the construction of a capacitor.**
EO 1.6	**DESCRIBE how a capacitor stores energy.**
EO 1.7	**DESCRIBE how a capacitor opposes a change in voltage.**
EO 1.8	**Given a circuit containing capacitors, CALCULATE total capacitance for series and parallel circuits.**
EO 1.9	**Given a circuit containing capacitors and resistors, CALCULATE the time constant of the circuit.**

Capacitor

Electrical devices that are constructed of two metal plates separated by an insulating material, called a *dielectric*, are known as capacitors (Figure 10a). Schematic symbols shown in Figures 10b and 10c apply to all capacitors.

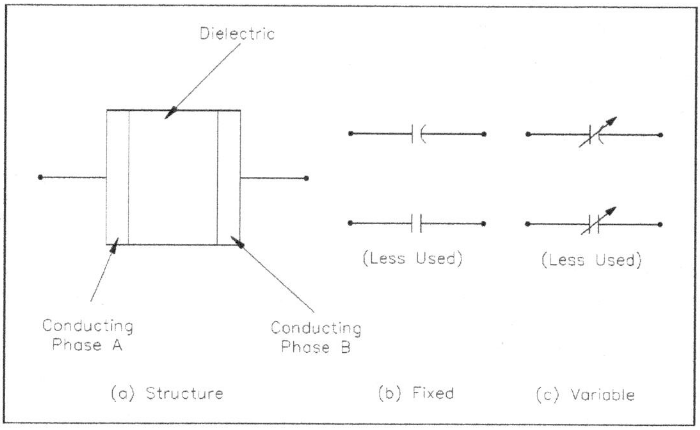

Figure 10 Capacitor and Symbols

The two conductor plates of the capacitor, shown in Figure 11a, are electrically neutral, because there are as many positive as negative charges on each plate. The capacitor, therefore, has no charge.

Now, we connect a battery across the plates (Figure 11b). When the switch is closed (Figure 11c), the negative charges on Plate A are attracted to the positive side of the battery, while the positive charges on Plate B are attracted to the negative side of the battery. This movement of charges will continue until the difference in charge between Plate A and Plate B is equal to the voltage of the battery. This is now a "charged capacitor." Capacitors store energy as an electric field between the two plates.

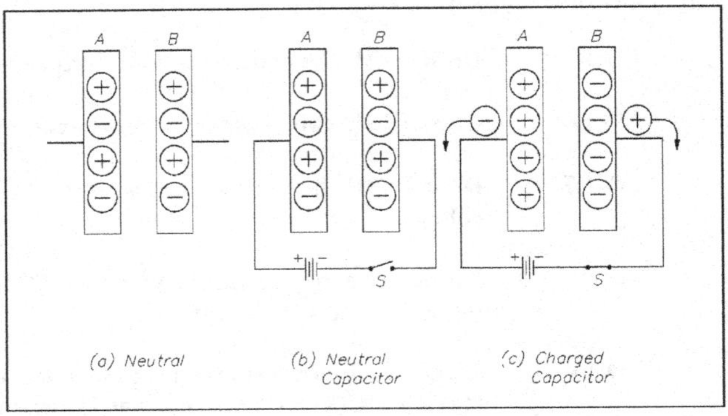

Figure 11 Charging a Capacitor

Because very few of the charges can cross between the plates, the capacitor will remain in the charged state even if the battery is removed. Because the charges on the opposing plates are attracted by one another, they will tend to oppose any changes in charge. In this manner, a capacitor will oppose any change in voltage felt across it.

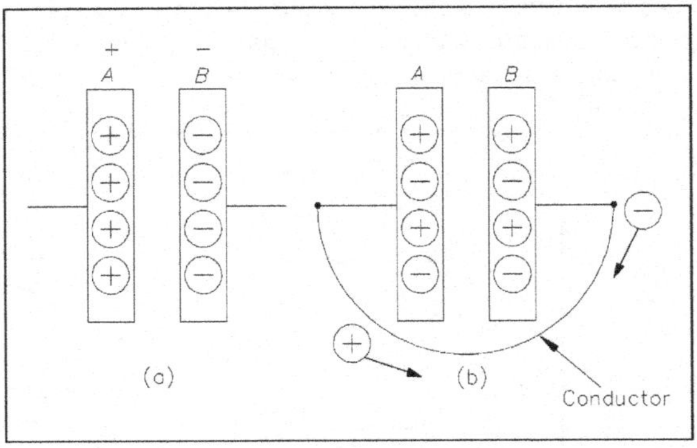

Figure 12 Discharging a Capacitor

If we place a conductor across the plates, electrons will find a path back to Plate A, and the charges will be neutralized again. This is now a "discharged" capacitor (Figure 12).

Capacitance

Capacitance is the ability to store an electrical charge. Capacitance is equal to the amount of charge that can be stored divided by the applied voltage, as shown in Equation (3-7).

$$C = \frac{Q}{V} \qquad (3\text{-}7)$$

where

C = capacitance (F)
Q = amount of charge (C)
V = voltage (V)

The unit of capacitance is the farad (F). A farad is the capacitance that will store one coulomb of charge when one volt is applied across the plates of the capacitor.

The dielectric constant (K) describes the ability of the dielectric to store electrical energy. Air is used as a reference and is given a dielectric constant of 1. Therefore, the dielectric constant is unitless. Some other dielectric materials are paper, teflon, bakelite, mica, and ceramic.

The capacitance of a capacitor depends on three things.

1. Area of conductor plates

2. Separation between the plates

3. Dielectric constant of insulation material

Equation (3-8) illustrates the formula to find the capacitance of a capacitor with two parallel plates.

$$C = K \frac{A}{d} (8.85 \times 10^{-12}) \qquad (3\text{-}8)$$

where

C = capacitance
K = dielectric constant
A = area
d = distance between the plates
8.85×10^{-12} = constant of proportionality

Example 1: Find the capacitance of a capacitor that stores 8 C of charge at 4 V.

$$C = \frac{Q}{V}$$

$$C = \frac{8}{4}$$

$$C = 2F$$

Example 2: What is the charge taken on by a 5F capacitor at 2 volts?

$$Q = C\ V$$

$$Q = (5F)\ (2V)$$

$$Q = 10C$$

Example 3: What is the capacitance if the area of a two plate mica capacitor is 0.0050 m^2 and the separation between the plates is 0.04 m? The dielectric constant for mica is 7.

$$C = K\ \frac{A}{d}\ (8.85 \times 10^{-12})$$

$$C = 7\ \frac{0.0050}{0.04}\ (8.85 \times 10^{-12})$$

$$C = 7.74 \times 10^{-12} F$$

$$C = 7.74\ pF$$

Types of Capacitors

All commercial capacitors are named according to their dielectrics. The most common are air, mica, paper, and ceramic capacitors, plus the electrolytic type. These types of capacitors are compared in Table 1.

TABLE 1
Types of Capacitors

Dielectric	Construction	Capacitance Range
Air	Meshed plates	10 - 400 pF
Mica	Stacked Sheets	10 - 5000 pF
Paper	Rolled foil	0.001 - 1 μF
Ceramic	Tubular	0.5 - 1600 pF
Disk	Tubular	0.002 - 0.1 μF
Electrolytic	Aluminum	5 - 1000 μF
Tantalum	Aluminum	0.01 - 300 μF

Capacitors in Series and Parallel

Capacitors in series are combined like resistors in parallel. The total capacitance, C_T, of capacitors connected in series (Figure 13), is shown in Equation (3-9).

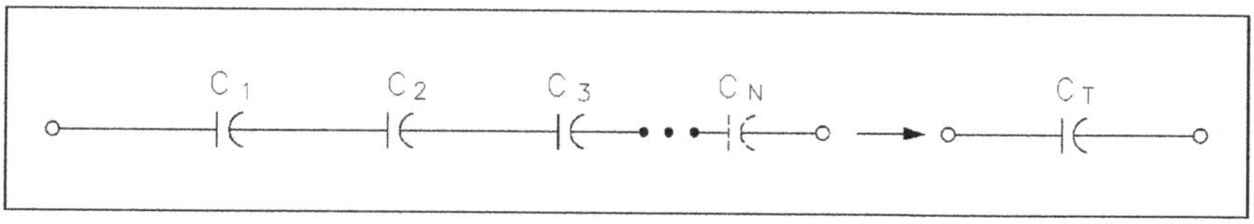

Figure 13 Capacitors Connected in Series

$$\frac{1}{C_T} = \frac{1}{C_1} + \frac{1}{C_2} + \frac{1}{C_3} + \ldots \frac{1}{C_N} \tag{3-9}$$

When only two capacitors are in series, Equation (3-9) may be simplified as given in Equation (3-10). As shown in Equation (3-10), this is valid when there are only two capacitors in series.

$$C_T = \frac{C_1 C_2}{C_1 + C_2} \tag{3-10}$$

When all the capacitors in series are the same value, the total capacitance can be found by dividing the capacitor's value by the number of capacitors in series as given in Equation (3-11).

$$C_T = \frac{C}{N} \tag{3-11}$$

where

C = value of any capacitor in series

N = the number of capacitors in series with the same value.

Capacitors in parallel are combined like resistors in series. When capacitors are connected in parallel (Figure 14), the total capacitance, C_T, is the sum of the individual capacitances as given in Equation (3-12).

$$C_T = C_1 + C_2 + C_3 + \ldots + C_N \tag{3-12}$$

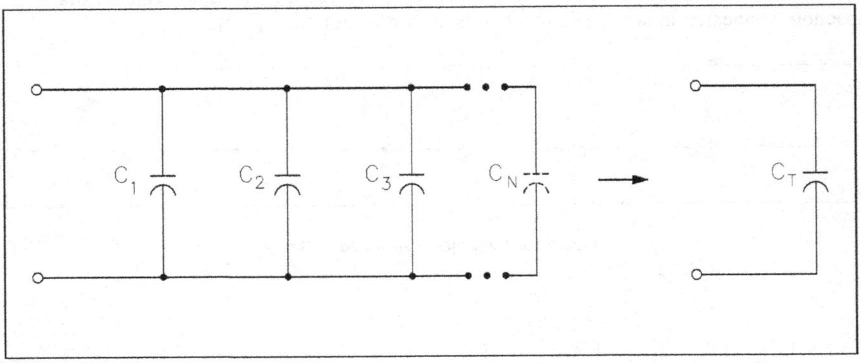

Figure 14 Capacitors Connected in Parallel

Example 1: Find the total capacitance of 3μF, 6μF, and 12μF capacitors connected in series (Figure 15).

$$\frac{1}{C_T} = \frac{1}{C_1} + \frac{1}{C_2} + \frac{1}{C_3}$$

$$= \frac{1}{3} + \frac{1}{6} + \frac{1}{12}$$

$$= \frac{4}{12} + \frac{2}{12} + \frac{1}{12}$$

$$= \frac{7}{12}$$

$$C_T = \frac{12}{7} = 1.7\mu f$$

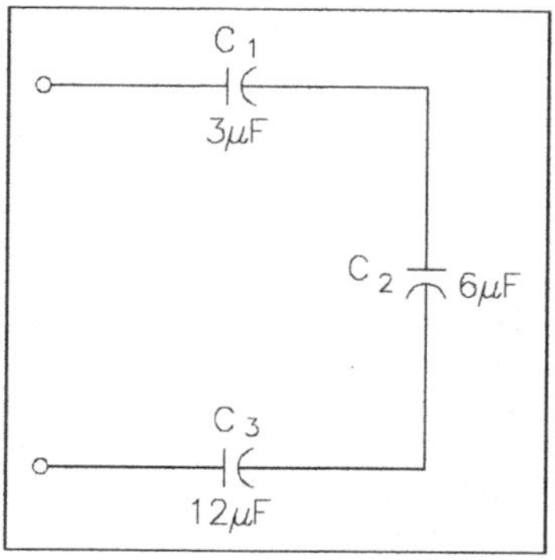

Figure 15 Example 1 - Capacitors Connected in Series

Example 2: Find the total capacitance and working voltage of two capacitors in series, when both have a value of 150 μF, 120 V (Figure 16).

$$C_T = \frac{C}{N}$$

$$= \frac{150}{2}$$

$$C_T = 75\mu f$$

Total voltage that can be applied across a group of capacitors in series is equal to the sum of the working voltages of the individual capacitors.

working voltage = 120 V + 120 V = 240 volts

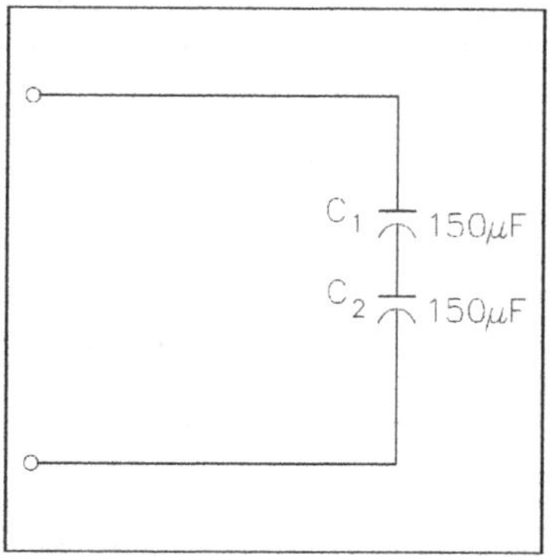

Figure 16 Example 2 - Capacitors Connected in Series

Example 3: Find the total capacitance of three capacitors in parallel, if the values are 15 μF-50 V, 10 μF-100 V, and 3 μF-150 V (Figure 17). What would be the working voltage?

$$C_T = C_1 + C_2 + C_3$$

$$= 15\mu F + 10\mu F + 3\mu F$$

$$C_T = 28\mu F$$

The working voltage of a group of capacitors in parallel is only as high as the lowest working voltage of an individual capacitor. Therefore, the working voltage of this combination is only 50 volts.

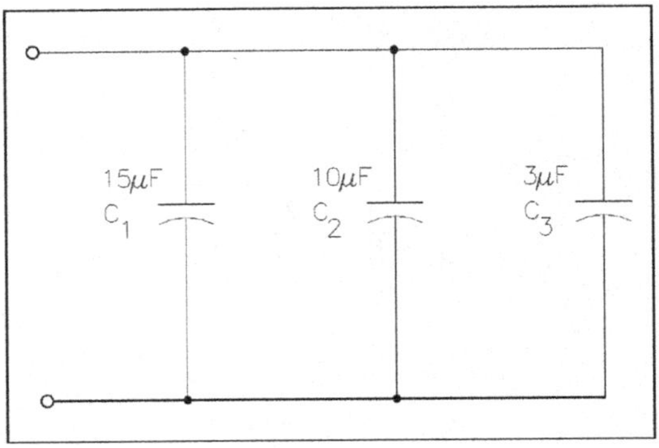

Figure 17 Example 3 - Capacitors Connected in Parallel

Capacitive Time Constant

When a capacitor is connected to a DC voltage source, it charges very rapidly. If no resistance was present in the charging circuit, the capacitor would become charged almost instantaneously. Resistance in a circuit will cause a delay in the time for charging a capacitor. The exact time required to charge a capacitor depends on the resistance (R) and the capacitance (C) in the charging circuit. Equation (3-13) illustrates this relationship.

$$T_C = RC \qquad (3\text{-}13)$$

where

T_C = capacitive time constant (sec)
R = resistance (ohms)
C = capacitance (farad)

The capacitive time constant is the time required for the capacitor to charge to 63.2 percent of its fully charged voltage. In the following time constants, the capacitor will charge an additional 63.2 percent of the remaining voltage. The capacitor is considered fully charged after a period of five time constants (Figure 18).

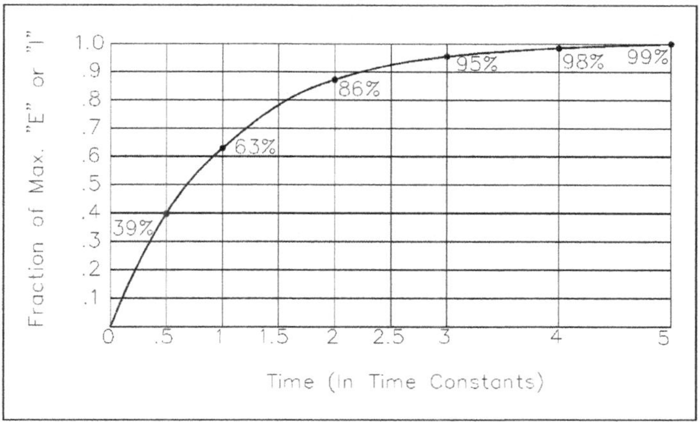

Figure 18 Capacitive Time Constant for Charging Capacitor

The capacitive time constant also shows that it requires five time constants for the voltage across a discharging capacitor to drop to its minimum value (Figure 19).

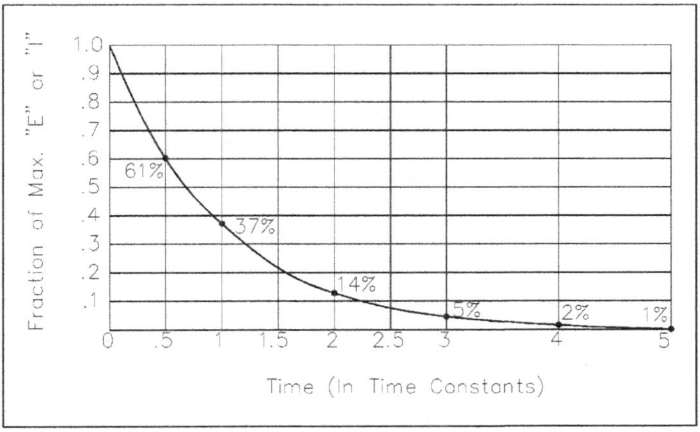

Figure 19 Capacitive Time Constant for Discharging Capacitor

Example: Find the time constant of a 100 μF capacitor in series with a 100Ω resistor (Figure 20).

$$T_C = RC$$

$$T_C = (100\Omega)(100\,\mu F)$$

$$T_C = 0.01 \text{ seconds}$$

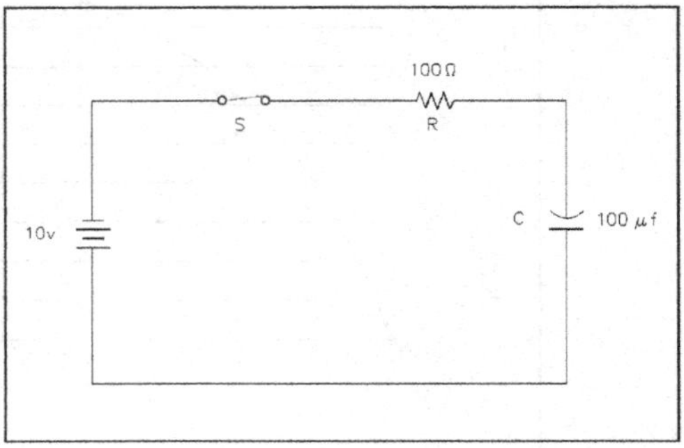

Figure 20 Example - Capacitive Time Constant

Summary

The important information on capacitors is summarized below.

Capacitance Summary

- A capacitor is constructed of two conductors (plates) separated by a dielectric.
- A capacitor will store energy in the form of an electric field caused by the attraction of the positively-charged particles in one plate to the negatively-charged particles in the other plate.
- The attraction of charges in the opposite plates of a capacitor opposes a change in voltage across the capacitor.
- Capacitors in series are combined like resistors in parallel.
- Capacitors in parallel are combined like resistors in series.
- The capacitive time constant is the time required for the capacitor to charge (or discharge) to 63.2 percent of its fully charged voltage.

Department of Energy
Fundamentals Handbook

ELECTRICAL SCIENCE
Module 4
Batteries

TABLE OF CONTENTS

LIST OF FIGURES iii

LIST OF TABLES iv

REFERENCES v

OBJECTIVES vi

BATTERY TERMINOLOGY 1

Voltaic Cell 1
Battery 1
Electrode 1
Electrolyte 1
Specific Gravity 1
Ampere-Hour 2
Summary 3

BATTERY THEORY 4

Batteries 4
Discharge and Charging of Lead-Acid Battery 6
Summary 9

BATTERY OPERATIONS 10

Series Cells 10
Parallel Cells 11
Primary Cell 11
Secondary Cells 11
Capacity 11
Internal Resistance 12
Shelf Life 13
Charge and Discharge 13
Summary 13

TABLE OF CONTENTS (Cont.)

TYPES OF BATTERIES 14

Wet and Dry Cells 14
Carbon-Zinc Cell 14
Alkaline Cell 14
Nickel-Cadmium Cell 15
Edison Cell 15
Mercury Cell 15
Summary 16

BATTERY HAZARDS 17

Shorted Cell 17
Gas Generation 17
Battery Temperature 18
Summary 19

LIST OF FIGURES

Figure 1 Simple Hydrometer 2

Figure 2 Basic Chemical Production of Electrical Power 5

Figure 3 Electron Flow Through a Battery 5

Figure 4 Chemical Action During Discharge 6

Figure 5 Chemical Action During Charging 7

Figure 6 Voltage and Specific Gravity During Charge and Discharge 8

Figure 7 Cells Connected in Series 10

Figure 8 Cells Connected in Parallel 11

Figure 9 Internal Resistance in a Chemical Cell 12

Figure 10 Internal Voltage Drop 12

LIST OF TABLES

NONE

REFERENCES

- Gussow, Milton, Schaum's Outline Series, Basic Electricity, McGraw-Hill.
- Academic Program for Nuclear Power Plant Personnel, Volume IV, Columbia, MD: General Physics Corporation, Library of Congress Card #A 326517, 1982.
- Academic Program for Nuclear Power Plant Personnel, Volume II, Columbia, MD: General Physics Corporation, Library of Congress Card #A 326517, 1982.
- Nasar and Unnewehr, Electromechanics and Electric Machines, John Wiley and Sons.
- Van Valkenburgh, Nooger, and Neville, Basic Electricity, Vol. 5, Hayden Book Company.
- Lister, Eugene C., Electric Circuits and Machines, 5th Edition, McGraw-Hill.
- Croft, Carr, Watt, and Summers, American Electricians Handbook, 10th Edition, McGraw-Hill.
- Mileaf, Harry, Electricity One - Seven, Revised 2nd Edition, Hayden Book Company.
- Buban and Schmitt, Understanding Electricity and Electronics, 3rd Edition, McGraw-Hill.
- Kidwell, Walter, Electrical Instruments and Measurements, McGraw-Hill.

TERMINAL OBJECTIVE

1.0 **DESCRIBE** the operating characteristics of a lead-acid battery to include methods of voltage production, state of charge, and hazards associated with storage batteries.

ENABLING OBJECTIVES

1.1 **DEFINE** the following terms as they relate to batteries and voltaic cells:
   a. Voltaic cell
   b. Battery
   c. Electrode
   d. Electrolyte
   e. Specific gravity
   f. Ampere-Hour

1.2 **STATE** the purpose of a battery.

1.3 **DESCRIBE** the operation of a simple voltaic cell.

1.4 **STATE** the chemical equation for the reaction that occurs when a lead-acid battery is being charged or discharged.

1.5 **EXPLAIN** the relationship between specific gravity and state of charge of a lead-acid battery.

1.6 **DESCRIBE** the relationship between total battery voltage and cell voltage for a series-connected battery.

1.7 **STATE** the advantage of connecting a battery in parallel with respect to current-carrying capability.

1.8 **STATE** the difference between primary and secondary cells with respect to recharge capability.

ENABLING OBJECTIVES (Cont.)

1.9 **STATE** the advantage of each of the following types of batteries:
   a. Carbon-zinc cell
   b. Alkaline cell
   c. Nickel-cadmium cell
   d. Edison cell
   e. Mercury cell

1.10 **EXPLAIN** the adverse effects of a shorted cell.

1.11 **EXPLAIN** how gas generation is minimized for a lead-acid battery.

1.12 **EXPLAIN** how heat is generated in a lead-acid battery.

Intentionally Left Blank

BATTERY TERMINOLOGY

Batteries are used for a wide variety of services throughout technology today. To begin to study battery operation and characteristics, a few terms that are used with batteries must be understood.

EO 1.1 DEFINE the following terms as they relate to batteries and voltaic cells:

a. Voltaic cell
b. Battery
c. Electrode
d. Electrolyte
e. Specific gravity
f. Ampere-Hour

Voltaic Cell

The term *voltaic cell* is defined as a combination of materials used to convert chemical energy into electrical energy. A voltaic or chemical cell consists of two electrodes made of different types of metals or metallic compounds placed in an electrolyte solution.

Battery

A *battery* is a group of two or more connected voltaic cells.

Electrode

An *electrode* is a metallic compound, or metal, which has an abundance of electrons (negative electrode) or an abundance of positive charges (positive electrode).

Electrolyte

An *electrolyte* is a solution which is capable of conducting an electric current. The electrolyte of a cell may be a liquid or a paste. If the electrolyte is a paste, the cell is referred to as a dry cell; if the electrolyte is a solution, it is called a wet cell.

Specific Gravity

Specific gravity is defined as the ratio comparing the weight of any liquid to the weight of an equal volume of water. The specific gravity of pure water is 1.000. Lead-acid batteries use an electrolyte which contains sulfuric acid. Pure sulfuric acid has a specific gravity of 1.835, since it weighs 1.835 times as much as pure water per unit volume.

Since the electrolyte of a lead-acid battery consists of a mixture of water and sulfuric acid, the specific gravity of the electrolyte will fall between 1.000 and 1.835. Normally, the electrolyte for a battery is mixed such that the specific gravity is less than 1.350.

Specific gravity is measured with a hydrometer. A simple hydrometer consists of a glass float inside a glass tube, as shown in Figure 1. The hydrometer float is weighted at one end and sealed at both ends. A scale calibrated in specific gravity is positioned lengthwise along the body of the float. The float is placed inside the glass tube, and the fluid to be tested is drawn into the tube. As the fluid is drawn into the tube, the hydrometer float will sink to a certain level in the fluid. The extent to which the hydrometer float protrudes above the level of the fluid depends on the specific gravity of the fluid. The reading on the float scale at the surface of the fluid is the specific gravity of the fluid.

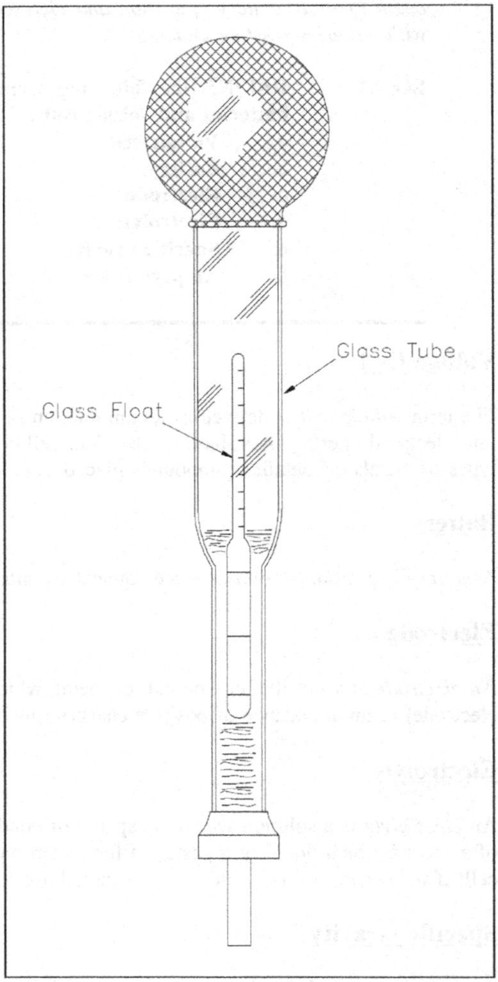

Figure 1 Simple Hydrometer

Ampere-Hour

An *ampere-hour* is defined as a current of one ampere flowing for one hour. If you multiply the current in amperes by the time of flow in hours, the result is the total number of ampere-hours. Ampere-hours are normally used to indicate the amount of energy a storage battery can deliver.

Summary

Battery terms are summarized below.

Battery Terminology Summary

- A voltaic cell is a combination of materials used to convert chemical energy into electrical energy.
- A battery is a group of two or more connected voltaic cells.
- An electrode is a metallic compound, or metal, which has an abundance of electrons (negative electrode) or an abundance of positive charges (positive electrode).
- An electrolyte is a solution which is capable of conducting an electric current.
- Specific gravity is defined as the ratio comparing the weight of any liquid to the weight of an equal volume of water.
- An ampere-hour is defined as a current of one ampere flowing for one hour.

BATTERY THEORY

A battery converts chemical energy to electrical energy. This conversion enables electrical power to be stored.

EO 1.2 **STATE the purpose of a battery.**

EO 1.3 **DESCRIBE the operation of a simple voltaic cell.**

EO 1.4 **STATE the chemical equation for the reaction that occurs when a lead-acid battery is being charged or discharged.**

EO 1.5 **EXPLAIN the relationship between specific gravity and state of charge of a lead-acid battery.**

Batteries

The purpose of a battery is to store chemical energy and to convert this chemical energy into electrical energy when the need arises.

As described in previous chapters, a chemical cell (or voltaic cell) consists of two electrodes of different types of metals or metallic compounds and an electrolyte solution which is capable of conducting an electric current.

A good example of a voltaic cell is one that contains zinc and copper electrodes. The zinc electrode contains an abundance of negatively charged atoms, and the copper electrode contains an abundance of positively charged atoms. When these electrodes are immersed in an electrolyte, chemical action begins. The zinc electrode will accumulate a much larger negative charge because it dissolves into the electrolyte. The atoms, which leave the zinc electrode, are positively charged and are attracted by the negatively charged ions of the electrolyte; the atoms repel the positively charged ions of the electrolyte toward the copper electrode (Figure 2).

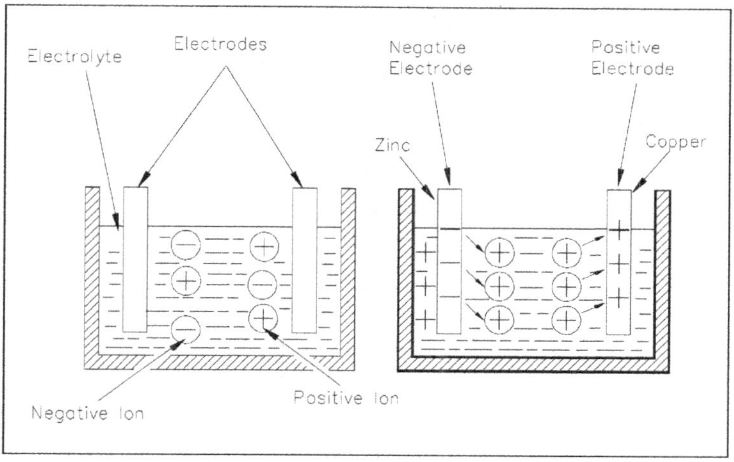

Figure 2 Basic Chemical Production of Electrical Power

This action causes electrons to be removed from the copper electrode, leaving it with an excess of positive charge. If a load is connected across the electrodes, the forces of attraction and repulsion will cause the free electrons in the negative zinc electrode to move through the connecting wire and load, and toward the positive copper electrode (Figure 3).

The potential difference that results allows the cell to function as a source of applied voltage.

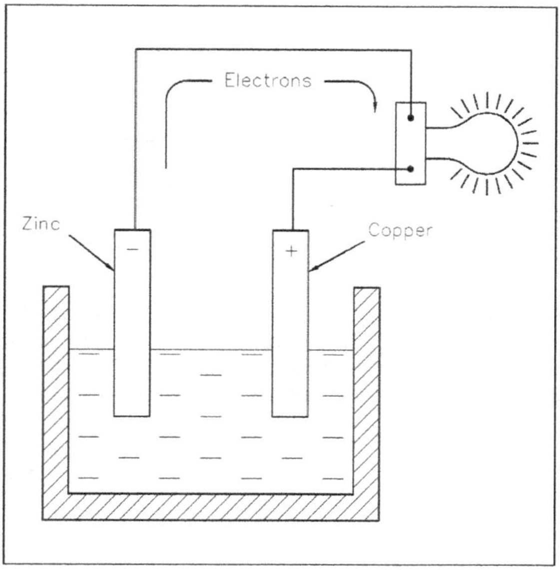

Figure 3 Electron Flow Through a Battery

Discharge and Charging of Lead-Acid Battery

In a lead-acid battery, two types of lead are acted upon electro-chemically by an electrolytic solution of diluted sulfuric acid (H_2SO_4). The positive plate consists of lead peroxide (PbO_2), and the negative plate is sponge lead (Pb), shown in Figure 4.

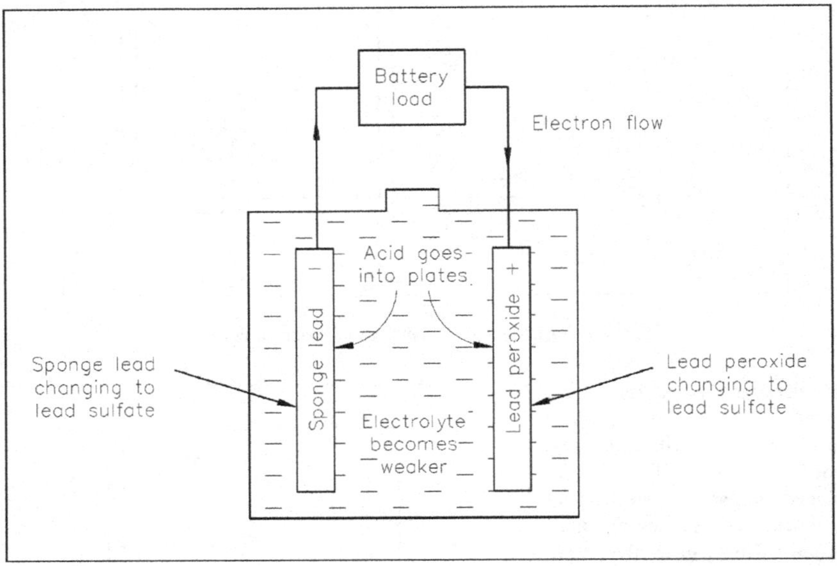

Figure 4 Chemical Action During Discharge

When a lead-acid battery is discharged, the electrolyte divides into H_2 and SO_4. The H_2 will combine with some of the oxygen that is formed on the positive plate to produce water (H_2O), and thereby reduces the amount of acid in the electrolyte. The sulfate (SO_4) combines with the lead (Pb) of both plates, forming lead sulphate ($PbSO_4$), as shown in Equation (4-1).

$$PbO_2 + Pb + 2H_2SO_4 \xrightarrow{\text{discharge}} 2PbSO_4 + 2H_2O \qquad (4\text{-}1)$$

As a lead-acid battery is charged in the reverse direction, the action described in the discharge is reversed. The lead sulphate ($PbSO_4$) is driven out and back into the electrolyte (H_2SO_4). The return of acid to the electrolyte will reduce the sulphate in the plates and increase the specific gravity. This will continue to happen until all of the acid is driven from the plates and back into the electrolyte, as shown in Equation (4-2) and Figure 5.

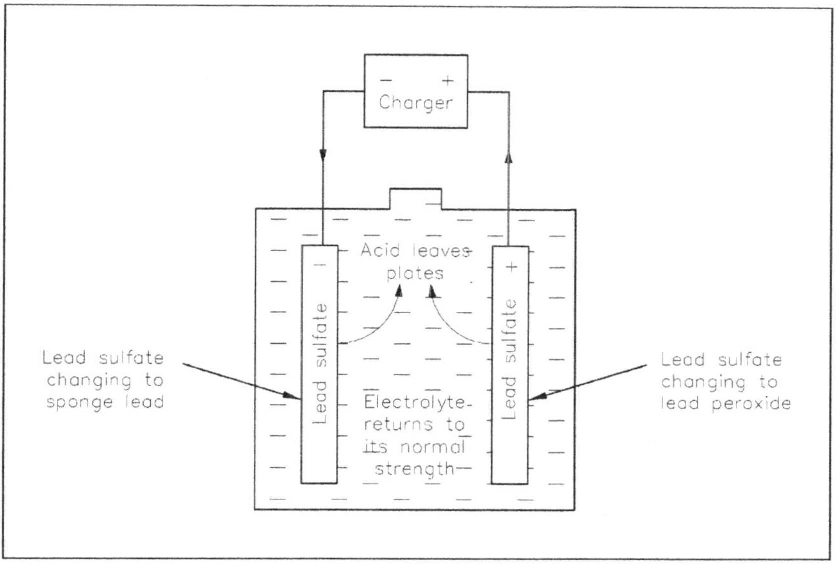

Figure 5 Chemical Action During Charging

$$PbO_2 + Pb + 2H_2SO_4 \xleftarrow{\text{charge}} 2PbSO_4 + 2H_2O \qquad (4\text{-}2)$$

As a lead-acid battery charge nears completion, hydrogen (H_2) gas is liberated at the negative plate, and oxygen (O_2) gas is liberated at the positive plate. This action occurs since the charging current is usually greater than the current necessary to reduce the remaining amount of lead sulfate on the plates. The excess current ionizes the water (H_2O) in the electrolyte. Since hydrogen is highly explosive, it is necessary to provide adequate ventilation to the battery whenever charging is in progress. Also, no smoking, electric sparks, or open flames are allowed near a charging battery.

The decrease in specific gravity on discharge is proportional to the ampere-hours discharged. While charging a lead-acid battery, the rise in specific gravity is not uniform, or proportional, to the amount of ampere-hours charged (Figure 6).

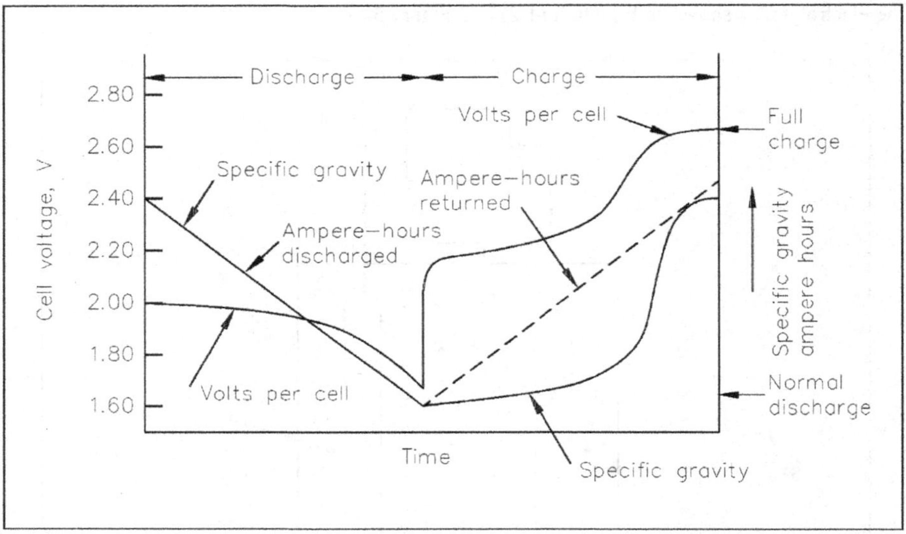

Figure 6 Voltage and Specific Gravity During Charge and Discharge

The electrolyte in a lead-acid battery plays a direct role in the chemical reaction. The specific gravity decreases as the battery discharges and increases to its normal, original value as it is charged. Since specific gravity of a lead-acid battery decreases proportionally during discharge, the value of specific gravity at any given time is an approximate indication of the battery's state of charge. To determine the state of charge, compare the specific gravity, as read using a hydrometer, with the full charge value and the manufacturer's published specific gravity drop, which is the decrease from full to nominal charge value.

Example: A lead-acid battery reads 1.175 specific gravity. Its average full charge specific gravity is 1.260 and has a normal gravity drop of 120 points (or.120) at an 8 hour discharge rate.

Solution:

Fully charged - 1.260
Present charge - 1.175

The battery is 85 points below its fully charged state. It is therefore about 85/120, or 71%, discharged.

Summary

Battery theory is summarized below.

Battery Theory Summary

- The purpose of a battery is to store chemical energy and to convert this chemical energy into electrical energy when the need arises.

- A voltaic cell develops a potential difference when electrodes of two different metals are immersed in an electrolyte. One electrode accumulates a positive charge. The potential difference is due to the difference in charge between the two electrodes.

- The chemical equation for a lead-acid battery during discharge is:

$$PbO_2 + Pb + 2H_2SO_4 \xrightarrow{\text{discharge}} 2PbSO_4 + 2H_2O.$$

- The chemical equation for a lead-acid battery during charge is:

$$PbO_2 + Pb + 2H_2SO_4 \xleftarrow{\text{charge}} 2PbSO_4 + 2H_2O.$$

- When a lead-acid battery is discharged, electrolyte and the active material on the plates of the battery are consumed to produce water and lead sulphate.

- When a lead-acid battery is charged, electrical energy is added to the battery, causing the water and lead sulphate to be consumed and produce electrolyte and active material.

- Since specific gravity of a lead-acid battery decreases proportionally during discharge, the value of specific gravity at any given time is an approximate indication of the battery's state of charge.

BATTERY OPERATIONS

Once the basic theory behind the operation of batteries is understood, we can apply these concepts to better understand the way batteries are utilized.

EO 1.6 **DESCRIBE the relationship between total battery voltage and cell voltage for a series-connected battery.**

EO 1.7 **STATE the advantage of connecting a battery in parallel with respect to current-carrying capability.**

EO 1.8 **STATE the difference between primary and secondary cells with respect to recharge capability.**

Series Cells

When several cells are connected in series (Figure 7), the total voltage output of the battery is equal to the sum of the individual cell voltages. In the example of the battery in Figure 7, the four 1.5V cells provide a total of 6 volts. When we connect cells in series, the positive terminal of one cell is connected to the negative terminal of the next cell. The current flow through a battery connected in series is the same as for one cell.

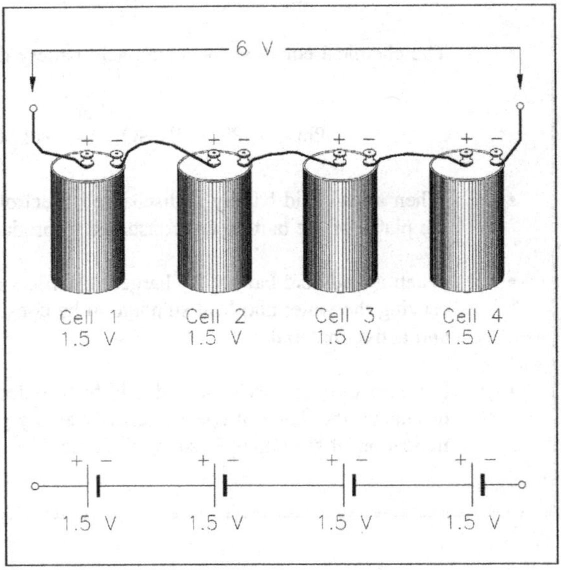

Figure 7 Cells Connected in Series

Parallel Cells

Cells connected in parallel (Figure 8), give the battery a greater current capacity. When cells are connected in parallel, all the positive terminals are connected together, and all the negative terminals are connected together. The total voltage output of a battery connected in parallel is the same as that of a single cell. Cells connected in parallel have the same effect as increasing the size of the electrodes and electrolyte in a single cell. The advantage of connecting cells in parallel is that it will increase the current-carrying capability of the battery.

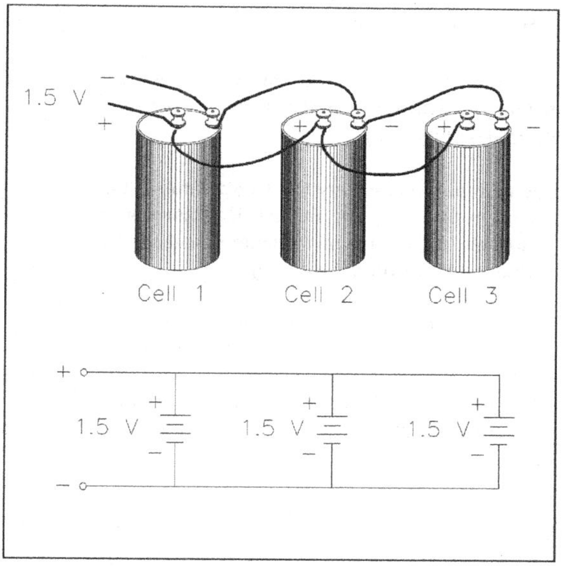

Figure 8 Cells Connected in Parallel

Primary Cell

Cells that cannot be returned to good condition, or recharged after their voltage output has dropped to a value that is not usable, are called *primary cells*. Dry cells that are used in flashlights and transistor radios (e.g., AA cells, C cells) are examples of primary cells.

Secondary Cells

Cells that can be recharged to nearly their original condition are called *secondary cells*. The most common example of a secondary, or rechargeable cell, is the lead-acid automobile battery.

Capacity

The capacity of a storage battery determines how long the storage battery will operate at a certain discharge rate and is rated in ampere-hours. For example, a 120 ampere-hour battery must be recharged after 12 hours if the discharge rate is 10 amps.

Internal Resistance

Internal resistance in a chemical cell is due mainly to the resistance of the electrolyte between electrodes (Figure 9).

Any current in the battery must flow through the internal resistance. The internal resistance is in series with the voltage of the battery, causing an internal voltage drop (Figure 10).

With no current flow, the voltage drop is zero; thus, the full battery voltage is developed across the output terminals (V_B). If a load is placed on the battery, load resistance (R_L) is in series with internal resistance (R_i).

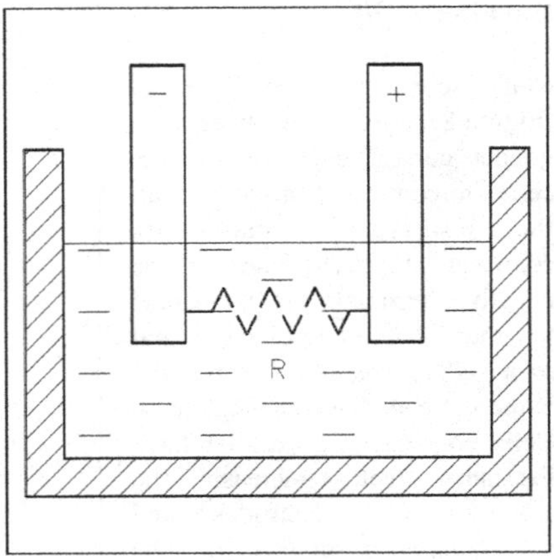

Figure 9 Internal Resistance in a Chemical Cell

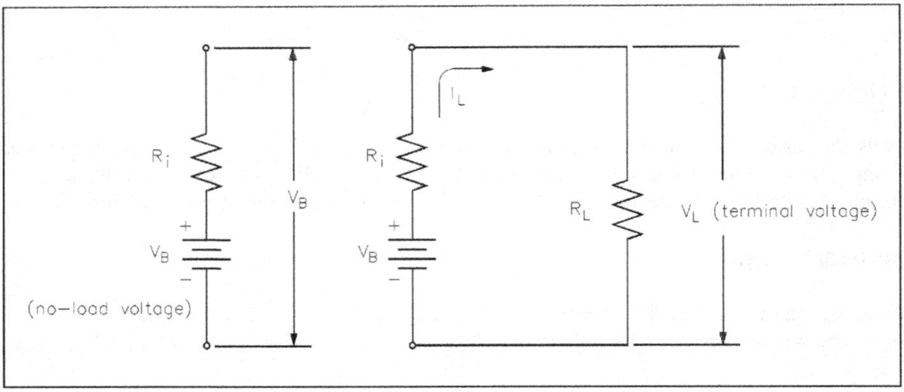

Figure 10 Internal Voltage Drop

When current flows in the circuit (I_L), the internal voltage drop (I_LR_i) drops the terminal voltage of the battery as shown in Equation (4-3). Thus, internal resistance reduces both the current and voltage available to the load.

$$V_L = V_B - I_LR_i \tag{4-3}$$

Shelf Life

The *shelf life* of a battery is the time which a battery may be stored and not lose more than 10 percent of its original capacity.

Charge and Discharge

The *charge* of a battery may refer to as one of two things: (1) the relative state of capacity of the battery, or (2) the actual act of applying current flow in the reverse direction to return the battery to a fully-charged state.

Discharge, simply stated, is the act of drawing current from a battery.

Summary

Battery operations are summarized below.

Battery Operations Summary

- The output voltage of a battery connected in series is equal to the sum of the cell voltages.
- A battery that is connected in parallel has the advantage of a greater current-carrying capability.
- Secondary cells can be recharged; primary cells cannot be recharged.
- The unit for battery capacity is the ampere-hour.
- Internal resistance in a battery will decrease the battery voltage when a load is placed on the battery.
- Shelf life is a term that is used to measure the time that a battery may sit idle and not lose more than 10 percent of its charge.
- The charge of a battery may refer to one of two things: (1) the relative state of capacity of the battery, or (2) the actual act of applying current flow in the reverse direction to restore the battery to a fully-charged condition.
- Discharge refers to the act of drawing current from a battery.

TYPES OF BATTERIES

The lead-acid battery is the most common type of battery in use today. There are other types of storage batteries, each having certain advantages.

EO 1.9 STATE the advantage of each of the following types of batteries:

a. Carbon-zinc cell
b. Alkaline cell
c. Nickel-cadmium cell
d. Edison cell
e. Mercury cell

Wet and Dry Cells

Wet and dry cells are classified by the type of electrolyte the battery uses. The electrolyte of a cell may be a liquid or a paste. If the electrolyte is a paste, the cell is referred to as a dry cell. If the electrolyte is a solution, the cell is called a wet cell.

Carbon-Zinc Cell

The carbon-zinc cell is one of the oldest and most widely used types of dry cells. The carbon in the battery is in the form of a rod in the center of the cell which acts as the positive terminal. The case is made from zinc and acts as the negative electrode. The electrolyte for this type of cell is a chemical paste-like mixture which is housed between the carbon electrode and the zinc case. The cell is then sealed to prevent any of the liquid in the paste from evaporating.

The advantage of a carbon-zinc battery is that it is durable and very inexpensive to produce. The cell voltage for this type of cell is about 1.5 volts.

Alkaline Cell

The alkaline cell is so called because it has an alkaline electrolyte of potassium hydroxide. The negative electrode is made from zinc, and the positive electrode is made of manganese dioxide. The typical alkaline cell generates 1.5 volts. The alkaline cell has the advantage of an extended life over that of a carbon-zinc cell of the same size; however, it is usually more expensive.

Nickel-Cadmium Cell

The nickel-cadmium cell is a secondary cell, and the electrolyte is potassium hydroxide. The negative electrode is made of nickel hydroxide, and the positive electrode is made of cadmium hydroxide. The nominal voltage of a nickel-cadmium cell is 1.25 volts. The nickel-cadmium battery has the advantage of being a dry cell that is a true storage battery with a reversible chemical reaction (i.e., it can be recharged). The nickel-cadmium battery is a rugged, dependable battery. It gives dependable service under extreme conditions of temperature, shock, and vibration. Due to its dependability, it is ideally suited for use in portable communications equipment.

Edison Cell

In an edison cell the positive plate consists of nickel and nickel hydrate, and the negative plate is made of iron. The electrolyte is an alkaline. Typical voltage output is 1.4 volts, and it should be recharged when it reaches 1.0 volts. The edison cell has the advantage of being a lighter and more rugged secondary cell than a lead-acid storage battery.

Mercury Cell

Mercury cells come in two types; one is a flat cell that is shaped like a button, while the other is a cylindrical cell that looks like a regular flashlight battery. Each cell produces about 1.35 volts. These cells are very rugged and have a relatively long shelf life. The mercury cell has the advantage of maintaining a fairly constant output under varying load conditions. For this reason, they are used in products such as electric watches, hearing aids, cameras, and test instruments.

Summary

Battery types are summarized below.

Battery Types Summary
• If the electrolyte is a paste, the cell is referred to as a dry cell. If the electrolyte is a solution, the cell is called a wet cell.
• The advantage of a carbon-zinc battery is that it is durable and very inexpensive to produce.
• The alkaline cell has the advantage of an extended life over that of a carbon-zinc cell of the same size.
• The nickel-cadmium battery has the advantage of being a dry cell that is a true storage battery with a reversible chemical reaction.
• The edison cell has the advantage of being a lighter and more rugged secondary cell than a lead-acid storage battery.
• The mercury cell has the advantage of maintaining a fairly constant output under varying load conditions.

BATTERY HAZARDS

Because batteries store large amounts of energy, there are certain hazards that are associated with battery operation. These hazards must be fully understood to ensure safe operation of batteries.

EO 1.10 EXPLAIN the adverse effects of a shorted cell.

EO 1.11 EXPLAIN how gas generation is minimized for a lead-acid battery.

EO 1.12 EXPLAIN how heat is generated in a lead-acid battery.

Shorted Cell

Cell short circuits can be caused by several conditions, which include the following: faulty separators; lead particles or other metals forming a circuit between the positive and negative plates; buckling of the plates; or excessive sediments in the bottom of the jar. The primary cause of some of these occurrences is overcharging and overdischarging of the battery, which causes sediment to build up due to flaking of active material and buckling of cell plates.

Overcharging and overdischarging should be avoided at all costs. Short circuits cause a great reduction in battery capacity. With each shorted cell, battery capacity is reduced by a percentage equal to one over the total number of cells.

Gas Generation

A lead-acid battery cannot absorb all the energy from the charging source when the battery is nearing the completion of the charge. This excess energy dissociates water by way of electrolysis into hydrogen and oxygen. Oxygen is produced by the positive plate, and hydrogen is produced by the negative plate. This process is known as gassing.

Gassing is first noticed when cell voltage reaches 2.30-2.35 volts per cell and increases as the charge progresses. At full charge, the amount of hydrogen produced is about one cubic foot per cell for each 63 ampere-hours input. If gassing occurs and the gases are allowed to collect, an explosive mixture of hydrogen and oxygen can be readily produced. It is necessary, therefore, to ensure that the area is well ventilated and that it remains free of any open flames or spark-producing equipment.

As long as battery voltage is greater than 2.30 volts per cell, gassing will occur and cannot be prevented entirely. To reduce the amount of gassing, charging voltages above 2.30 volts per cell should be minimized (e.g., 13.8 volts for a 12 volt battery).

Battery Temperature

The operating temperature of a battery should preferably be maintained in the nominal band of 60-80°F. Whenever the battery is charged, the current flowing through the battery will cause heat to be generated by the electrolysis of water. The current flowing through the battery (I) will also cause heat to be generated (P) during charge and discharge as it passes through the internal resistance (R_i), as illustrated using the formula for power in Equation (4-4).

$$P = I^2 R_i \tag{4-4}$$

Higher temperatures will give some additional capacity, but they will eventually reduce the life of the battery. Very high temperatures, 125°F and higher, can actually do damage to the battery and cause early failure.

Low temperatures will lower battery capacity but also prolong battery life under floating (i.e., slightly charging) operation or storage. Extremely low temperatures can freeze the electrolyte, but only if the battery is low in specific gravity.

Summary

Battery hazards are summarized below.

Battery Hazards Summary

- Short circuits cause a great reduction in battery capacity.
- To prevent short circuits in a battery, overcharging and overdischarging should be avoided at all costs.
- The adverse effect of gassing is that if gassing occurs and the gases are allowed to collect, an explosive mixture of hydrogen and oxygen can be readily produced.
- To reduce the amount of gassing, charging voltages above 2.30 volts per cell should be minimized.
- Whenever the battery is charged, the current flowing through the battery will cause heat to be generated by the electrolysis of water and by I^2R_i power generation.
- Higher temperatures will give some additional capacity, but they will eventually reduce the life of the battery. Very high temperatures, 125°F and higher, can actually do damage to the battery and cause early failure.

Intentionally Left Blank

Department of Energy
Fundamentals Handbook

ELECTRICAL SCIENCE
Module 5
DC Generators

TABLE OF CONTENTS

LIST OF FIGURES . . . iii

LIST OF TABLES . . . iv

REFERENCES . . . v

OBJECTIVES . . . vi

DC EQUIPMENT TERMINOLOGY . . . 1

Terminal Voltage . . . 1
Counter-Electromotive Force . . . 1
Applied Voltage . . . 1
Commutation . . . 1
Summary . . . 2

DC EQUIPMENT CONSTRUCTION . . . 3

Armature . . . 3
Rotor . . . 3
Stator . . . 4
Field . . . 4
Summary . . . 4

DC GENERATOR THEORY . . . 5

Voltage Production . . . 5
Theory of Operation . . . 5
Commutator Action . . . 7
Field Excitation . . . 9
Terminal Voltage . . . 9
DC Generator Ratings . . . 10
Internal Losses . . . 10
Copper Losses . . . 10
Eddy-Current Losses . . . 10
Hysteresis Losses . . . 11
Mechanical Losses . . . 11
Summary . . . 12

TABLE OF CONTENTS (Cont.)

DC GENERATOR CONSTRUCTION 13

Shunt-Wound DC Generators 13
Series-Wound DC Generators 15
Compound Generators 16
Summary 18

LIST OF FIGURES

Figure 1 AC to DC Conversion with a Commutator 2

Figure 2 Basic DC Machine 3

Figure 3 Basic Operation of a DC Generator 6

Figure 4 Left-Hand Rule for Generators 7

Figure 5 Commutator Segments and Brushes 7

Figure 6 Commutation in a DC Generator 8

Figure 7 Varying Generator Terminal Voltage 9

Figure 8 Shunt-Wound DC Generator 13

Figure 9 Output Voltage-vs-Load Current for Shunt-Wound DC Generator 14

Figure 10 Series-Wound DC Generator 15

Figure 11 Output Voltage-vs-Load Current for Series-Wound DC Generator 15

Figure 12 Compounded DC Generator 16

Figure 13 Voltage-vs-Current for a Compounded DC Generator 17

LIST OF TABLES

NONE

REFERENCES

- Gussow, Milton, Schaum's Outline Series, Basic Electricity, McGraw-Hill.
- Academic Program for Nuclear Power Plant Personnel, Volume IV, Columbia, MD: General Physics Corporation, Library of Congress Card #A 326517, 1982.
- Academic Program for Nuclear Power Plant Personnel, Volume II, Columbia, MD: General Physics Corporation, Library of Congress Card #A 326517 1982.
- Nasar and Unnewehr, Electromechanics and Electric Machines, John Wiley and Sons.
- Van Valkenburgh, Nooger, and Neville, Basic Electricity, Vol. 5, Hayden Book Company.
- Lister, Eugene C., Electric Circuits and Machines, 5th Edition, McGraw-Hill.
- Croft, Carr, Watt, and Summers, American Electricians Handbook, 10th Edition, McGraw-Hill.
- Mileaf, Harry, Electricity One - Seven, Revised 2nd Edition, Hayden Book Company.
- Buban and Schmitt, Understanding Electricity and Electronics, 3rd Edition, McGraw-Hill.
- Kidwell, Walter, Electrical Instruments and Measurements, McGraw-Hill.

TERMINAL OBJECTIVE

1.0 Given the type and application of a direct current (DC) generator, **DESCRIBE** the operating characteristics of that generator including methods of voltage production, advantages of each type, and voltage-vs-load characteristics.

ENABLING OBJECTIVES

1.1 **DEFINE** terminal voltage as it applies to DC generators.

1.2 **DEFINE** counter-electromotive force (CEMF) as it applies to a DC machine.

1.3 **DESCRIBE** the effects of commutation in a DC generator.

1.4 **STATE** the purpose of each of the following components of a DC machine:
- a. Armature
- b. Rotor
- c. Stator
- d. Field

1.5 **LIST** the three conditions necessary to induce a voltage into a conductor.

1.6 Using the left-hand rule of generators, **DETERMINE** the direction of the magnetic field, the motion of the conductor, or the direction of current induced into a conductor.

1.7 **DESCRIBE** how terminal voltage of a DC generator is adjusted.

1.8 **STATE** the basis behind each of the four DC generator ratings.

1.9 **LIST** the four internal losses found in a DC generator.

1.10 **DESCRIBE** the differences in construction between a shunt-wound and a series-wound DC generator with respect to the relationship between the field and the armature.

1.11 **DESCRIBE** the relationship between the shunt and series fields for cumulatively-compounded and differentially-compounded DC generators.

ENABLING OBJECTIVES (Cont.)

1.12 **DESCRIBE** the voltage-vs-load current characteristics for a flat-compounded, over-compounded, and under-compounded DC generator.

Intentionally Left Blank

DC EQUIPMENT TERMINOLOGY

Direct current devices are used frequently in today's technology. Before the construction and operation of these devices can be introduced, a few common terms must be understood.

EO 1.1	**DEFINE terminal voltage as it applies to DC generators.**
EO 1.2	**DEFINE counter-electromotive force (CEMF) as it applies to a DC machine.**
EO 1.3	**DESCRIBE the effects of commutation in a DC generator.**

Terminal Voltage

Terminal voltage, as applied to DC generators, is defined as the voltage that can be measured at the output of the generator.

Counter-Electromotive Force (CEMF)

In a generator using a rotating armature, the conductors cut the magnetic lines of force in the magnetic field. Voltage is induced in the armature conductors. This induced voltage opposes the applied voltage; it counteracts some of the applied voltage, which reduces the current flow through the armature. This induced voltage acts counter to applied voltage; therefore, it is called *counter-electromotive force* (CEMF).

Applied Voltage

Applied voltage is defined as the voltage that is delivered across the load. This voltage should be the same as terminal voltage; however, various circuit faults and losses may reduce the terminal voltage.

Commutation

Commutation is the positioning of the DC generator brushes so that the commutator segments change brushes at the same time the armature current changes direction. More simply stated, commutation is the mechanical conversion from AC to DC at the brushes of a DC machine, as shown in Figure 1.

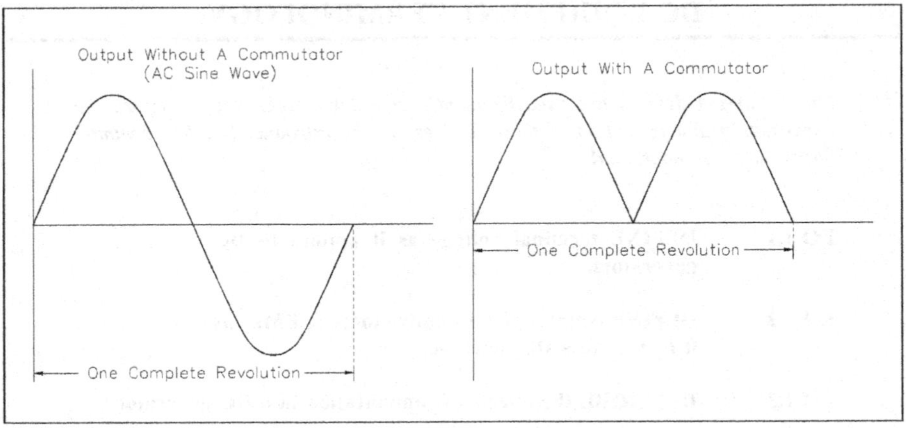

Figure 1 AC to DC Conversion with a Commutator

In a DC generator, commutation provides for the conversion of AC to a DC output that is generated in the armature windings. Commutation will be discussed in greater detail in subsequent chapters.

Summary

DC equipment terms are summarized below.

DC Equipment Terminology Summary

- Terminal voltage, as applied to DC generators, is defined as the voltage that can be measured at the output of the generator.
- Counter-electromotive force (CEMF) is defined as the induced voltage that acts to counter the applied voltage in a DC motor or a DC generator.
- Applied voltage is defined as the voltage that is delivered across the load.
- Commutation is the positioning of the DC generator brushes so that the commutator segments change brushes at the same time the armature current changes direction.
- In a DC generator, commutation provides for the conversion of AC to a DC output that is generated in the armature windings.

DC EQUIPMENT CONSTRUCTION

Direct current machines are energy transfer devices. These machines can function as either a motor or a generator. DC motors and generators have the same basic construction, differing primarily in the energy conversion. To better understand the operation and construction of DC machines, a few basic terms must be understood.

EO 1.4 STATE the purpose of each of the following components of a DC machine:
- **a. Armature**
- **b. Rotor**
- **c. Stator**
- **d. Field**

Armature

The purpose of the *armature* is to provide the energy conversion in a DC machine (refer to Figure 2).

In a DC generator, the armature is rotated by an external mechanical force, such as a steam turbine. This rotation induces a voltage and current flow in the armature. Thus, the armature converts mechanical energy to electrical energy.

In a DC motor, the armature receives voltage from an outside electrical source and converts electrical energy into mechanical energy in the form of torque.

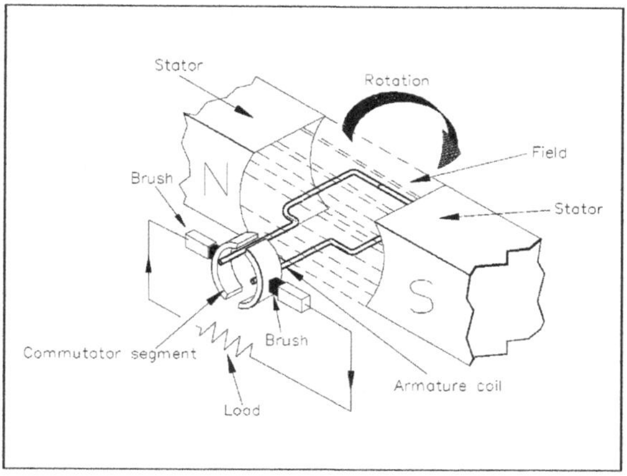

Figure 2 Basic DC Machine

Rotor

The purpose of the *rotor* is to provide the rotating element in a DC machine (refer to Figure 2). In a DC generator, the rotor is the component that is rotated by an external force. In a DC motor, the rotor is the component that turns a piece of equipment. In both types of DC machines, the rotor is the armature.

Stator

The *stator* is the part of a motor or generator that is stationary (refer to Figure 2). In DC machines, the purpose of the stator is to provide the magnetic field. The stator in Figure 2 is provided by a permanent magnet.

Field

The purpose of the *field* in a DC machine is to provide a magnetic field for producing either a voltage (generator) or a torque (motor) (refer to Figure 2). The field in a DC machine is produced by either a permanent magnet or an electromagnet. Normally, electromagnets are used because they have an increased magnetic strength, and the magnetic strength is more easily varied using external devices. In Figure 2, the field is provided by the stator.

Summary

The construction of DC equipment is summarized below.

DC Equipment Construction Summary

- The purpose of the armature is to provide the energy conversion in a DC machine.
- The purpose of the rotor is to provide the rotating element in a DC machine.
- In DC machines, the purpose of the stator is to provide the field.
- The purpose of the field in a DC machine is to provide a magnetic field for producing either a voltage or a torque.

DC GENERATOR THEORY

DC generators are widely used to produce a DC voltage. The amount of voltage produced depends on a variety of factors.

EO 1.5 **LIST the three conditions necessary to induce a voltage into a conductor.**

EO 1.6 **Using the left-hand rule of generators, DETERMINE the direction of the magnetic field, the motion of the conductor, or the direction of current induced into a conductor.**

EO 1.7 **DESCRIBE how terminal voltage of a DC generator is adjusted.**

EO 1.8 **STATE the basis behind each of the four DC generator ratings.**

EO 1.9 **LIST the four internal losses found in a DC generator.**

Voltage Production

Recall from Module 3, DC Circuits, that there are three conditions necessary to induce a voltage into a conductor.

1. A magnetic field
2. A conductor
3. Relative motion between the two

A DC generator provides these three conditions to produce a DC voltage output.

Theory of Operation

A basic DC generator has four basic parts: (1) a magnetic field; (2) a single conductor, or loop; (3) a commutator; and (4) brushes (Figure 3). The magnetic field may be supplied by either a permanent magnet or an electromagnet. For now, we will use a permanent magnet to describe a basic DC generator.

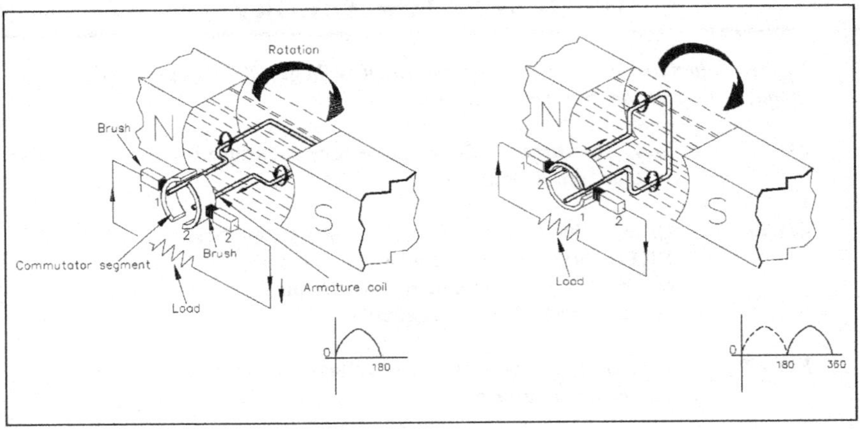

Figure 3 Basic Operation of a DC Generator

A single conductor, shaped in the form of a loop, is positioned between the magnetic poles. As long as the loop is stationary, the magnetic field has no effect (no relative motion). If we rotate the loop, the loop cuts through the magnetic field, and an EMF (voltage) is induced into the loop.

When we have relative motion between a magnetic field and a conductor in that magnetic field, and the direction of rotation is such that the conductor cuts the lines of flux, an EMF is induced into the conductor. The magnitude of the induced EMF depends on the field strength and the rate at which the flux lines are cut, as given in equation (5-1). The stronger the field or the more flux lines cut for a given period of time, the larger the induced EMF.

$$E_g = K\Phi N \qquad (5\text{-}1)$$

where

E_g = generated voltage
K = fixed constant
Φ = magnetic flux strength
N = speed in RPM

The direction of the induced current flow can be determined using the "left-hand rule" for generators. This rule states that if you point the index finger of your left hand in the direction of the magnetic field (from North to South) and point the thumb in the direction of motion of the conductor, the middle finger will point in the direction of current flow (Figure 4). In the generator shown in Figure 4, for example, the conductor closest to the N pole is traveling upward across the field; therefore, the current flow is to the right, lower corner. Applying the left-hand rule to both sides of the loop will show that current flows in a counter-clockwise direction in the loop.

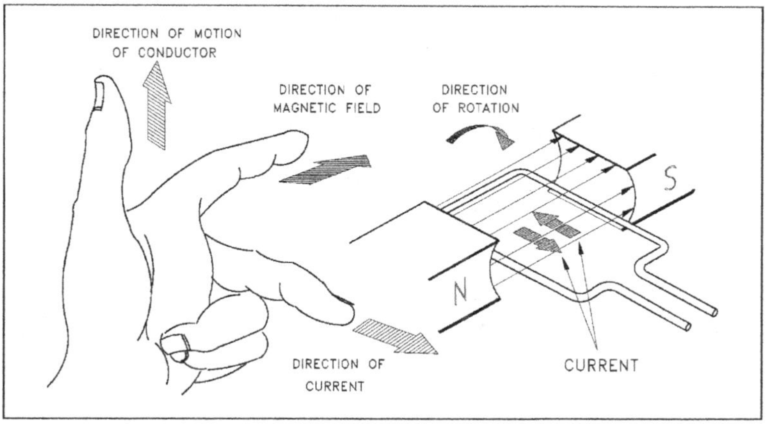

Figure 4 Left-Hand Rule for Generators

Commutator Action

The commutator converts the AC voltage generated in the rotating loop into a DC voltage. It also serves as a means of connecting the brushes to the rotating loop. The purpose of the brushes is to connect the generated voltage to an external circuit. In order to do this, each brush must make contact with one of the ends of the loop. Since the loop or armature rotates, a direct connection is impractical. Instead, the brushes are connected to the ends of the loop through the commutator.

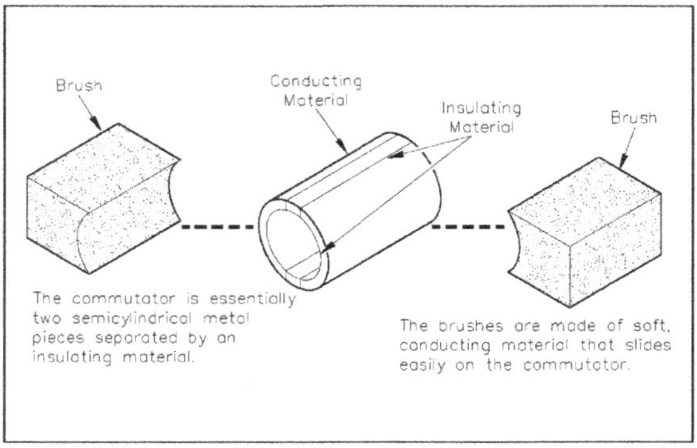

Figure 5 Commutator Segments and Brushes

In a simple one-loop generator, the commutator is made up of two semicylindrical pieces of a smooth conducting material, usually copper, separated by an insulating material, as shown in Figure 5. Each half of the commutator segments is permanently attached to one end of the rotating loop, and the commutator rotates with the loop. The brushes, usually made of carbon, rest against the commutator and slide along the commutator as it rotates. This is the means by which the brushes make contact with each end of the loop.

Each brush slides along one half of the commutator and then along the other half. The brushes are positioned on opposite sides of the commutator; they will pass from one commutator half to the other at the instant the loop reaches the point of rotation, at which point the voltage that was induced reverses the polarity. Every time the ends of the loop reverse polarity, the brushes switch from one commutator segment to the next. This means that one brush is always positive with respect to another. The voltage between the brushes fluctuates in amplitude (size or magnitude) between zero and some maximum value, but is always of the same polarity (Figure 6). In this manner, commutation is accomplished in a DC generator.

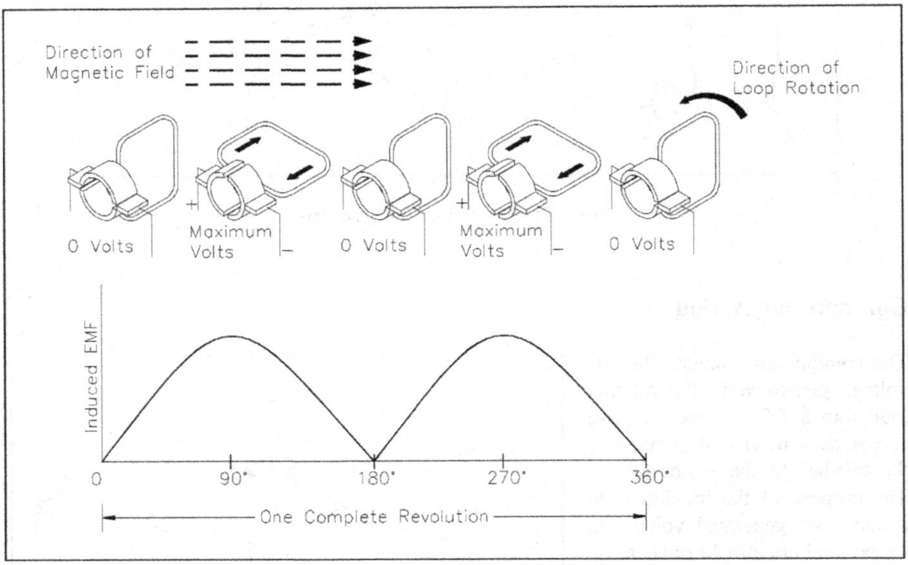

Figure 6 Commutation in a DC Generator

One important point to note is that, as the brushes pass from one segment to the other, there is an instant when the brushes contact both segments at the same time. The induced voltage at this point is zero. If the induced voltage at this point were not zero, extremely high currents would be produced due to the brushes shorting the ends of the loop together. The point at which the brushes contact both commutator segments, when the induced voltage is zero, is called the "neutral plane."

Field Excitation

The magnetic fields in DC generators are most commonly provided by electromagnets. A current must flow through the electromagnet conductors to produce a magnetic field. In order for a DC generator to operate properly, the magnetic field must always be in the same direction. Therefore, the current through the field winding must be direct current. This current is known as the *field excitation current* and can be supplied to the field winding in one of two ways. It can come from a separate DC source external to the generator (e.g., a separately excited generator) or it can come directly from the output of the generator, in which case it is called a *self-excited generator*.

In a self-excited generator, the field winding is connected directly to the generator output. The field may be connected in series with the output, in parallel with the output, or a combination of the two.

Separate excitation requires an external source, such as a battery or another DC source. It is generally more expensive than a self-excited generator. Separately excited generators are, therefore, used only where self-excitation is not satisfactory. They would be used in cases where the generator must respond quickly to an external control source or where the generated voltage must be varied over a wide range during normal operations.

Terminal Voltage

DC generator output voltage is dependent on three factors (recall equation 5-1): (1) the number of conductor loops in series in the armature, (2) armature speed, and (3) magnetic field strength. In order to change the generator output, one of these three factors must be varied. The number of conductors in the armature cannot be changed in a normally operating generator, and it is usually impractical to change the speed at which the armature rotates. The strength of the magnetic field, however, can be changed quite easily by varying the current through the field winding. This is the most widely used method for regulating the output voltage of a DC generator (Figure 7).

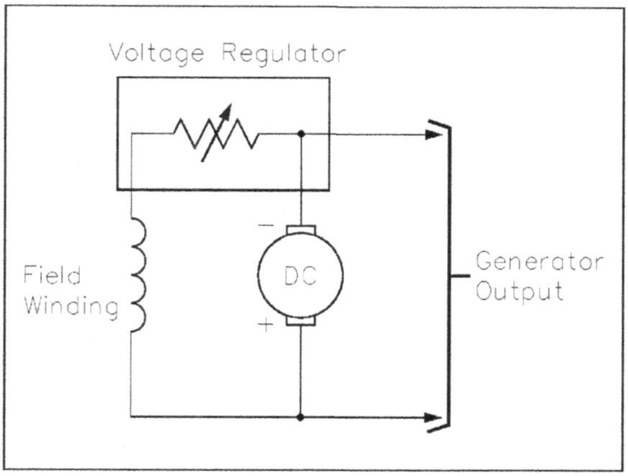

Figure 7 Varying Generator Terminal Voltage

DC Generator Ratings

A DC generator contains four ratings.

Voltage: Voltage rating of a machine is based on the insulation type and design of the machine.

Current: The current rating is based on the size of the conductor and the amount of heat that can be dissipated in the generator.

Power: The power rating is based on the mechanical limitations of the device that is used to turn the generator and on the thermal limits of conductors, bearings, and other components of the generator.

Speed: Speed rating, at the upper limit, is determined by the speed at which mechanical damage is done to the machine. The lower speed rating is based on the limit for field current (as speed increases, a higher field current is necessary to produce the same voltage).

Internal Losses

There are four internal losses that contribute to lower efficiency of a DC generator.

- Copper losses
- Eddy-current losses
- Hysteresis losses
- Mechanical losses

Each of these is described in the paragraphs that follow.

Copper Losses

Copper loss is the power lost as heat in the windings; it is caused by the flow of current through the coils of the DC armature or DC field. This loss varies directly with the square of the current in the armature or field and the resistance of the armature or field coils.

Armature: $I_a^2 R_a$
Field: $I_f^2 R_f$

Eddy-Current Losses

As the armature rotates within the field, it cuts the lines of flux at the same time that the copper coils of wire that are wound on the armature cut the lines of flux. Since the armature is made of iron, an EMF is induced in the iron, which causes a current to flow. These circulating currents within the iron core are called *eddy-currents*.

To reduce eddy-currents, the armature and field cores are constructed from laminated (layered) steel sheets. The laminated sheets are insulated from one another so that current cannot flow from one sheet to the other.

Hysteresis Losses

Hysteresis losses occur when the armature rotates in a magnetic field. The magnetic domains of the armature are held in alignment with the field in varying numbers, dependent upon field strength. The magnetic domains rotate, with respect to the particles not held in alignment, by one complete turn during each rotation of the armature. This rotation of magnetic domains in the iron causes friction and heat. The heat produced by this friction is called magnetic hysteresis loss.

To reduce hysteresis losses, most DC armatures are constructed of heat-treated silicon steel, which has an inherently low hysteresis loss. After the heat-treated silicon steel is formed to the desired shape, the laminations are heated to a dull red and then allowed to cool. This process, known as annealing, reduces hysteresis losses to a very low value.

Mechanical Losses

Rotational or *mechanical losses* can be caused by bearing friction, brush friction on the commutator, or air friction (called windage), which is caused by the air turbulence due to armature rotation. Careful maintenance can be instrumental in keeping bearing friction to a minimum. Clean bearings and proper lubrication are essential to the reduction of bearing friction. Brush friction is reduced by assuring proper brush seating, using proper brushes, and maintaining proper brush tension. A smooth and clean commutator also aids in the reduction of brush friction.

Summary

DC generator theory is summarized below.

DC Generator Theory Summary

- The three conditions necessary to induce a voltage into a conductor are:
  - Magnetic field
  - Conductor
  - Relative motion between the two

- The left-hand rule states that if you point the index finger of the left hand in the direction of the magnetic field and point the thumb in the direction of motion of the conductor, the middle finger will point in the direction of current flow.

- The terminal voltage of a DC generator is adjusted by varying the field strength.
- The voltage rating of a DC generator is based on the insulation type and design of the machine.

- The current rating of a DC generator is based on the size of the conductor and the amount of heat that can be dissipated in the generator.

- The power rating of a DC generator is based on the mechanical limitation of the device that is used to turn the generator.

- The upper speed rating of a DC generator is determined by the speed at which mechanical damage is done to the machine. The lower speed rating is based on the limit for field current.

- There are four internal losses that contribute to lower efficiency of a DC generator.
  - Copper losses
  - Eddy-current losses
  - Hysteresis losses
  - Mechanical losses

DC GENERATOR CONSTRUCTION

A DC generator may be constructed in a variety of ways depending upon the relationship and location of each of the fields. Each type of construction contains certain advantages.

EO 1.10 **DESCRIBE the differences in construction between a shunt-wound and a series-wound DC generator with respect to the relationship between the field and the armature.**

EO 1.11 **DESCRIBE the relationship between the shunt and series fields for cumulatively-compounded and differentially-compounded DC generators.**

EO 1.12 **DESCRIBE the voltage-vs-load current characteristics for a flat-compounded, over-compounded, and under-compounded DC generator.**

Shunt-Wound DC Generators

When the field winding of a generator is connected in parallel with the generator armature, the generator is called a shunt-wound generator (Figure 8).

The excitation current in a shunt-wound generator is dependent upon the output voltage and the field resistance. Normally, field excitation is maintained between 0.5 and 5 percent of the total current output of the generator.

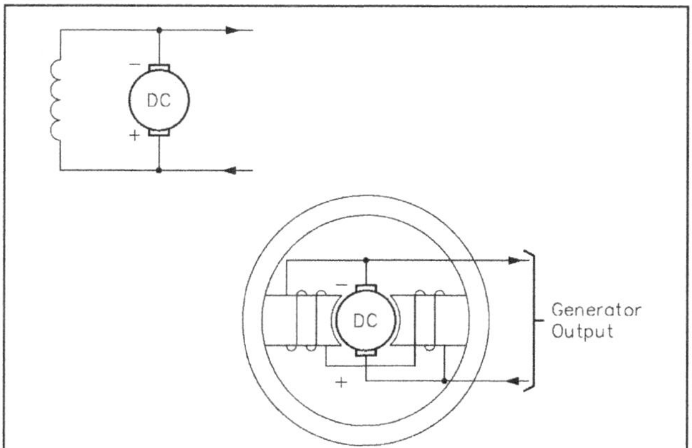

Figure 8 Shunt-Wound DC Generator

The shunt-wound generator, running at a constant speed under varying load conditions, has a much more stable voltage output than does a series-wound generator. Some change in output voltage does take place. This change is caused by the fact that, as the load current increases, the voltage drop (I_aR_a) across the armature coil increases, causing output voltage to decrease. As a result, the current through the field decreases, reducing the magnetic field and causing voltage to decrease even more. If load current is much higher than the design of the generator, the drop in output voltage is severe. For load current within the design range of the generator, the drop in output voltage is minimal (Figure 9).

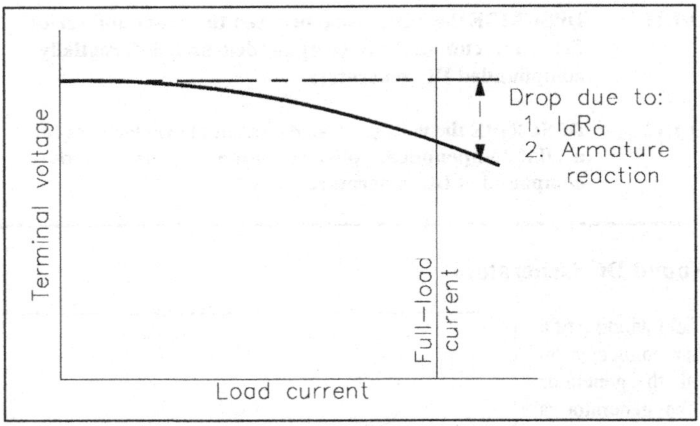

Figure 9 Output Voltage-vs-Load Current for Shunt-Wound DC Generator

Series-Wound DC Generators

When the field winding of a DC generator is connected in series with the armature, the generator is called a series-wound generator (Figure 10).

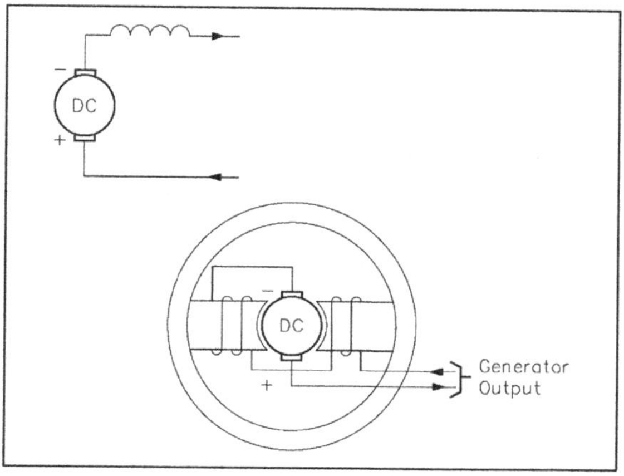

Figure 10 Series-Wound DC Generator

The excitation current in a series-wound generator is the same as the current the generator delivers to the load. If the load has a high resistance and only draws a small amount of current, the excitation current is also small. Therefore, the magnetic field of the series field winding is weak, making the generated voltage low. Conversely, if the load draws a large current, the excitation current is also high. Therefore, the magnetic field of the series field winding is very strong, and the generated voltage is high.

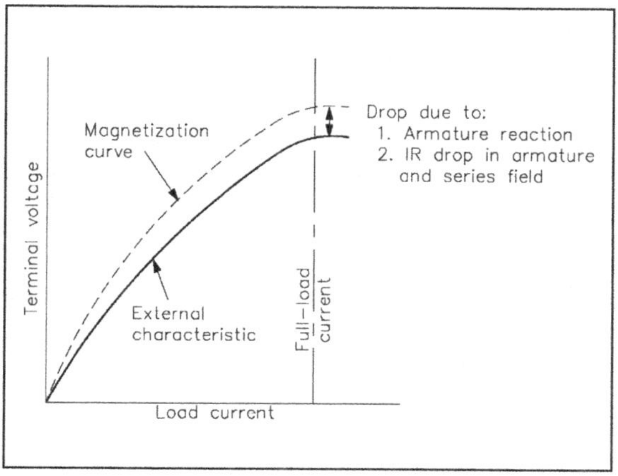

Figure 11 Output Voltage-vs-Load Current for Series-Wound DC Generator

As you can see in Figure 11, in a series generator, changes in load current drastically affect the generator output voltage. A series generator has poor voltage regulation, and, as a result, series generators are not used for fluctuating loads. As is the case for the shunt-wound generator, a series-wound generator also exhibits some losses due to the resistance of the windings and armature reaction. These losses cause a lower terminal voltage than that for an ideal magnetization curve.

Compound Generators

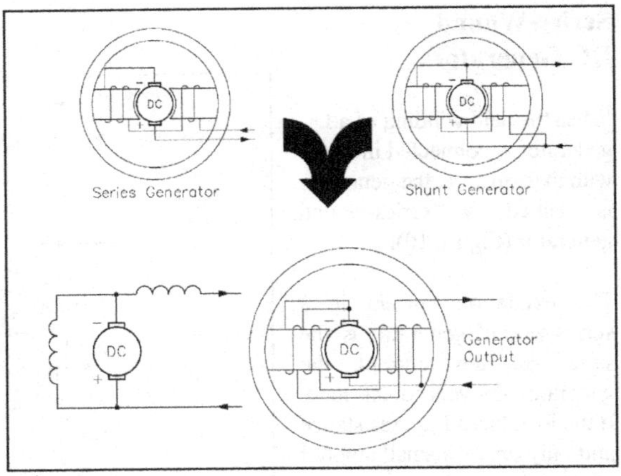

Figure 12 Compounded DC Generator

Series-wound and shunt-wound generators have a disadvantage in that changes in load current cause changes in generator output voltage. Many applications in which generators are used require a more stable output voltage than can be supplied by a series-wound or shunt-wound generator. One means of supplying a stable output voltage is by using a compound generator.

The compound generator has a field winding in parallel with the generator armature (the same as a shunt-wound generator) and a field winding in series with the generator armature (the same as a series-wound generator) (Figure 12).

The two windings of the compounded generator are made such that their magnetic fields will either aid or oppose one another.

If the two fields are wound so that their flux fields oppose one another, the generator is said to be *differentially-compounded.* Due to the nature of this type of generator, it is used only in special cases and will not be discussed further in this text.

If the two fields of a compound generator are wound so that their magnetic fields aid one another, the generator is said to be *cumulatively-compounded.* As the load current increases, the current through the series field winding increases, increasing the overall magnetic field strength and causing an increase in the output voltage of the generator. With proper design, the increase in the magnetic field strength of the series winding will compensate for the decrease in shunt field strength. Therefore, the overall strength of the combined magnetic fields remains almost unchanged, so the output voltage will remain constant. In reality, the two fields cannot be made so that their magnetic field strengths compensate for each other completely. There will be some change in output voltage from the no-load to full-load conditions.

In practical compounded generators, the change in output voltage from no-load to full-load is less than 5 percent. A generator with this characteristic is said to be *flat-compounded* (Figure 13).

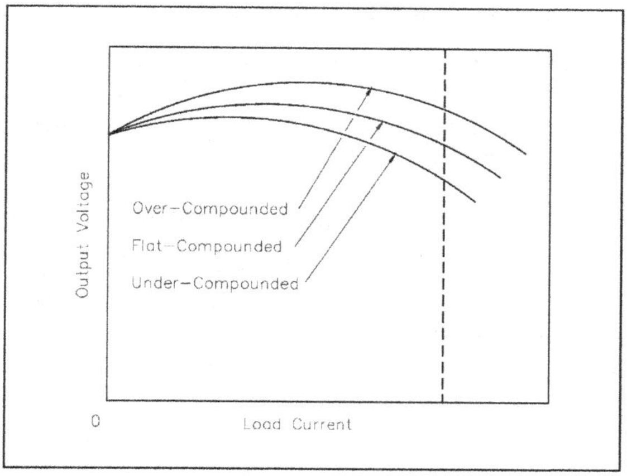

Figure 13 Voltage-vs-Current for a Compounded DC Generator

For some applications, the series winding is wound so that it overcompensates for a change in the shunt field. The output gradually rises with increasing load current over the normal operating range of the machine. This type of generator is called an *over-compounded* generator. The series winding can also be wound so that it undercompensates for the change in shunt field strength. The output voltage decreases gradually with an increase in load current. This type of generator is called an *under-compounded* generator.

Summary

DC generator construction is summarized below.

DC Generator Construction Summary

- A shunt-wound DC generator is constructed so that the field winding is in parallel with the armature winding.
- The voltage of a shunt-wound DC generator decreases with an increase in load current.
- A series-wound DC generator is constructed so that the field winding is in series with the armature winding.
- The voltage of a series-wound DC generator increases sharply with an increase in load.
- In a cumulatively-compounded DC generator, the series and shunt fields aid one another.
- In a differentially-compounded DC generator, the series and shunt fields oppose one another.
- The voltage of a flat-compounded DC generator changes less than 5 percent from no-load to full-load.
- The voltage of an over-compounded DC generator gradually rises with an increasing load.

Department of Energy
Fundamentals Handbook

ELECTRICAL SCIENCE
Module 6
DC Motors

www.KnowledgePublications.com

TABLE OF CONTENTS

LIST OF FIGURES ii

LIST OF TABLES iii

REFERENCES iv

OBJECTIVES v

DC MOTOR THEORY 1

- Inducing a Force on a Conductor 1
- Theory of Operation 2
- Torque 4
- Generator Action in a Motor 5
- DC Motor Speed 6
- Summary 7

TYPES OF DC MOTORS 8

- DC Motor Connections 8
- Shunt-Wound Motor Operation 9
- Shunt-Wound Motor Applications 9
- Series-Wound Motor 10
- Series-Wound Motor Applications 10
- Compounded Motor 10
- Summary 11

DC MOTOR OPERATION 12

- Starting of DC Motors 12
- DC Motor Ratings 13
- Summary 14

LIST OF FIGURES

Figure 1 Left-Hand Rule for Current-Carrying Conductors 2

Figure 2 Current-Carrying Conductor in a Magnetic Field 2

Figure 3 Motor Action 3

Figure 4 Right-Hand Rule for Motors 3

Figure 5 Armature Current in a Basic DC Motor 4

Figure 6 Counterelectromotive Force (CEMF) 5

Figure 7 DC Motor Connections 8

Figure 8 Torque-vs-Speed for a Shunt-Wound DC Motor 9

Figure 9 Torque-vs-Speed for a Series-Wound Motor 10

LIST OF TABLES

NONE

REFERENCES

- Gussow, Milton, Schaum's Outline Series, Basic Electricity, McGraw-Hill.
- Academic Program for Nuclear Power Plant Personnel, Volume IV, Columbia, MD: General Physics Corporation, Library of Congress Card #A 326517, 1982.
- Academic Program for Nuclear Power Plant Personnel, Volume II, Columbia, MD: General Physics Corporation, Library of Congress Card #A 326517, 1982.
- Nasar and Unnewehr, Electromechanics and Electric Machines, John Wiley and Sons.
- Van Valkenburgh, Nooger, and Neville, Basic Electricity, Vol. 5, Hayden Book Company.
- Lister, Eugene C., Electric Circuits and Machines, 5th Edition, McGraw-Hill.
- Croft, Carr, Watt, and Summers, American Electricians Handbook, 10th Edition, McGraw-Hill.
- Mileaf, Harry, Electricity One - Seven, Revised 2nd Edition, Hayden Book Company.
- Buban and Schmitt, Understanding Electricity and Electronics, 3rd Edition, McGraw-Hill.
- Kidwell, Walter, Electrical Instruments and Measurements, McGraw-Hill.

TERMINAL OBJECTIVE

1.0 Given the type and application of a DC motor, **DESCRIBE** the operating characteristics of that motor to include methods of speed variation, advantages of each type, and torque vs speed characteristics.

ENABLING OBJECTIVES

1.1 Using the right-hand rule for motors, **DETERMINE** the direction of the magnetic field, direction of current flow, or force on a conductor.

1.2 **STATE** the function of torque in a direct current motor and how it is developed.

1.3 **DESCRIBE** how Counterelectromotive Force (CEMF) is developed in a DC motor.

1.4 **DESCRIBE** the relationship between field current and magnetic field size in a DC motor.

1.5 **STATE** the function of the CEMF that is developed in a DC motor.

1.6 **DESCRIBE** how the speed of a DC motor is adjusted.

1.7 **DESCRIBE** the relationship between armature current and torque produced in a DC motor.

1.8 **DESCRIBE** the differences in construction between a shunt-wound and a series-wound DC motor with respect to the relationship between the field and the armature windings.

1.9 **DESCRIBE** the construction of a compounded DC motor.

1.10 **DESCRIBE** the torque-vs-speed characteristics for a shunt-wound and a series-wound DC motor.

1.11 **EXPLAIN** why starting resistors are necessary for large DC motors.

1.12 **LIST** the four nameplate ratings for a DC motor.

Intentionally Left Blank

DC MOTOR THEORY

DC motors are widely used to drive various equipment. The speed and torque produced in a DC motor depends on a variety of factors.

EO 1.1 **Using the right-hand rule for motors, DETERMINE the direction of the magnetic field, direction of current flow, or force on a conductor.**

EO 1.2 **STATE the function of torque in a direct current motor and how it is developed.**

EO 1.3 **DESCRIBE how Counterelectromotive Force (CEMF) is developed in a DC motor.**

EO 1.4 **DESCRIBE the relationship between field current and magnetic field size in a DC motor.**

EO 1.5 **STATE the function of the CEMF that is developed in a DC motor.**

EO 1.6 **DESCRIBE how the speed of a DC motor is adjusted.**

EO 1.7 **DESCRIBE the relationship between armature current and torque produced in a DC motor.**

Inducing a Force on a Conductor

There are two conditions which are necessary to produce a force on a conductor.

- The conductor must be carrying current.
- The conductor must be within a magnetic field.

When these two conditions exist, a force will be applied to the conductor, which will attempt to move the conductor in a direction perpendicular to the magnetic field. This is the basic theory by which all DC motors operate.

Theory of Operation

Every current-carrying conductor has a magnetic field around it. The direction of this magnetic field may be found by using the left-hand rule for current-carrying conductors. When the thumb points in the direction of current flow, the fingers will point in the direction of the magnetic field produced, as shown in Figure 1.

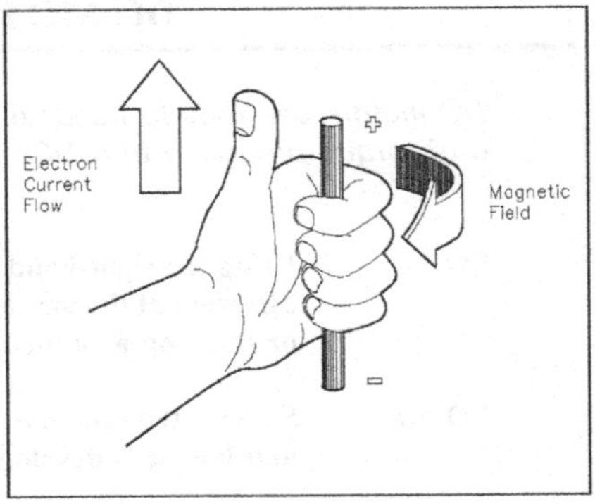

Figure 1 Left-Hand Rule for Current-Carrying Conductors

If a current-carrying conductor is placed in a magnetic field, the combined fields will be similar to those shown in Figure 2. The direction of current flow through the conductor is indicated with an "x" or a "·". The "x" indicates the current flow is away from the reader, or into the page. The "·" indicates the current flow is towards the reader, or out of the page.

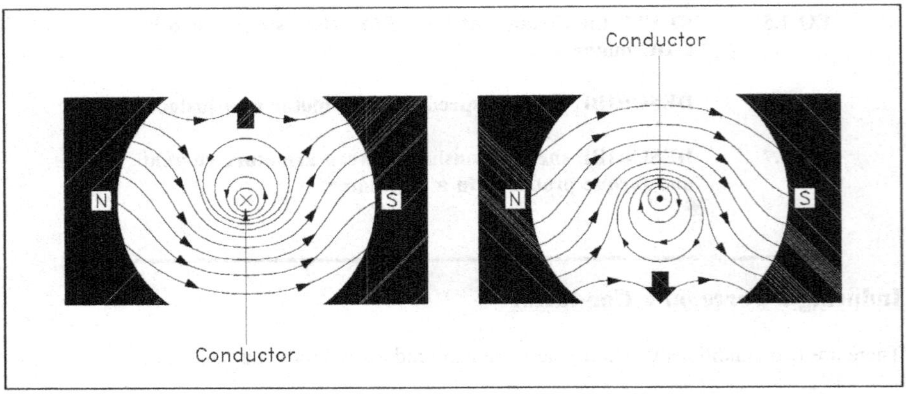

Figure 2 Current-Carrying Conductor in a Magnetic Field

Above the conductor on the left, the field caused by the conductor is in the opposite direction of the main field, and therefore, opposes the main field. Below the conductor on the left, the field caused by the conductor is in the same direction as the main field, and therefore, aids the main field. The net result is that above the conductor the main field is weakened, or flux density is decreased; below the conductor the field is strengthened, or flux density is increased. A force is developed on the conductor that moves the conductor in the direction of the weakened field (upward).

Above the conductor on the right, the field caused by the conductor is in the same direction as the main field, and therefore, aids the main field. Below the conductor on the right, the field caused by the conductor is in the opposite direction of the main field, and therefore, opposes the main field. The net result is that above the conductor the field is strengthened, or flux density is increased, and below the conductor, the field is weakened, or flux density is decreased. A force is developed on the conductor that moves the conductor in the direction of the weakened field (downward).

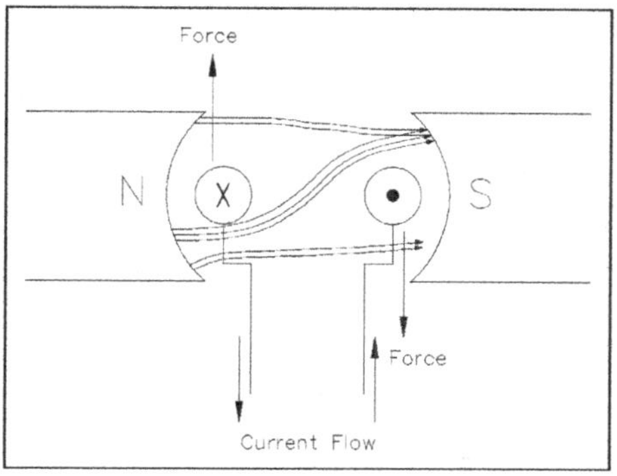

Figure 3 Motor Action

In a DC motor, the conductor will be formed in a loop such that two parts of the conductor are in the magnetic field at the same time, as shown in Figure 3.

This combines the effects of both conductors to distort the main magnetic field and produce a force on each part of the conductor. When the conductor is placed on a rotor, the force exerted on the conductors will cause the rotor to rotate clockwise, as shown on Figure 3.

You can think of these magnetic lines of force as rubber bands that are always trying to shorten themselves. The lines of force above the conductor exert a downward force due to the magnetic lines of force trying to straighten themselves.

The above explanation of how a force is developed is convenient; however, it is somewhat artificial. It is based on a fundamental principle of physics which may be stated as follows:

"A current-carrying conductor in a magnetic field tends to move at right angles to that field."

Another important way to show the relationship between the current-carrying conductor, magnetic field, and motion, is the right-hand rule for motors, as shown in Figure 4.

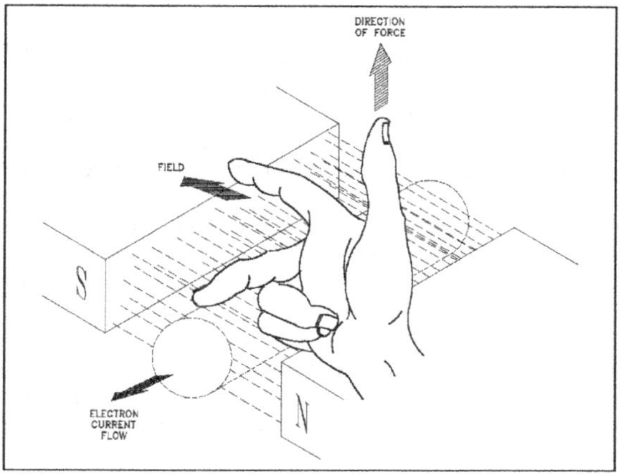

Figure 4 Right-Hand Rule for Motors

The right-hand rule for motors shows the direction in which a current-carrying conductor moves in a magnetic field. When the forefinger is pointed in the direction of the magnetic field lines, and the center finger is pointed in the direction of current flow, the thumb will point in the direction of force (motion).

Torque

Torque is defined as that force which tends to produce and maintain rotation. The function of torque in a DC motor is to provide the mechanical output or drive the piece of equipment that the DC motor is attached to.

When a voltage is applied to a motor, current will flow through the field winding, establishing a magnetic field. Current will also flow through the armature winding, from the negative brush to the positive brush as shown in Figure 5.

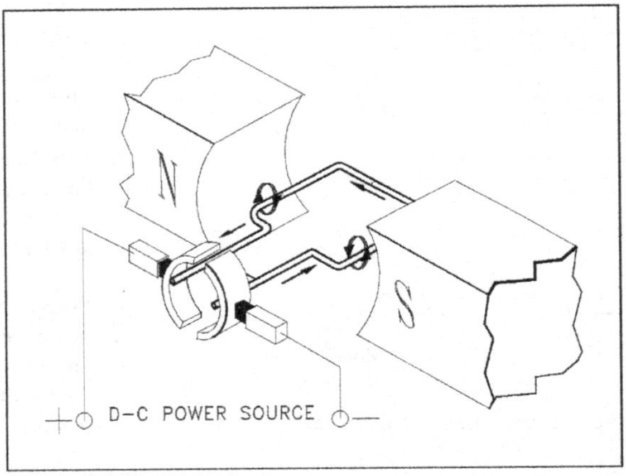

Figure 5 Armature Current in a Basic DC Motor

Since the armature is a current-carrying conductor in a magnetic field, the conductor has a force exerted on it, tending to move it at right angles to that field. Using the left-hand rule for current-carrying conductors, you will see that the magnetic field on one side is strengthened at the bottom, while it is weakened on the other side. Using the right-hand rule for motors, we can see that there is a force exerted on the armature which tends to turn the armature in the counter-clockwise direction. The sum of the forces, in pounds, multiplied by the radius of the armature, in feet, is equal to the torque developed by the motor in pound-feet (1b - ft).

It is evident from Figure 5 that if the armature current were reversed, but the field were the same, torque would be developed in the opposite direction. Likewise, if the field polarity were reversed and the armature remained the same, torque would also be developed in the opposite direction.

The force that is developed on a conductor of a motor armature is due to the combined action of the magnetic fields. The force developed is directly proportional to the strength of the main field flux and the strength of the field around the armature conductor. As we know, the field strength around each armature conductor depends on the amount of current flowing through the armature conductor. Therefore, the torque which is developed by the motor can be determined using Equation (6-1).

$$T = K\Phi I_a \tag{6-1}$$

where

T = torque, lb-ft
K = a constant depending on physical size of motor
Φ = field flux, number of lines of force per pole
I_a = armature current

Generator Action in a Motor

A generator action is developed in every motor. When a conductor cuts lines of force, an EMF is induced in that conductor.

Current to start the armature turning will flow in the direction determined by the applied DC power source. After rotation starts, the conductor cuts lines of force. By applying the left-hand rule for generators, the EMF that is induced in the armature will produce a current in the opposite direction. The induced EMF, as a result of motor operation, is called counterelectromotive force, or CEMF, as illustrated in Figure 6.

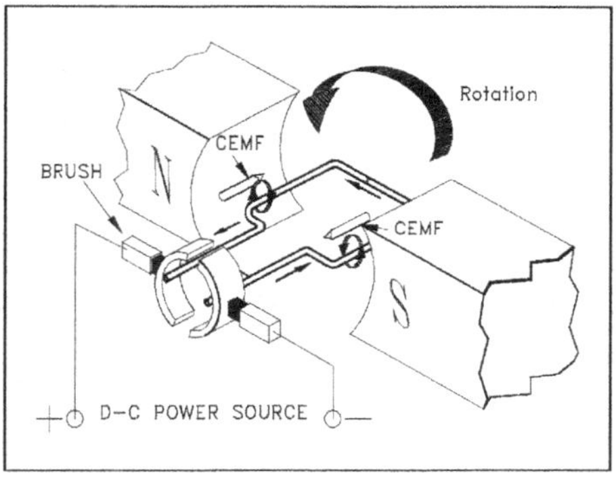

Figure 6 Counterelectromotive Force (CEMF)

Since the CEMF is generated by the action of the armature cutting lines of force, the value of CEMF will depend on field strength and armature speed, as shown in Equation (6-2).

$$E_{CEMF} = K\Phi N \tag{6-2}$$

where

E_{CEMF} = counter EMF
K = constant
Φ = field flux strength
N = speed of the armature

The CEMF opposes the applied voltage and functions to lower armature current. The effective voltage acting in the armature of a motor is the applied voltage, minus the counter EMF. Armature current can be found by using Ohm's law, as shown in Equation (6-3).

$$I_a = \frac{E_t - E_{CEMF}}{R_a} \tag{6-3}$$

where

I_a = armature current
E_t = terminal voltage
E_{CEMF} = counter EMF
R_a = armature resistance

DC Motor Speed

The field of a DC motor is varied using external devices, usually field resistors. For a constant applied voltage to the field (E), as the resistance of the field (R_f) is lowered, the amount of current flow through the field (I_f) increases as shown by Ohm's law in Equation (6-4).

$$\uparrow I_f = \frac{\overset{\leftrightarrow}{E}}{\downarrow R_f} \tag{6-4}$$

An increase in field current will cause field flux (Φ_f) to increase. Conversely, if the resistance of the field is increased, field flux will decrease. If the field flux of a DC motor is decreased, the motor speed will increase. The reduction of field strength reduces the CEMF of the motor, since fewer lines of flux are being cut by the armature conductors, as shown in Equation (6-5).

$$\downarrow E_{CEMF} = \vec{K}\,\overset{\downarrow}{\Phi_F}\,\vec{N} \tag{6-5}$$

A reduction of counter EMF allows an increase in armature current as shown in Equation (6-6).

$$\uparrow I_a = \frac{\overset{\rightarrow}{E_t} - \overset{\downarrow}{E_{CEMF}}}{\overset{\rightarrow}{R_a}} \tag{6-6}$$

This increase in armature current causes a larger torque to be developed; the increase in armature current more than offsets the decrease in field flux as shown in Equation (6-7).

$$\uparrow T = \vec{K}\,\overset{\downarrow}{\Phi_F}\,\overset{\uparrow}{I_a} \tag{6-7}$$

This increased torque causes the motor to increase in speed.

$$\uparrow T \propto N \uparrow$$

This increase in speed will then proportionately increase the CEMF. The speed and CEMF will continue to increase until the armature current and torque are reduced to values just large enough to supply the load at a new constant speed.

Summary

DC motor theory is summarized below.

DC Motor Theory Summary

- There are two conditions necessary to produce a force on a conductor:
  - The conductor must be carrying current.
  - The conductor must be within a magnetic field.
- The right-hand rule for motors states that when the forefinger is pointed in the direction of the magnetic field lines, and the center finger is pointed in the direction of current flow, the thumb will point in the direction of motion.
- The function of torque in a DC motor is to provide the mechanical output to drive the piece of equipment that the DC motor is attached to.
- Torque is developed in a DC motor by the armature (current-carrying conductor) being present in the motor field (magnetic field).
- CEMF is developed in a DC motor by the armature (conductor) rotating (relative motion) in the field of the motor (magnetic field).
- The function of the voltage that is developed in a DC motor (CEMF) opposes the applied voltage and results in the lowering of armature current.
- The speed of a DC motor may be changed by using resistors to vary the field current and, therefore, the field strength.

TYPES OF DC MOTORS

There are various types of DC motors found in industry today. Each type contains various characteristics that makes it desirable for certain applications.

EO 1.8 **DESCRIBE the differences in construction between a shunt-wound and a series-wound DC motor with respect to the relationship between the field and the armature windings.**

EO 1.9 **DESCRIBE the construction of a compounded DC motor.**

EO 1.10 **DESCRIBE the torque-vs-speed characteristics for a shunt-wound and a series-wound DC motor.**

DC Motor Connections

Figure 7 shows schematically the different methods of connecting the field and armature circuits in a DC motor. The circular symbol represents the armature circuit, and the squares at the side of the circle represent the brush commutator system. The direction of the arrows indicates the direction of the magnetic fields.

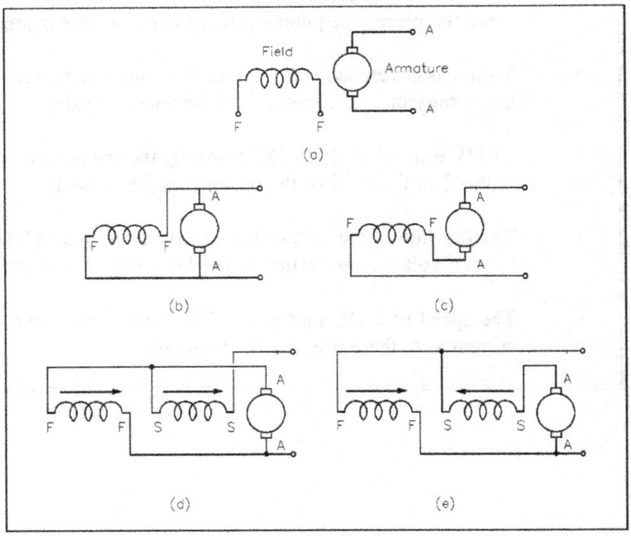

Figure 7 DC Motor Connections

- Figure 7a shows an externally-excited DC motor. This type of DC motor is constructed such that the field is not connected to the armature. This type of DC motor is not normally used.

- Figure 7b shows a shunt DC motor. The motor is called a "shunt" motor because the field is in parallel, or "shunts" the armature.

- Figure 7c shows a series DC motor. The motor field windings for a series motor are in series with the armature.

- Figures 7d and 7e show a compounded DC motor. A compounded DC motor is constructed so that it contains both a shunt and a series field. Figure 7d is called a "cumulatively-compounded" DC motor because the shunt and series fields are aiding one another. Figure 7e is called a "differentially-compounded" DC motor because the shunt and series field oppose one another.

Shunt-Wound Motor Operation

The speed-torque relationship for a typical shunt-wound motor is shown in Figure 8.

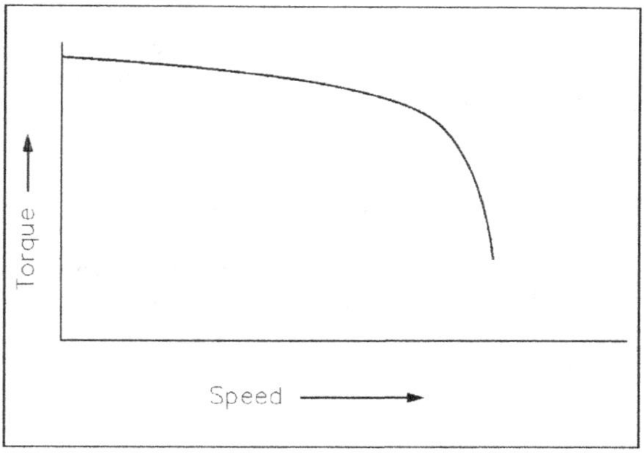

Figure 8 Torque-vs-Speed for a Shunt-Wound DC Motor

A shunt-wound DC motor has a decreasing torque when speed increases. The decreasing torque-vs-speed is caused by the armature resistance voltage drop and armature reaction. At a value of speed near 2.5 times the rated speed, armature reaction becomes excessive, causing a rapid decrease in field flux, and a rapid decline in torque until a stall condition is reached.

Shunt-Wound Motor Applications

The characteristics of a shunt-wound motor give it very good speed regulation, and it is classified as a constant speed motor, even though the speed does slightly decrease as load is increased. Shunt-wound motors are used in industrial and automotive applications where precise control of speed and torque are required.

Series-Wound Motor

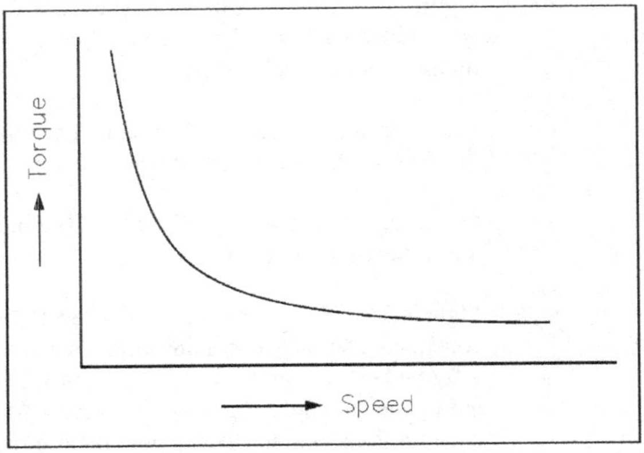

Figure 9 Torque-vs-Speed for a Series-Wound Motor

Since the armature and field in a series-wound motor are connected in series, the armature and field currents become identical, and the torque can be expressed as shown in Equation (6-8).

$$T = KI_a^2 \qquad (6\text{-}8)$$

The torque-vs-speed characteristics of a series-wound motor with a constant voltage source are shown in Figure 9. As the speed decreases, the torque for a series-wound motor increases sharply. As load is removed from a series motor, the speed will increase sharply. For these reasons, series-wound motors must have a load connected to prevent damage from high speed conditions.

Series-Wound Motor Applications

The advantage of a series-wound motor is that it develops a large torque and can be operated at low speed. It is a motor that is well-suited for starting heavy loads; it is often used for industrial cranes and winches where very heavy loads must be moved slowly and lighter loads moved more rapidly.

Compounded Motor

The compounded motor is desirable for a variety of applications because it combines the characteristics of a series-wound motor and a shunt-wound motor. The compounded motor has a greater torque than a shunt motor due to the series field; however, it has a fairly constant speed due to the shunt field winding. Loads such as presses, shears, and reciprocating machines are often driven by compounded motors.

Summary

The types of DC motors are summarized below.

Types of DC Motors Summary

- In a shunt-wound motor, the field is in parallel, or "shunts" the armature.
- In a series-wound motor, the field is in series with the armature.
- A compounded DC motor is constructed so that it contains both a shunt and a series field.
- A shunt-wound DC motor has a decreasing torque as speed increases.
- The characteristics of a shunt-wound motor give it very good speed regulation, and it is classified as a constant speed motor, even though the speed does slightly decrease as load is increased.
- A series-wound motor has a rapidly increasing torque when speed decreases. As load is removed from a series-wound motor, the speed will increase sharply.
- The advantages of a series-wound motor are that it develops a large torque and can be operated at low speed. It is a motor that is well-suited for starting heavy loads.

DC MOTOR OPERATION

DC motors require special starting resistors for operation due to their unique design. A knowledge of the operation of these starting resistors is necessary to understand DC motor operation.

EO 1.11 EXPLAIN why starting resistors are necessary for large DC motors.

EO 1.12 LIST the four nameplate ratings for a DC motor.

Starting of DC Motors

At the moment a DC motor is started the armature is stationary and there is no counter EMF being generated. The only component to limit starting current is the armature resistance, which, in most DC motors is a very low value (approximately one ohm or less), as shown in Equation (6-9).

$$I_a = \frac{E_t - E_{CEMF}}{R_a} \tag{6-9}$$

In order to reduce this very high starting current, an external resistance must be placed in series with the armature during the starting period. To show why this is essential, let us consider a 10-hp motor with an armature resistance of 0.4 ohms. If the motor were supplied by a 260 VDC source, the resulting current would be as shown in Equation (6-9).

$$I_a = \frac{E_t - E_{CEMF}}{R_a}$$

$$I_a = \frac{260\,\text{VDC} - 0}{0.4\Omega}$$

$$I_a = 650 \text{ amps}$$

This large current is approximately twelve times greater than actual full-load current for this motor. This high current would, in all probability, cause severe damage to the brushes, commutator, or windings. Starting resistors are usually incorporated into the motor design to limit starting current to 125 to 200 percent of full load current.

The amount of starting resistance necessary to limit starting current to a more desirable value is calculated using Equation (6-10.

$$R_s = \frac{E_t}{I_s} - R_a \tag{6-10}$$

where

R_s = starting resistance
E_t = terminal voltage
I_s = desired armature starting current
R_a = armature resistance

Example: If the full load current of the motor mentioned previously is 50 amps, and it is desired to limit starting current to 125% of this value, find the required resistance that must be added in series with the armature.

$$R_s = \frac{E_t}{I_s} - R_a$$

$$R_s = \frac{260\,\text{VDC}}{125\%(50\ \text{amps})} - 0.4\Omega$$

$$R_s = 3.76\Omega$$

The starting resistors are used in a DC motor by placing them in the starting circuit of the motor controller that is used to start the DC motor. Starting resistors are normally of variable resistances, with the value of resistance in the circuit at any time being either manually or automatically controlled. The maximum amount of resistance will always be inserted when the motor is first started. As the speed of the motor increases, counter EMF will begin to increase, decreasing armature current. The starting resistors may then be cut out, in successive steps, until the motor reaches full running speed.

DC Motor Ratings

The nameplate ratings of a DC motor refer to the conditions of voltage, current, speed, and power at which the motor is normally operated. The principal rating is known as the *continuous* rating, which is the rating described on the nameplate of a motor. The continuous power rating is a thermal rating. At this power, the motor can be operated for long periods of time without a large rise in temperature and beyond the limits of the conductor insulating material, bearings and other components, which are greatly affected by temperature.

The speed rating of a DC motor is often given on the nameplate. This speed is the upper limit at which a motor can be operated without mechanical damage occurring.

Summary

DC motor operation is summarized below.

DC Motor Operation Summary

- Starting resistors are necessary for large DC motors to prevent damage due to high currents while starting the motor.

- Starting resistors are placed in the starting circuits for the controllers that start the motor. When the motor reaches full speed, the starting resistors are cut out of the circuit.

- The four nameplate ratings for a DC motor include:

  - voltage
  - current
  - speed
  - power

DOE-HDBK-1011/3-92
JUNE 1992

DOE FUNDAMENTALS HANDBOOK
ELECTRICAL SCIENCE
Volume 3 of 4

U.S. Department of Energy
Washington, D.C. 20585

FSC-6910

Order No. DE92019787

ABSTRACT

The *Electrical Science Fundamentals Handbook* was developed to assist nuclear facility operating contractors provide operators, maintenance personnel, and the technical staff with the necessary fundamentals training to ensure a basic understanding of electrical theory, terminology, and application. The handbook includes information on alternating current (AC) and direct current (DC) theory, circuits, motors, and generators; AC power and reactive components; batteries; AC and DC voltage regulators; transformers; and electrical test instruments and measuring devices. This information will provide personnel with a foundation for understanding the basic operation of various types of DOE nuclear facility electrical equipment.

Key Words: Training Material, Magnetism, DC Theory, DC Circuits, Batteries, DC Generators, DC Motors, AC Theory, AC Power, AC Generators, Voltage Regulators, AC Motors, Transformers, Test Instruments, Electrical Distribution

FOREWORD

The *Department of Energy (DOE) Fundamentals Handbooks* consist of ten academic subjects, which include Mathematics; Classical Physics; Thermodynamics, Heat Transfer, and Fluid Flow; Instrumentation and Control; Electrical Science; Material Science; Mechanical Science; Chemistry; Engineering Symbology, Prints, and Drawings; and Nuclear Physics and Reactor Theory. The handbooks are provided as an aid to DOE nuclear facility contractors.

These handbooks were first published as Reactor Operator Fundamentals Manuals in 1985 for use by DOE category A reactors. The subject areas, subject matter content, and level of detail of the Reactor Opera tor Fundamentals Manuals were determined from several sources. DOE Category A reactor training managers determined which materials should be included, and served as a primary reference in the initial development phase. Training guidelines from the commercial nuclear power industry, results of job and task analyses, and independent input from contractors and operations-oriented personnel were all considered and included to some degree in developing the text material and learning objectives.

The *DOE Fundamentals Handbooks* represent the needs of various DOE nuclear facilities' fundamental training requirements. To increase their applicability to nonreactor nuclear facilities, the Reactor Operator Fundamentals Manual learning objectives were distributed to the Nuclear Facility Training Coordination Program Steering Committee for review and comment. To update their reactor-specific content, DOE Category A reactor training managers also reviewed and commented on the content. On the basis of feedback from these sources, information that applied to two or more DOE nuclear facilities was considered generic and was included. The final draft of each of the handbooks was then reviewed by these two groups. This approach has resulted in revised modular handbooks that contain sufficient detail such that each facility may adjust the content to fit their specific needs.

Each handbook contains an abstract, a foreword, an overview, learning objectives, and text material, and is divided into modules so that content and order may be modified by individual DOE contractors to suit their specific training needs. Each subject area is supported by a separate examination bank with an answer key.

The *DOE Fundamentals Handbooks* have been prepared for the Assistant Secretary for Nuclear Energy, Office of Nuclear Safety Policy and Standards, by the DOE Training Coordination Program. This program is managed by EG&G Idaho, Inc.

OVERVIEW

The *Department of Energy Fundamentals Handbook* entitled *Electrical Science* was prepared as an information resource for personnel who are responsible for the operation of the Department's nuclear facilities. A basic understanding of electricity and electrical systems is necessary for DOE nuclear facility operators, maintenance personnel, and the technical staff to safely operate and maintain the facility and facility support systems. The information in the handbook is presented to provide a foundation for applying engineering concepts to the job. This knowledge will help personnel more fully understand the impact that their actions may have on the safe and reliable operation of facility components and systems.

The *Electrical Science* handbook consists of fifteen modules that are contained in four volumes. The following is a brief description of the information presented in each module of the handbook.

Volume 1 of 4

Module 1 - Basic Electrical Theory

This module describes basic electrical concepts and introduces electrical terminology.

Module 2 - Basic DC Theory

This module describes the basic concepts of direct current (DC) electrical circuits and discusses the associated terminology.

Volume 2 of 4

Module 3 - DC Circuits

This module introduces the rules associated with the reactive components of inductance and capacitance and how they affect DC circuits.

Module 4 - Batteries

This module introduces batteries and describes the types of cells used, circuit arrangements, and associated hazards.

Module 5 - DC Generators

This module describes the types of DC generators and their application in terms of voltage production and load characteristics.

Module 6 - DC Motors

This module describes the types of DC motors and includes discussions of speed control, applications, and load characteristics.

Volume 3 of 4

Module 7 - Basic AC Theory

This module describes the basic concepts of alternating current (AC) electrical circuits and discusses the associated terminology.

Module 8 - AC Reactive Components

This module describes inductance and capacitance and their effects on AC circuits.

Module 9 - AC Power

This module presents power calculations for single-phase and three-phase AC circuits and includes the power triangle concept.

Module 10 - AC Generators

This module describes the operating characteristics of AC generators and includes terminology, methods of voltage production, and methods of paralleling AC generation sources.

Module 11 - Voltage Regulators

This module describes the basic operation and application of voltage regulators.

Volume 4 of 4

Module 12 - AC Motors

This module explains the theory of operation of AC motors and discusses the various types of AC motors and their application.

Module 13 - Transformers

This module introduces transformer theory and includes the types of transformers, voltage/current relationships, and application.

Module 14 - Test Instruments and Measuring Devices

This module describes electrical measuring and test equipment and includes the parameters measured and the principles of operation of common instruments.

Module 15 - Electrical Distribution Systems

This module describes basic electrical distribution systems and includes characteristics of system design to ensure personnel and equipment safety.

The information contained in this handbook is by no means all encompassing. An attempt to present the entire subject of electrical science would be impractical. However, the *Electrical Science* handbook does present enough information to provide the reader with a fundamental knowledge level sufficient to understand the advanced theoretical concepts presented in other subject areas, and to better understand basic system and equipment operations.

Department of Energy
Fundamentals Handbook

ELECTRICAL SCIENCE
Module 7
Basic AC Theory

www.KnowledgePublications.com

TABLE OF CONTENTS

LIST OF FIGURES ii

LIST OF TABLES iii

REFERENCES iv

OBJECTIVES v

AC GENERATION 1

Development of a Sine-Wave Output 1
Summary 3

AC GENERATION ANALYSIS 4

Effective Values 4
Phase Angle 7
Voltage Calculations 8
Current Calculations 9
Frequency Calculations 9
Summary 10

LIST OF FIGURES

Figure 1 Simple AC Generator 1

Figure 2 Developing a Sine-Wave Voltage 2

Figure 3 Voltage Sine Wave 4

Figure 4 Effective Value of Current 6

Figure 5 Phase Relationship 7

LIST OF TABLES

NONE

REFERENCES

- Gussow, Milton, Schaum's Outline Series, Basic Electricity, McGraw-Hill.
- Academic Program for Nuclear Power Plant Personnel, Volume IV, Columbia, MD: General Physics Corporation, Library of Congress Card #A 326517, 1982.
- Sienko and Plane, Chemical Principles and Properties, 2nd Edition, McGraw-Hill.
- Academic Program for Nuclear Power Plant Personnel, Volume II, Columbia, MD: General Physics Corporation, Library of Congress Card #A 326517, 1981.
- Nasar and Unnewehr, Electromechanics and Electric Machines, John Wiley and Sons.
- Van Valkenburgh, Nooger, and Neville, Basic Electricity, Vol. 5, Hayden Book Company.
- Exide Industrial Marketing Division, The Storage Battery, Lead-Acid Type, The Electric Storage Battery Company.
- Lister, Eugene C., Electric Circuits and Machines, 5th Edition, McGraw-Hill.
- Croft, Carr, Watt, and Summers, American Electricians Handbook, 10th Edition, McGraw-Hill.
- Mason, C. Russel, The Art and Science of Protective Relaying, John Wiley and Sons.
- Mileaf, Harry, Electricity One - Seven, Revised 2nd Edition, Hayden Book Company.
- Buban and Schmitt, Understanding Electricity and Electronics, 3rd Edition, McGraw-Hill.
- Kidwell, Walter, Electrical Instruments and Measurements, McGraw-Hill.

TERMINAL OBJECTIVE

1.0 Given an alternating current (AC) waveform, **DESCRIBE** the relationship between average and RMS values of voltage and current, and the angular velocity within that waveform.

ENABLING OBJECTIVES

1.1 **DESCRIBE** the construction and operation of a simple AC generator.

1.2 **EXPLAIN** the development of a sine-wave output in an AC generator.

1.3 **DEFINE** the following terms in relation to AC generation:
   a. Radians/second
   b. Hertz
   c. Period

1.4 **DEFINE** effective value of an AC current relative to DC current.

1.5 Given a maximum value, **CALCULATE** the effective (RMS) and average values of AC voltage.

1.6 Given a diagram of two sine waves, **DESCRIBE** the phase relationship between the two waves.

Intentionally Left Blank

AC GENERATION

An understanding of how an AC generator develops an AC output will help the student analyze the AC power generation process.

EO 1.1 DESCRIBE the construction and operation of a simple AC generator.

EO 1.2 EXPLAIN the development of a sine-wave output in an AC generator.

The elementary AC generator (Figure 1) consists of a conductor, or loop of wire in a magnetic field that is produced by an electromagnet. The two ends of the loop are connected to slip rings, and they are in contact with two brushes. When the loop rotates it cuts magnetic lines of force, first in one direction and then the other.

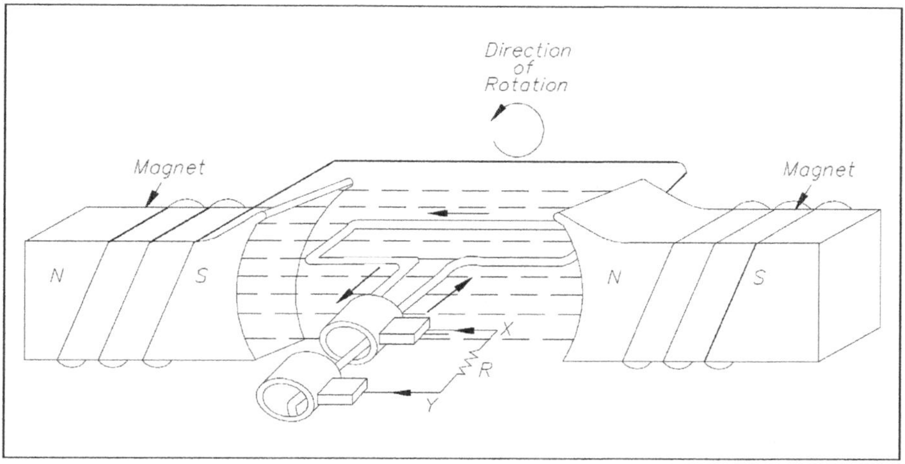

Figure 1 Simple AC Generator

Development of a Sine-Wave Output

At the instant the loop is in the vertical position (Figure 2, 0°), the coil sides are moving parallel to the field and do not cut magnetic lines of force. In this instant, there is no voltage induced in the loop. As the coil rotates in a counter-clockwise direction, the coil sides will cut the magnetic lines of force in opposite directions. The direction of the induced voltages depends on the direction of movement of the coil.

The induced voltages add in series, making slip ring X (Figure 1) positive (+) and slip ring Y (Figure 1) negative (-). The potential across resistor R will cause a current to flow from Y to X through the resistor. This current will increase until it reaches a maximum value when the coil is horizontal to the magnetic lines of force (Figure 2, 90°). The horizontal coil is moving perpendicular to the field and is cutting the greatest number of magnetic lines of force. As the coil continues to turn, the voltage and current induced decrease until they reach zero, where the coil is again in the vertical position (Figure 2, 180°). In the other half revolution, an equal voltage is produced except that the polarity is reversed (Figure 2, 270°, 360°). The current flow through R is now from X to Y (Figure 1).

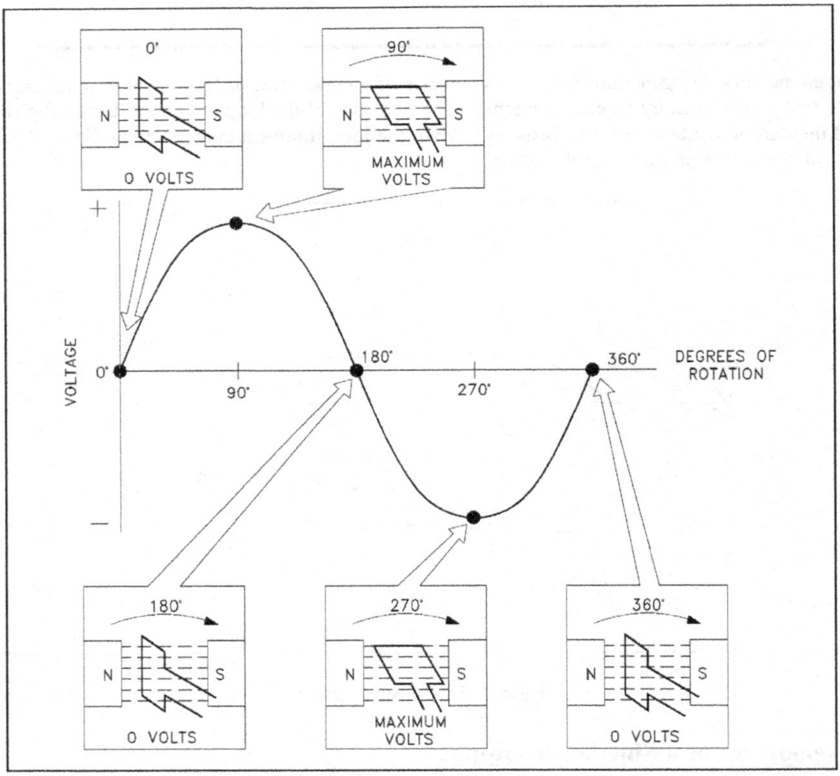

Figure 2 Developing a Sine-Wave Voltage

The periodic reversal of polarity results in the generation of a voltage, as shown in Figure 2. The rotation of the coil through 360° results in an AC sine wave output.

Summary

AC generation is summarized below.

AC Generation Summary

- A simple generator consists of a conductor loop turning in a magnetic field, cutting across the magnetic lines of force.
- The sine wave output is the result of one side of the generator loop cutting lines of force. In the first half turn of rotation this produces a positive current and in the second half of rotation produces a negative current. This completes one cycle of AC generation.

AC GENERATION ANALYSIS

Analysis of the AC power generation process and of the alternating current we use in almost every aspect of our lives is necessary to better understand how AC power is used in today's technology.

EO 1.3 **DEFINE the following terms in relation to AC generation:**
- **a. Radians/second**
- **b. Hertz**
- **c. Period**

EO 1.4 **DEFINE effective value of an AC current relative to DC current.**

EO 1.5 **Given a maximum value, CALCULATE the effective (RMS) and average values of AC voltage.**

EO 1.6 **Given a diagram of two sine waves, DESCRIBE the phase relationship between the two waves.**

Effective Values

The output voltage of an AC generator can be expressed in two ways. One is graphically by use of a sine wave (Figure 3). The second way is algebraically by the equation $e = E_{max} \sin \omega t$, which will be covered later in the text.

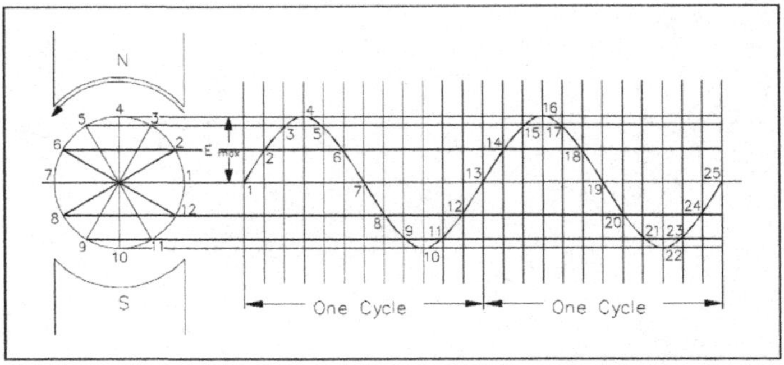

Figure 3 Voltage Sine Wave

When a voltage is produced by an AC generator, the resulting current varies in step with the voltage. As the generator coil rotates 360°, the output voltage goes through one complete cycle. In one cycle, the voltage increases from zero to E_{max} in one direction, decreases to zero, increases to E_{max} in the opposite direction (negative E_{max}), and then decreases to zero again. The value of E_{max} occurs at 90° and is referred to as peak voltage. The time it takes for the generator to complete one cycle is called the *period*, and the number of cycles per second is called the *frequency* (measured in *hertz*).

One way to refer to AC voltage or current is by peak voltage (E_p) or peak current (I_p). This is the maximum voltage or current for an AC sine wave.

Another value, the peak-to-peak value (E_{p-p} or I_{p-p}), is the magnitude of voltage, or current range, spanned by the sine wave. However, the value most commonly used for AC is *effective value.* Effective value of AC is the amount of AC that produces the same heating effect as an equal amount of DC. In simpler terms, one ampere effective value of AC will produce the same amount of heat in a conductor, in a given time, as one ampere of DC. The heating effect of a given AC current is proportional to the square of the current. Effective value of AC can be calculated by squaring all the amplitudes of the sine wave over one period, taking the average of these values, and then taking the square root. The effective value, being the root of the mean (average) square of the currents, is known as the root-mean-square, or RMS value. In order to understand the meaning of effective current applied to a sine wave, refer to Figure 4.

The values of I are plotted on the upper curve, and the corresponding values of I^2 are plotted on the lower curve. The I^2 curve has twice the frequency of I and varies above and below a new axis. The new axis is the average of the I^2 values, and the square root of that value is the RMS, or effective value, of current. The average value is $\frac{1}{2} I_{max}^2$. The RMS value is then $\frac{\sqrt{2 I_{max}^2}}{2}$ OR $\frac{\sqrt{2}}{2} I_{max}$, which is equal to 0.707 I_{max}.

There are six basic equations that are used to convert a value of AC voltage or current to another value, as listed below.

Average value = peak value x 0.637 (7-1)

Effective value (RMS) = peak value x 0.707 (7-2)

Peak value = average value x 1.57 (7-3)

Effective value (RMS) = average value x 1.11 (7-4)

Peak value = effective value (RMS) x 1.414 (7-5)

Average value = effective (RMS) x 0.9 (7-6)

The values of current (I) and voltage (E) that are normally encountered are assumed to be RMS values; therefore, no subscript is used.

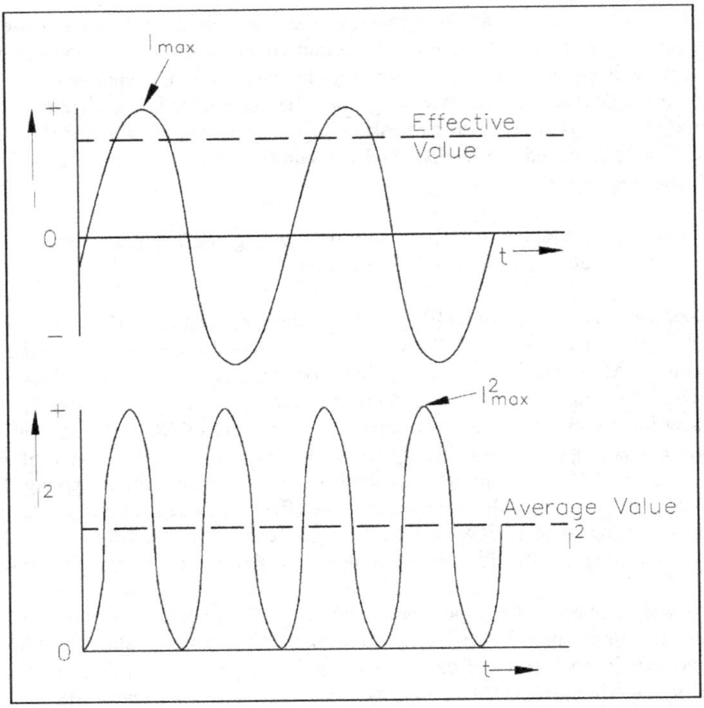

Figure 4 Effective Value of Current

Another useful value is the average value of the amplitude during the positive half of the cycle. Equation (7-7) is the mathematical relationship between I_{av}, I_{max}, and I.

$$I_{av} = 0.637\ I_{max} = 0.90\ I \tag{7-7}$$

Equation (7-8) is the mathematical relationship between E_{av}, E_{max}, and E.

$$E_{av} = 0.637\ E_{max} = 0.90\ E \tag{7-8}$$

Example 1: The peak value of voltage in an AC circuit is 200 V. What is the RMS value of the voltage?

$E = 0.707\ E_{max}$
$E = 0.707\ (200\ V)$
$E = 141.4\ V$

Example 2: The peak current in an AC circuit is 10 amps. What is the average value of current in the circuit?

$$I_{av} = 0.637\ I_{max}$$
$$I_{av} = 0.637\ (10\ \text{amps})$$
$$I_{av} = 6.37\ \text{amps}$$

Phase Angle

Phase angle is the fraction of a cycle, in degrees, that has gone by since a voltage or current has passed through a given value. The given value is normally zero. Referring back to Figure 3, take point 1 as the starting point or zero phase. The phase at Point 2 is 30°, Point 3 is 60°, Point 4 is 90°, and so on, until Point 13 where the phase is 360°, or zero. A term more commonly used is phase difference. The phase difference can be used to describe two different voltages that have the same frequency, which pass through zero values in the same direction at different times. In Figure 5, the angles along the axis indicate the phases of voltages e_1 and e_2 at any point in time. At 120°, e_1 passes through the zero value, which is 60° ahead of e_2 (e_2 equals zero at 180°). The voltage e_1 is said to lead e_2 by 60 electrical degrees, or it can be said that e_2 lags e_1 by 60 electrical degrees.

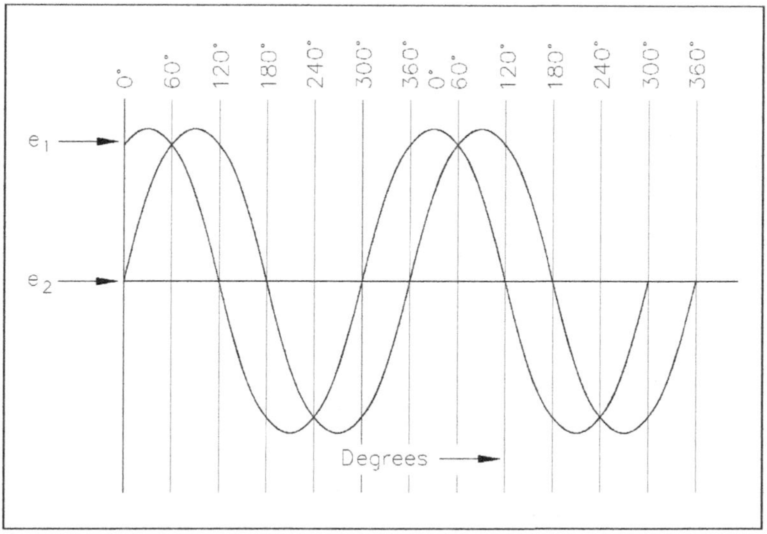

Figure 5 Phase Relationship

Phase difference is also used to compare two different currents or a current and a voltage. If the phase difference between two currents, two voltages, or a voltage and a current is zero degrees, they are said to be "in-phase." If the phase difference is an amount other than zero, they are said to be "out-of-phase."

Voltage Calculations

Equation (7-9) is a mathematical representation of the voltage associated with any particular orientation of a coil (inductor).

$$e = E_{max} \sin\theta \qquad (7\text{-}9)$$

where

e = induced EMF
E_{max} = maximum induced EMF
θ = angle from reference (degrees or radians)

Example 1: What is the induced EMF in a coil producing a maximum EMF of 120 V when the angle from reference is 45°?

$e = E_{max} \sin \theta$
$e = 120 \text{ V} (\sin 45°)$
$e = 84.84 \text{ V}$

The maximum induced voltage can also be called peak voltage E_p. If (t) is the time in which the coil turns through the angle (θ), then the angular velocity (ω) of the coil is equal to θ/t and is expressed in units of radians/sec. Equation (7-10) is the mathematical representation of the angular velocity.

$$\theta = \omega t \qquad (7\text{-}10)$$

where

ω = angular velocity (radians/sec)
t = time to turn through the angle from reference (sec)
θ = angle from reference (radians)

Using substitution laws, a relationship between the voltage induced, the maximum induced voltage, and the angular velocity can be expressed. Equation (7-11) is the mathematical representation of the relationship between the voltage induced, the maximum voltage, and the angular velocity, and is equal to the output of an AC Generator.

$$e = E_{max} \sin(\omega t) \qquad (7\text{-}11)$$

where

e	=	induced EMF (volts)
E_{max}	=	maximum induced EMF (volts)
ω	=	angular velocity (radians/sec)
t	=	time to turn through the angle from reference (sec)

Current Calculations

Maximum induced current is calculated in a similar fashion. Equation (7-12) is a mathematical representation of the relationship between the maximum induced current and the angular velocity.

$$i = I_{max} \sin(\omega t) \tag{7-12}$$

where

i = induced current (amps)
I_{max} = maximum induced current (amps)
ω = angular velocity (radians/sec)
t = time to turn through the angle from reference (sec)

Frequency Calculations

The frequency of an alternating voltage or current can be related directly to the angular velocity of a rotating coil. The units of angular velocity are ***radians per second***, and 2π radians is a full revolution. A radian is an angle that subtends an arc equal to the radius of a circle. One radian equals 57.3 degrees. One cycle of the sine wave is generated when the coil rotates 2π radians. Equation (7-13) is the mathematical relationship between frequency (f) and the angular velocity (ω) in an AC circuit.

$$\omega = 2\pi f \tag{7-13}$$

where

ω = angular velocity (radians/sec)
f = frequency (HZ)

Example 1: The frequency of a 120 V AC circuit is 60 Hz. Find the following:
1. Angular velocity
2. Angle from reference at 1 msec
3. Induced EMF at that point

Solution:

1. $\omega = 2\pi f$
$= 2\ (3.14)\ (60\ Hz)$
$= 376.8$ radians/sec

2. $\theta = \omega t$
$= (376.8$ radian/sec$)\ (.001$ sec$)$
$= 0.3768$ radians

3. $e = E_{max} \sin\theta$
$= (120\ V)\ (\sin 0.3768$ radians$)$
$= (120\ V)\ (0.3679)$
$= 44.15\ V$

Summary

AC generation analysis is summarized below.

Voltage, Current, and Frequency Summary

- The following terms relate to the AC cycle: radians/second, the velocity the loop turns; hertz, the number of cycles in one second; period, the time to complete one cycle.
- Effective value of AC equals effective value of DC.
- Root mean square (RMS) values equate AC to DC equivalents:
  - $I = 0.707\ I_{max}$ = Effective Current
  - $E = 0.707\ E_{max}$ = Effective Voltage
  - $I_{av} = 0.636\ I_{max}$ = 0.9 I = Average Current
  - $E_{av} = 0.636\ E_{max}$ = 0.9 E = Average Voltage
- Phase angle is used to compare two wave forms. It references the start, or zero point, of each wave. It compares differences by degrees of rotation. Wave forms with the same start point are "in-phase" while wave forms "out-of-phase" either lead or lag.

Department of Energy
Fundamentals Handbook

ELECTRICAL SCIENCE
Module 8
Basic AC Reactive Components

TABLE OF CONTENTS

LIST OF FIGURES ii

LIST OF TABLES iii

REFERENCES iv

OBJECTIVES v

INDUCTANCE 1

Inductive Reactance 1
Voltage and Current Phase Relationships in an Inductive Circuit 2
Summary 4

CAPACITANCE 5

Capacitors 5
Capacitive Reactance 6
Summary 8

IMPEDANCE 9

Impedance 9
Impedance in R-L Circuits 11
Impedance in R-C Circuits 12
Impedance in R-C-L Circuits 13
Summary 18

RESONANCE 19

Resonant Frequency 19
Series Resonance 19
Parallel Resonance 20
Summary 20

LIST OF FIGURES

Figure 1 Current, Self-Induced EMF, and Applied Voltage in Inductive Circuit 2

Figure 2 Coil Circuit and Phasor Diagram 3

Figure 3 Voltage, Charge, and Current in a Capacitor 5

Figure 4 Circuit and Phasor Diagram 7

Figure 5 Relationship Between Resistance, Reactance, and Impedance 10

Figure 6 Simple R-L Circuit 12

Figure 7 Simple R-C Circuit 13

Figure 8 Series R-C-L Impedance-Phasor 14

Figure 9 Simple R-C-L Circuit 15

Figure 10 Simple Parallel R-C-L Circuit 16

LIST OF TABLES

NONE

REFERENCES

- Gussow, Milton, Schaum's Outline Series, Basic Electricity, McGraw-Hill.
- Academic Program for Nuclear Power Plant Personnel, Volume IV, Columbia, MD: General Physics Corporation, Library of Congress Card #A 326517, 1982.
- Sienko and Plane, Chemical Principles and Properties, 2nd Edition, McGraw-Hill.
- Academic Program for Nuclear Power Plant Personnel, Volume II, Columbia, MD: General Physics Corporation, Library of Congress Card #A 326517, 1982.
- Nasar and Unnewehr, Electromechanics and Electric Machines, John Wiley and Sons.
- Van Valkenburgh, Nooger, and Neville, Basic Electricity, Vol. 5, Hayden Book Company.
- Exide Industrial Marketing Division, The Storage Battery, Lead-Acid Type, The Electric Storage Battery Company.
- Lister, Eugene C., Electric Circuits and Machines, 5th Edition, McGraw-Hill.
- Croft, Carr, Watt, and Summers, American Electricians Handbook, 10th Edition, McGraw-Hill.
- Mason, C. Russel, The Art and Science of Protective Relaying, John Wiley and Sons.
- Mileaf, Harry, Electricity One - Seven, Revised 2nd Edition, Hayden Book Company.
- Buban and Schmitt, Understanding Electricity and Electronics, 3rd Edition, McGraw-Hill.
- Kidwell, Walter, Electrical Instruments and Measurements, McGraw-Hill.

TERMINAL OBJECTIVE

1.0 Using the rules associated with inductors and capacitors, **DESCRIBE** the characteristics of these elements when they are placed in an AC circuit.

ENABLING OBJECTIVES

1.1 **DESCRIBE** inductive reactance (X_L).

1.2 Given the operation frequency (f) and the value of inductance (L), **CALCULATE** the inductive reactance (X_L) of a simple circuit.

1.3 **DESCRIBE** the effect of the phase relationship between current and voltage in an inductive circuit.

1.4 **DRAW** a simple phasor diagram representing AC current (I) and voltage (E) in an inductive circuit.

1.5 **DEFINE** capacitive reactance (X_C).

1.6 Given the operating frequency (f) and the value of capacitance (C), **CALCULATE** the capacitive reactance (X_C) of a simple AC circuit.

1.7 **DESCRIBE** the effect on phase relationship between current (I) and voltage (E) in a capacitive circuit.

1.8 **DRAW** a simple phasor diagram representing AC current (I) and voltage (E) in a capacitive circuit.

1.9 **DEFINE** impedance (Z).

1.10 Given the values for resistance (R) and inductance (L) and a simple R-L series AC circuit, **CALCULATE** the impedance (Z) for that circuit.

1.11 Given the values for resistance (R) and capacitance (C) and a simple R-C series AC circuit, **CALCULATE** the impedance (Z) for that circuit.

1.12 Given a simple R-C-L series AC circuit and the values for resistance (R), inductive reactance (X_L), and capacitive reactance (X_C), **CALCULATE** the impedance (Z) for that circuit.

ENABLING OBJECTIVES (Cont.)

1.13 **STATE** the formula for calculating total current (I_T) in a simple parallel R-C-L AC circuit.

1.14 Given a simple R-C-L parallel AC circuit and the values for voltage (V_T), resistance (R), inductive reactance (X_L), and capacitive reactance (X_C), **CALCULATE** the impedance (Z) for that circuit.

1.15 **DEFINE** resonance.

1.16 Given the values of capacitance (C) and inductance (L), **CALCULATE** the resonant frequency.

1.17 Given a series R-C-L circuit at resonance, **DESCRIBE** the net reactance of the circuit.

1.18 Given a parallel R-C-L circuit at resonance, **DESCRIBE** the circuit output relative to current (I).

INDUCTANCE

Any device relying on magnetism or magnetic fields to operate is a form of inductor. Motors, generators, transformers, and coils are inductors. The use of an inductor in a circuit can cause current and voltage to become out-of-phase and inefficient unless corrected.

EO 1.1 **DESCRIBE inductive reactance (X_L).**

EO 1.2 **Given the operation frequency (f) and the value of inductance (L), CALCULATE the inductive reactance (X_L) of a simple circuit.**

EO 1.3 **DESCRIBE the effect of the phase relationship between current and voltage in an inductive circuit.**

EO 1.4 **DRAW a simple phasor diagram representing AC current (I) and voltage (E) in an inductive circuit.**

Inductive Reactance

In an inductive AC circuit, the current is continually changing and is continuously inducing an EMF. Because this EMF opposes the continuous change in the flowing current, its effect is measured in ohms. This opposition of the inductance to the flow of an alternating current is called *inductive reactance* (X_L). Equation (8-1) is the mathematical representation of the current flowing in a circuit that contains only inductive reactance.

$$I = \frac{E}{X_L} \tag{8-1}$$

where

I = effective current (A)
X_L = inductive reactance (Ω)
E = effective voltage across the reactance (V)

The value of X_L in any circuit is dependent on the inductance of the circuit and on the rate at which the current is changing through the circuit. This rate of change depends on the frequency of the applied voltage. Equation (8-2) is the mathematical representation for X_L.

$$X_L = 2\pi fL \tag{8-2}$$

where

π = ~3.14
f = frequency (Hertz)
L = inductance (Henries)

The magnitude of an induced EMF in a circuit depends on how fast the flux that links the circuit is changing. In the case of self-induced EMF (such as in a coil), a counter EMF is induced in the coil due to a change in current and flux in the coil. This CEMF opposes any change in current, and its value at any time will depend on the rate at which the current and flux are changing at that time. In a purely inductive circuit, the resistance is negligible in comparison to the inductive reactance. The voltage applied to the circuit must always be equal and opposite to the EMF of self-induction.

Voltage and Current Phase Relationships in an Inductive Circuit

As previously stated, any change in current in a coil (either a rise or a fall) causes a corresponding change of the magnetic flux around the coil. Because the current changes at its maximum rate when it is going through its zero value at 90° (point b on Figure 1) and 270° (point d), the flux change is also the greatest at those times. Consequently, the self-induced EMF in the coil is at its maximum (or minimum) value at these points, as shown in Figure 1. Because the current is not changing at the point when it is going through its peak value at 0° (point a), 180° (point c), and 360° (point e), the flux change is zero at those times. Therefore, the self-induced EMF in the coil is at its zero value at these points.

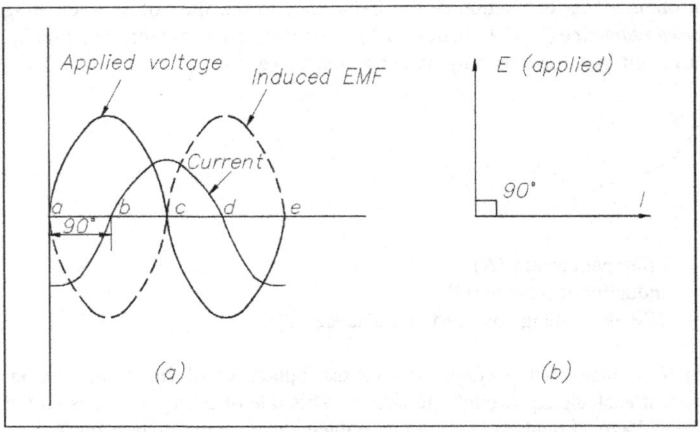

Figure 1 Current, Self-Induced EMF, and Applied Voltage in an Inductive Circuit

According to Lenz's Law (refer to Module 1, Basic Electrical Theory), the induced voltage always opposes the change in current. Referring to Figure 1, with the current at its maximum negative value (point a), the induced EMF is at a zero value and falling. Thus, when the current rises in a positive direction (point a to point c), the induced EMF is of opposite polarity to the applied voltage and opposes the rise in current. Notice that as the current passes through its zero value (point b) the induced voltage reaches its maximum negative value. With the current now at its maximum positive value (point c), the induced EMF is at a zero value and rising. As the current is falling toward its zero value at 180° (point c to point d), the induced EMF is of the same polarity as the current and tends to keep the current from falling. When the current reaches a zero value, the induced EMF is at its maximum positive value. Later, when the current is increasing from zero to its maximum negative value at 360° (point d to point e), the induced voltage is of the opposite polarity as the current and tends to keep the current from increasing in the negative direction. Thus, the induced EMF can be seen to lag the current by 90°.

The value of the self-induced EMF varies as a sine wave and lags the current by 90°, as shown in Figure 1. The applied voltage must be equal and opposite to the self-induced EMF at all times; therefore, the current lags the applied voltage by 90° in a purely inductive circuit.

If the applied voltage (E) is represented by a vector rotating in a counterclockwise direction (Figure 1b), then the current can be expressed as a vector that is lagging the applied voltage by 90°. Diagrams of this type are referred to as *phasor diagrams*.

Example: A 0.4 H coil with negligible resistance is connected to a 115V, 60 Hz power source (see Figure 2). Find the inductive reactance of the coil and the current through the circuit. Draw a phasor diagram showing the phase relationship between current and applied voltage.

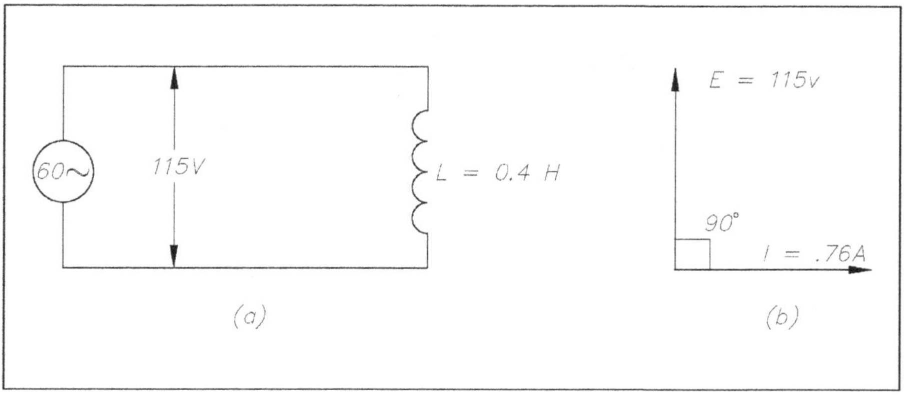

Figure 2 Coil Circuit and Phasor Diagram

Solution:

1. Inductive reactance of the coil

$$X_L = 2\pi fL$$

$$= (2)(3.14)(60)(0.4)$$

$$X_L = 150.7\Omega$$

2. Current through the circuit

$$I = \frac{E}{X_L}$$

$$= \frac{115}{150.7}$$

$$I = 0.76 \text{ amps}$$

3. Draw a phasor diagram showing the phase relationship between current and applied voltage.

   Phasor diagram showing the current lagging voltage by 90° is drawn in Figure 2b.

Summary

Inductive reactance is summarized below.

Inductive Reactance Summary

- Opposition to the flow of alternating current caused by inductance is called Inductive Reactance (X_L).
- The formula for calculating X_L is:

  $$X_L = 2\pi fL$$

- Current (I) lags applied voltage (E) in a purely inductive circuit by 90° phase angle.
- The phasor diagram shows the applied voltage (E) vector leading (above) the current (I) vector by the amount of the phase angle differential due to the relationship between voltage and current in an inductive circuit.

CAPACITANCE

There are many natural causes of capacitance in AC power circuits, such as transmission lines, fluorescent lighting, and computer monitors. Normally, these are counteracted by the inductors previously discussed. However, where capacitors greatly outnumber inductive devices, we must calculate the amount of capacitance to add or subtract from an AC circuit by artificial means.

EO 1.5 **DEFINE capacitive reactance (X_C).**

EO 1.6 **Given the operating frequency (f) and the value of capacitance (C), CALCULATE the capacitive reactance (X_C) of a simple AC circuit.**

EO 1.7 **DESCRIBE the effect on phase relationship between current (I) and voltage (E) in a capacitive circuit.**

EO 1.8 **DRAW a simple phasor diagram representing AC current (I) and voltage (E) in a capacitive circuit.**

Capacitors

The variation of an alternating voltage applied to a capacitor, the charge on the capacitor, and the current flowing through the capacitor are represented by Figure 3.

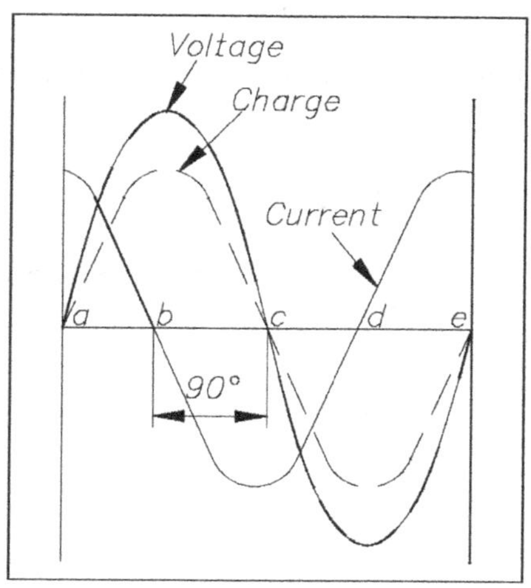

Figure 3 Voltage, Charge, and Current in a Capacitor

The current flow in a circuit containing capacitance depends on the rate at which the voltage changes. The current flow in Figure 3 is greatest at points a, c, and e. At these points, the voltage is changing at its maximum rate (i.e., passing through zero). Between points a and b, the voltage and charge are increasing, and the current flow is into the capacitor, but decreasing in value. At point b, the capacitor is fully charged, and the current is zero. From points b to c, the voltage and charge are decreasing as the capacitor discharges, and its current flows in a direction opposite to the voltage. From points c to d, the capacitor begins to charge in the opposite direction, and the voltage and current are again in the same direction.

At point d, the capacitor is fully charged, and the current flow is again zero. From points d to e, the capacitor discharges, and the flow of current is opposite to the voltage. Figure 3 shows the current leading the applied voltage by 90°. In any purely capacitive circuit, current leads applied voltage by 90°.

Capacitive Reactance

Capacitive reactance is the opposition by a capacitor or a capacitive circuit to the flow of current. The current flowing in a capacitive circuit is directly proportional to the capacitance and to the rate at which the applied voltage is changing. The rate at which the applied voltage is changing is determined by the frequency of the supply; therefore, if the frequency of the capacitance of a given circuit is increased, the current flow will increase. It can also be said that if the frequency or capacitance is increased, the opposition to current flow decreases; therefore, capacitive reactance, which is the opposition to current flow, is inversely proportional to frequency and capacitance. Capacitive reactance X_C, is measured in ohms, as is inductive reactance. Equation (8-3) is a mathematical representation for capacitive reactance.

$$X_C = \frac{1}{2\pi fC} \tag{8-3}$$

where

f = frequency (Hz)
π = ~3.14
C = capacitance (farads)

Equation (8-4) is the mathematical representation of capacitive reactance when capacitance is expressed in microfarads (μF).

$$X_C = \frac{1{,}000{,}000}{2\pi fC} \tag{8-4}$$

Equation (8-5) is the mathematical representation for the current that flows in a circuit with only capacitive reactance.

$$I = \frac{E}{X_C} \tag{8-5}$$

where

I = effective current (A)
E = effective voltage across the capacitive reactance (V)
X_C = capacitive reactance (Ω)

Example: A 10μF capacitor is connected to a 120V, 60Hz power source (see Figure 4). Find the capacitive reactance and the current flowing in the circuit. Draw the phasor diagram.

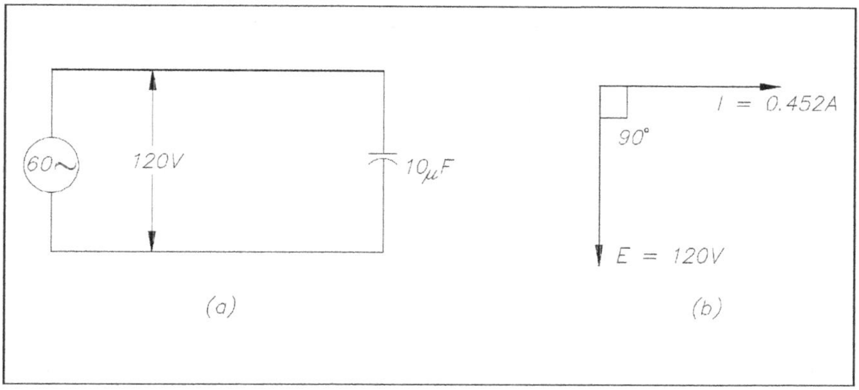

Figure 4 Circuit and Phasor Diagram

Solution:

1. Capacitive reactance

$$X_C = \frac{1{,}000{,}000}{2\pi fC}$$

$$= \frac{1{,}000{,}000}{(2)(3.14)(60)(10)}$$

$$= \frac{1{,}000{,}000}{3768}$$

$$X_C = 265.4\Omega$$

2. Current flowing in the circuit

$$I = \frac{E}{X_C}$$

$$= \frac{120}{265.4}$$

$$I = 0.452 \text{ amps}$$

3. Phasor diagram showing current leading voltage by 90° is drawn in Figure 4b.

Summary

Capacitive reactance is summarized below.

Capacitive Reactance Summary

- Opposition to the flow of alternating current caused by capacitance is called capacitive reactance (X_C).
- The formula for calculating X_C is:

$$X_C = \frac{1}{2\pi fC}$$

- Current (I) leads applied voltage by 90° in a purely capacitive circuit.
- The phasor diagram shows the applied voltage (E) vector leading (below) the current (I) vector by the amount of the phase angle differential due to the relationship between voltage and current in a capacitive circuit.

IMPEDANCE

Whenever inductive and capacitive components are used in an AC circuit, the calculation of their effects on the flow of current is important.

EO 1.9 **DEFINE impedance (Z).**

EO 1.10 **Given the values for resistance (R) and inductance (L) and a simple R-L series AC circuit, CALCULATE the impedance (Z) for that circuit.**

EO 1.11 **Given the values for resistance (R) and capacitance (C) and a simple R-C series AC circuit, CALCULATE the impedance (Z) for that circuit.**

EO 1.12 **Given a simple R-C-L series AC circuit and the values for resistance (R), inductive reactance (X_L), and capacitive reactance (X_C), CALCULATE the impedance (Z) for that circuit.**

EO 1.13 **STATE the formula for calculating total current (I_T) in a simple parallel R-C-L AC circuit.**

EO 1.14 **Given a simple R-C-L parallel AC circuit and the values for voltage (V_T), resistance (R), inductive reactance (X_L), and capacitive reactance (X_C), CALCULATE the impedance (Z) for that circuit.**

Impedance

No circuit is without some resistance, whether desired or not. Resistive and reactive components in an AC circuit oppose current flow. The total opposition to current flow in a circuit depends on its resistance, its reactance, and the phase relationships between them. *Impedance* is defined as the total opposition to current flow in a circuit. Equation (8-6) is the mathematical representation for the magnitude of impedance in an AC circuit.

$$Z = \sqrt{R^2 + X^2} \qquad (8\text{-}6)$$

where

Z = impedance (Ω)
R = resistance (Ω)
X = net reactance (Ω)

The relationship between resistance, reactance, and impedance is shown in Figure 5.

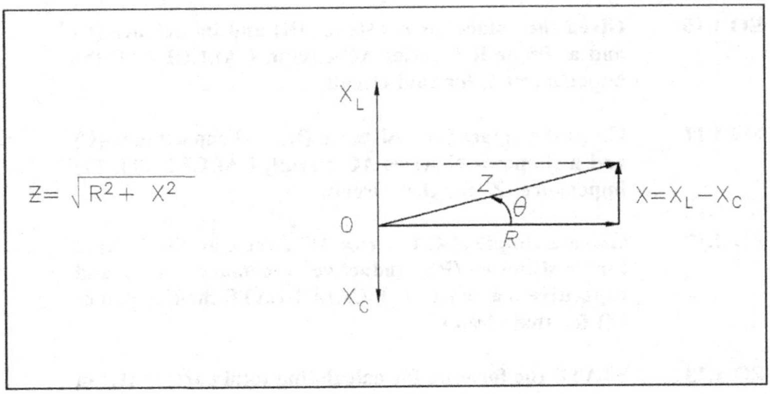

Figure 5 Relationship Between Resistance, Reactance, and Impedance

The current through a certain resistance is always in phase with the applied voltage. Resistance is shown on the zero axis. The current through an inductor lags applied voltage by 90°; inductive reactance is shown along the 90° axis. Current through a capacitor leads applied voltage by 90°; capacitive reactance is shown along the -90° axis. Net reactance in an AC circuit is the difference between inductive and capacitive reactance. Equation (8-7) is the mathematical representation for the calculation of net reactance when X_L is greater than X_C.

$$X = X_L - X_C \qquad (8\text{-}7)$$

where

X = net reactance (Ω)
X_L = inductive reactance (Ω)
X_C = capacitive reactance (Ω)

Equation (8-8) is the mathematical representation for the calculation of net reactance when X_C is greater than X_L.

$$X = X_C - X_L \tag{8-8}$$

Impedance is the vector sum of the resistance and net reactance (X) in a circuit, as shown in Figure 5. The angle θ is the phase angle and gives the phase relationship between the applied voltage and the current. Impedance in an AC circuit corresponds to the resistance of a DC circuit. The voltage drop across an AC circuit element equals the current times the impedance. Equation (8-9) is the mathematical representation of the voltage drop across an AC circuit.

$$V = IZ \tag{8-9}$$

where

V = voltage drop (V)
I = current (A)
Z = impedance (Ω)

The phase angle θ gives the phase relationship between current and the voltage.

Impedance in R-L Circuits

Impedance is the resultant of phasor addition of R and X_L. The symbol for impedance is Z. Impedance is the total opposition to the flow of current and is expressed in ohms. Equation (8-10) is the mathematical representation of the impedance in an RL circuit.

$$Z = \sqrt{R^2 + X_L^2} \tag{8-10}$$

Example: If a 100 Ω resistor and a 60 Ω X_L are in series with a 115V applied voltage (Figure 6), what is the circuit impedance?

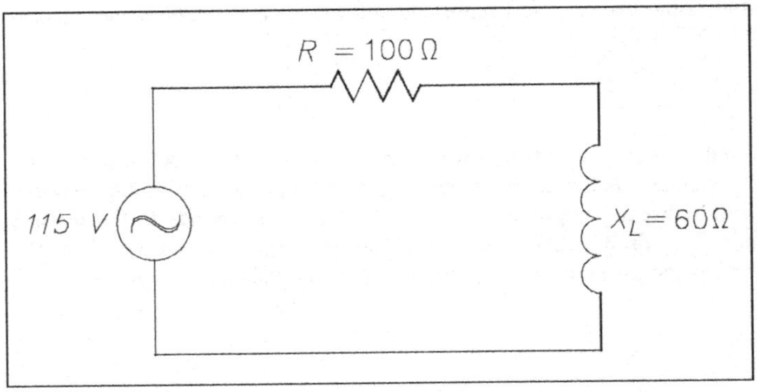

Figure 6 Simple R-L Circuit

Solution:

$$Z = \sqrt{R^2 + X_L^2}$$

$$= \sqrt{100^2 + 60^2}$$

$$= \sqrt{10,000 + 3600}$$

$$= \sqrt{13,600}$$

$$Z = 116.6\Omega$$

Impedance in R-C Circuits

In a capacitive circuit, as in an inductive circuit, impedance is the resultant of phasor addition of R and X_C. Equation (8-11) is the mathematical representation for impedance in an R-C circuit.

$$Z = \sqrt{R^2 + X_C^2} \quad (8\text{-}11)$$

Example: A 50 Ω X_C and a 60 Ω resistance are in series across a 110V source (Figure 7). Calculate the impedance.

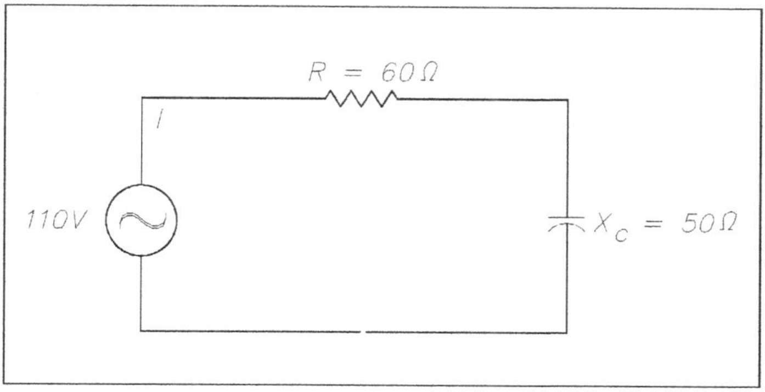

Figure 7 Simple R-C Circuit

Solution:

$$Z = \sqrt{R^2 + X_C^2}$$

$$= \sqrt{60^2 + 50^2}$$

$$= \sqrt{3600 + 2500}$$

$$= \sqrt{6100}$$

$$Z = 78.1\Omega$$

Impedance in R-C-L Circuits

Impedance in an R-C-L series circuit is equal to the phasor sum of resistance, inductive reactance, and capacitive reactance (Figure 8).

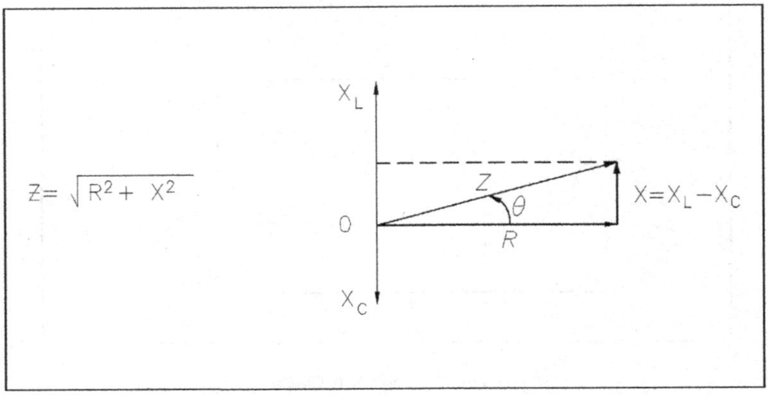

Figure 8 Series R-C-L Impedance-Phasor

Equations (8-12) and (8-13) are the mathematical representations of impedance in an R-C-L circuit. Because the difference between X_L and X_C is squared, the order in which the quantities are subtracted does not affect the answer.

$$Z = \sqrt{R^2 + (X_L - X_C)^2} \tag{8-12}$$

$$Z = \sqrt{R^2 + (X_C - X_L)^2} \tag{8-13}$$

Example: Find the impedance of a series R-C-L circuit, when R = 6 Ω, X_L = 20 Ω, and X_C = 10 Ω (Figure 9).

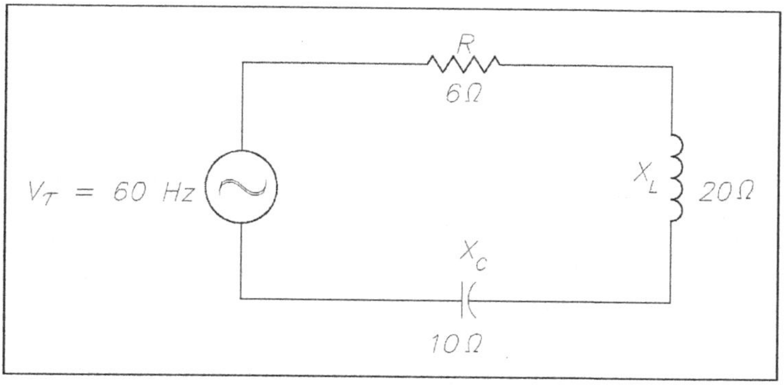

Figure 9 Simple R-C-L Circuit

Solution:

$$Z = \sqrt{R^2 + (X_L - X_C)^2}$$

$$= \sqrt{6^2 + (20 - 10)^2}$$

$$= \sqrt{6^2 + 10^2}$$

$$= \sqrt{36 + 100}$$

$$= \sqrt{136}$$

$$Z = 11.66\Omega$$

Impedance in a parallel R-C-L circuit equals the voltage divided by the total current. Equation (8-14) is the mathematical representation of the impedance in a parallel R-C-L circuit.

$$Z_T = \frac{V_T}{I_T} \tag{8-14}$$

where

Z_T = total impedance (Ω)
V_T = total voltage (V)
I_T = total current (A)

Total current in a parallel R-C-L circuit is equal to the square root of the sum of the squares of the current flows through the resistance, inductive reactance, and capacitive reactance branches of the circuit. Equations (8-15) and (8-16) are the mathematical representations of total current in a parallel R-C-L circuit. Because the difference between I_L and I_C is squared, the order in which the quantities are subtracted does not affect the answer.

$$I_T = \sqrt{I_R^2 + (I_C - I_L)^2} \tag{8-15}$$

$$I_T = \sqrt{I_R^2 + (I_L - I_C)^2} \tag{8-16}$$

where

I_T = total current (A)
I_R = current through resistance leg of circuit (A)
I_C = current through capacitive reactance leg of circuit (A)
I_L = current through inductive reactance leg of circuit (A)

Example: A 200 Ω resistor, a 100 Ω X_L, and an 80 Ω X_C are placed in parallel across a 120V AC source (Figure 10). Find: (1) the branch currents, (2) the total current, and (3) the impedance.

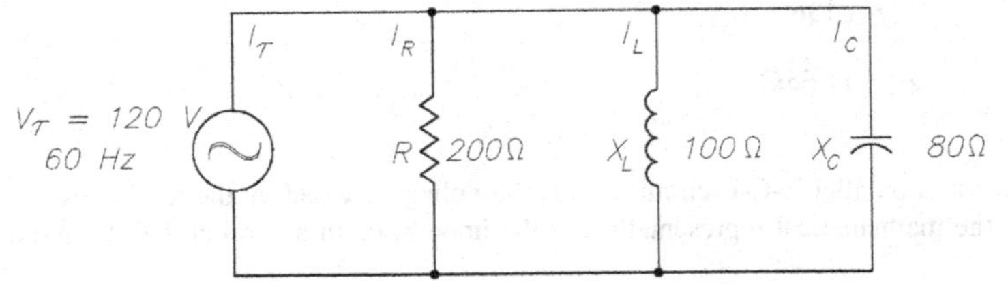

(a) Parallel RLC circuit

Figure 10 Simple Parallel R-C-L Circuit

Solution:

1. Branch currents

$$I_R = \frac{V_T}{R} \qquad I_L = \frac{V_T}{X_L} \qquad I_C = \frac{V_T}{X_C}$$

$$= \frac{120}{200} \qquad = \frac{120}{100} \qquad = \frac{120}{80}$$

$$I_R = 0.6\ \text{A} \qquad I_L = 1.2\ \text{A} \qquad I_C = 1.5\ \text{A}$$

2. Total current

$$I_T = \sqrt{I_R^2 + (I_C - I_L)^2}$$

$$= \sqrt{(0.6)^2 + (1.5 - 1.2)^2}$$

$$= \sqrt{(0.6)^2 + (0.3)^2}$$

$$= \sqrt{0.36 + 0.09}$$

$$= \sqrt{0.45}$$

$$I_T = 0.671\ \text{A}$$

3. Impedance

$$Z = \frac{V_T}{I_T}$$

$$= \frac{120}{0.671}$$

$$Z = 178.8\Omega$$

Summary

Impedance is summarized below.

Impedance Summary

- Impedance (Z) is the total opposition to current flow in an AC circuit.
- The formula for impedance in a series AC circuit is:

  $$Z = \sqrt{R^2 + X^2}$$

- The formula for impedance in a parallel R-C-L circuit is:

  $$Z = \sqrt{R^2 + (X_C - X_L)^2}$$

- The formulas for finding total current (I_T) in a parallel R-C-L circuit are:

  where $I_C > I_L$, $I_T = \sqrt{I_R^2 + (I_C - I_L)^2}$

  where $I_L > I_C$, $I_T = \sqrt{I_R^2 + (I_L - I_C)^2}$

RESONANCE

In the chapters on inductance and capacitance we have learned that both conditions are reactive and can provide opposition to current flow, but for opposite reasons. Therefore, it is important to find the point where inductance and capacitance cancel one another to achieve efficient operation of AC circuits.

EO 1.15 DEFINE resonance.

EO 1.16 Given the values of capacitance (C) and inductance (L), CALCULATE the resonant frequency.

EO 1.17 Given a series R-C-L circuit at resonance, DESCRIBE the net reactance of the circuit.

EO 1.18 Given a parallel R-C-L circuit at resonance, DESCRIBE the circuit output relative to current (I).

Resonant Frequency

Resonance occurs in an AC circuit when inductive reactance and capacitive reactance are equal to one another: $X_L = X_C$. When this occurs, the total reactance, $X = X_L - X_C$ becomes zero and the impendence is totally resistive. Because inductive reactance and capacitive reactance are both dependent on frequency, it is possible to bring a circuit to resonance by adjusting the frequency of the applied voltage. Resonant frequency (f_{Res}) is the frequency at which resonance occurs, or where $X_L = X_C$. Equation (8-14) is the mathematical representation for resonant frequency.

$$f_{Res} = \frac{1}{2\pi\sqrt{LC}} \qquad (8\text{-}14)$$

where

f_{Res} = resonant frequency (Hz)
L = inductance (H)
C = capacitance (f)

Series Resonance

In a series R-C-L circuit, as in Figure 9, at resonance the net reactance of the circuit is zero, and the impedance is equal to the circuit resistance; therefore, the current output of a series resonant circuit is at a maximum value for that circuit and is determined by the value of the resistance. (Z=R)

$$I = \frac{V_T}{Z_T} = \frac{V_T}{R}$$

Parallel Resonance

Resonance in a parallel R-C-L circuit will occur when the reactive current in the inductive branches is equal to the reactive current in the capacitive branches (or when $X_L = X_C$). Because inductive and capacitive reactance currents are equal and opposite in phase, they cancel one another at parallel resonance.

If a capacitor and an inductor, each with negligible resistance, are connected in parallel and the frequency is adjusted such that reactances are exactly equal, current will flow in the inductor and the capacitor, but the total current will be negligible. The parallel C-L circuit will present an almost infinite impedance. The capacitor will alternately charge and discharge through the inductor. Thus, in a parallel R-C-L, as in Figure 10, the net current flow through the circuit is at minimum because of the high impendence presented by X_L and X_C in parallel.

Summary

Resonance is summarized below.

Resonance Summary

- Resonance is a state in which the inductive reactance equals the capacitive reactance ($X_L = X_C$) at a specified frequency (f_{Res}).
- Resonant frequency is:

$$f_{Res} = \frac{1}{2\pi\sqrt{LC}}$$

- R-C-L series circuit at resonance is when net reactance is zero and circuit current output is determined by the series resistance of the circuit.
- R-C-L parallel circuit at resonance is when net reactance is maximum and circuit current output is at minimum.

Department of Energy
Fundamentals Handbook

ELECTRICAL SCIENCE
Module 9
Basic AC Power

TABLE OF CONTENTS

LIST OF FIGURES ii

LIST OF TABLES iii

REFERENCES iv

OBJECTIVES v

POWER TRIANGLE 1

- Power Triangle 1
- Apparent Power 2
- True Power 3
- Reactive Power 3
- Total Power 4
- Power Factor 4
- Power in Series R-L Circuit 5
- Power in Parallel R-L Circuit 6
- Power in Series R-C Circuit 8
- Power in Parallel R-C Circuit 10
- Power in Series R-C-L Circuit 12
- Power in Parallel R-C-L Circuit 14
- Summary 16

THREE-PHASE CIRCUITS 17

- Three-Phase Systems 17
- Power in Balanced 3ϕ Loads 19
- Unbalanced 3ϕ Loads 23
- Summary 26

LIST OF FIGURES

Figure 1 Power Triangle 2

Figure 2 Lagging Power Factor 4

Figure 3 Leading Power Factor 4

Figure 4 Series R-L Circuit 5

Figure 5 Parallel R-L Circuit 6

Figure 6 Series R-C Circuit 8

Figure 7 Parallel R-C Circuit 10

Figure 8 Series R-C-L Circuit 12

Figure 9 Parallel R-C-L Circuit 14

Figure 10 Three-Phase AC 18

Figure 11 3ϕ AC Power Connections 19

Figure 12 3ϕ Balanced Loads 19

Figure 13 3ϕ Power Triangle 21

Figure 14 Three-Phase Delta Generator 21

Figure 15 Three-Phase Wye Generator 22

Figure 16 3ϕ Unbalanced Load 24

LIST OF TABLES

NONE

REFERENCES

- Gussow, Milton, Schaum's Outline Series, Basic Electricity, McGraw-Hill.
- Academic Program for Nuclear Power Plant Personnel, Volume IV, Columbia, MD: General Physics Corporation, Library of Congress Card #A 326517, 1982.
- Academic Program for Nuclear Power Plant Personnel, Volume II, Columbia, MD: General Physics Corporation, Library of Congress Card #A 326517, 1982.
- Nasar and Unnewehr, Electromechanics and Electric Machines, John Wiley and Sons.
- Van Valkenburgh, Nooger, and Neville, Basic Electricity, Vol. 5, Hayden Book Company.
- Lister, Eugene C., Electric Circuits and Machines, 5th Edition, McGraw-Hill.
- Croft, Carr, Watt, and Summers, American Electricians Handbook, 10th Edition, McGraw-Hill.
- Mason, C. Russel, The Art and Science of Protective Relaying, John Wiley and Sons.
- Mileaf, Harry, Electricity One - Seven, Revised 2nd Edition, Hayden Book Company.
- Buban and Schmitt, Understanding Electricity and Electronics, 3rd Edition, McGraw-Hill.
- Kidwell, Walter, Electrical Instruments and Measurements, McGraw-Hill.

TERMINAL OBJECTIVE

1.0 Given an AC single-phase or three-phase circuit, **DESCRIBE** the power characteristics in the circuit.

ENABLING OBJECTIVES

1.1 **DESCRIBE** the relationship between apparent, true, and reactive power by definition or by using a power triangle.

1.2 **DEFINE** power factor as it relates to true power and apparent power.

1.3 Given the necessary values for voltage (E), resistance (R), reactance (X), impedance (Z), and/or current (I), **CALCULATE** the following power components for an AC circuit:
- a. True power (P)
- b. Apparent power (S)
- c. Reactive power (Q)
- d. Power factor (pf)

1.4 **DEFINE** the following terms:
- a. Leading power factor
- b. Lagging power factor

1.5 **STATE** the reasons that three-phase power systems are used in the industry.

1.6 Given values for current, voltage, and power factor in a three-phase system, **CALCULATE** the following:
- a. Real power
- b. Reactive power
- c. Apparent power

1.7 Given a diagram of a wye- or delta-connected three-phase system, **DESCRIBE** the voltage/current relationships of the circuit.

1.8 **STATE** the indications of an unbalanced load in a three-phase power system.

Intentionally Left Blank

POWER TRIANGLE

While direct current has one form of power, alternating current has three different forms of power that are related in a unique relationship. In this chapter, you will learn that power in AC circuits cannot be calculated in the same manner as in DC circuits.

EO 1.1 **DESCRIBE the relationship between apparent, true, and reactive power by definition or by using a power triangle.**

EO 1.2 **DEFINE power factor as it relates to true power and apparent power.**

EO 1.3 **Given the necessary values for voltage (E), resistance (R), reactance (X), impedance (Z), and/or current (I), CALCULATE the following power components for an AC circuit:**

- **a. True power (P)**
- **b. Apparent power (S)**
- **c. Reactive power (Q)**
- **d. Power factor (pf)**

EO 1.4 **DEFINE the following terms:**

- **a. Leading power factor**
- **b. Lagging power factor**

Power Triangle

In AC circuits, current and voltage are normally out of phase and, as a result, not all the power produced by the generator can be used to accomplish work. By the same token, power cannot be calculated in AC circuits in the same manner as in DC circuits. The power triangle, shown in Figure 1, equates AC power to DC power by showing the relationship between generator output (apparent power - S) in volt-amperes (VA), usable power (true power - P) in watts, and wasted or stored power (reactive power - Q) in volt-amperes-reactive (VAR). The phase angle (θ) represents the inefficiency of the AC circuit and corresponds to the total reactive impedance (Z) to the current flow in the circuit.

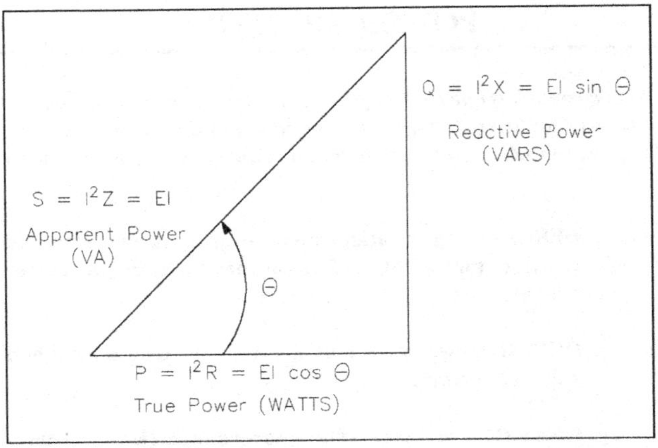

Figure 1 Power Triangle

The power triangle represents comparable values that can be used directly to find the efficiency level of generated power to usable power, which is expressed as the power factor (discussed later). Apparent power, reactive power, and true power can be calculated by using the DC equivalent (RMS value) of the AC voltage and current components along with the power factor.

Apparent Power

Apparent power (S) is the power delivered to an electrical circuit. Equation (9-1) is a mathematical representation of apparent power. The measurement of apparent power is in volt-amperes (VA).

$$S = I^2Z = I_TE \qquad (9\text{-}1)$$

where

S = apparent power (VA)
I = RMS current (A)
E = RMS voltage (V)
Z = impedance (Ω)

True Power

True power (P) is the power consumed by the resistive loads in an electrical circuit. Equation (9-2) is a mathematical representation of true power. The measurement of true power is in watts.

$$P = I^2R = EI\cos\theta \qquad (9\text{-}2)$$

where

P = true power (watts)
I = RMS current (A)
E = RMS voltage (V)
R = resistance (Ω)
θ = angle between E and I sine waves

Reactive Power

Reactive power (Q) is the power consumed in an AC circuit because of the expansion and collapse of magnetic (inductive) and electrostatic (capacitive) fields. Reactive power is expressed in volt-amperes-reactive (VAR). Equation (9-3) is a mathematical representation for reactive power.

$$Q = I^2X = EI\sin\theta \qquad (9\text{-}3)$$

where

Q = reactive power (VAR)
I = RMS current (A)
X = net reactance (Ω)
E = RMS voltage (V)
θ = angle between the E and I sine waves

Unlike true power, reactive power is not useful power because it is stored in the circuit itself. This power is stored by inductors, because they expand and collapse their magnetic fields in an attempt to keep current constant, and by capacitors, because they charge and discharge in an attempt to keep voltage constant. Circuit inductance and capacitance consume and give back reactive power. Reactive power is a function of a system's amperage. The power delivered to the inductance is stored in the magnetic field when the field is expanding and returned to the source when the field collapses. The power delivered to the capacitance is stored in the electrostatic field when the capacitor is charging and returned to the source when the capacitor discharges. None of the power delivered to the circuit by the source is consumed. It is all returned to the source. The true power, which is the power consumed, is thus zero. We know that alternating current constantly changes; thus, the cycle of expansion and collapse of the magnetic and electrostatic fields constantly occurs.

Total Power

The *total power* delivered by the source is the apparent power. Part of this apparent power, called true power, is dissipated by the circuit resistance in the form of heat. The rest of the apparent power is returned to the source by the circuit inductance and capacitance.

Power Factor

Power factor (pf) is the ratio between true power and apparent power. True power is the power consumed by an AC circuit, and reactive power is the power that is stored in an AC circuit. Cosθ is called the power factor (pf) of an AC circuit. It is the ratio of true power to apparent power, where θ is the phase angle between the applied voltage and current sine waves and also between P and S on a power triangle (Figure1). Equation (9-4) is a mathematical representation of power factor.

$$\cos\theta = \frac{P}{S} \quad (9\text{-}4)$$

where

$\cos\theta$ = power factor (pf)
P = true power (watts)
S = apparent power (VA)

Power factor also determines what part of the apparent power is real power. It can vary from 1, when the phase angle is 0°, to 0, when the phase angle is 90°. In an inductive circuit, the current lags the voltage and is said to have a lagging power factor, as shown in Figure 2.

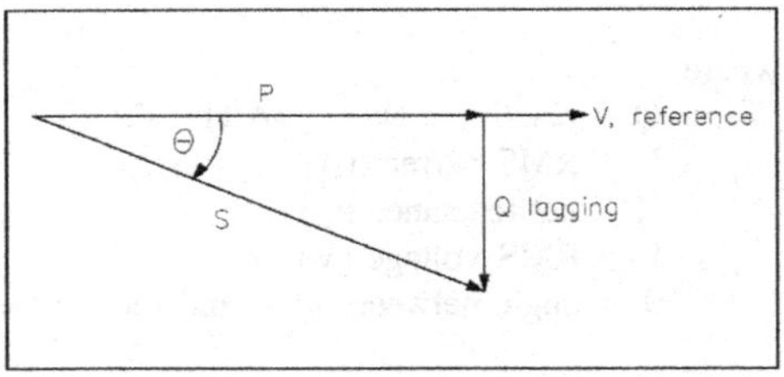

Figure 2 Lagging Power Factor

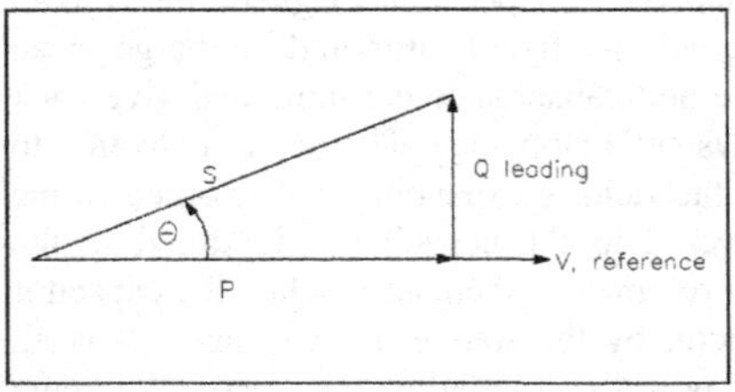

Figure 3 Leading Power Factor

In a capacitive circuit, the current leads the voltage and is said to have a leading power factor, as shown in Figure 3.

A mnemonic memory device, "ELI the ICE man," can be used to remember the voltage/current relationship in AC circuits. ELI refers to an inductive circuit (L) where current (I) lags voltage (E). ICE refers to a capacitive circuit (C) where current (I) leads voltage (E).

Power in Series R-L Circuit

Example: A 200 Ω resistor and a 50 Ω X_L are placed in series with a voltage source, and the total current flow is 2 amps, as shown in Figure 4.

Find:
1. pf
2. applied voltage, V
3. P
4. Q
5. S

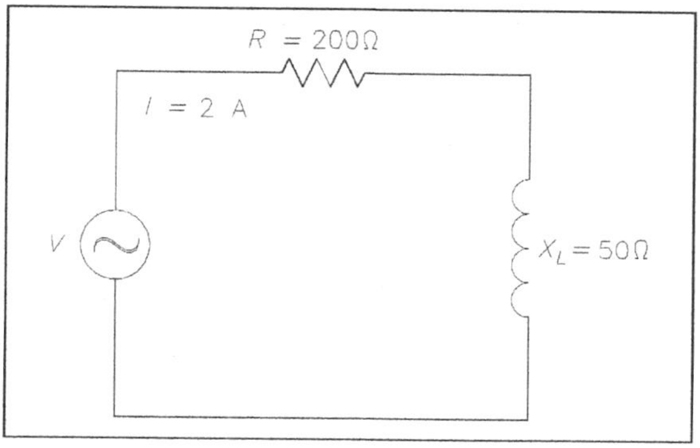

Figure 4 Series R-L Circuit

Solution:

1. $$\text{pf} = \cos\theta \qquad \theta = \arctan\left(\frac{X_L}{R}\right)$$

$$= \cos\left(\arctan\left(\frac{X_L}{R}\right)\right)$$

$$= \cos\left(\arctan\left(\frac{50}{200}\right)\right)$$

$$= \cos\ (14°)$$

$$\text{pf} = 0.097$$

2. $$V = IZ \qquad Z = \sqrt{R^2 + X_L^2}$$

$$= I\sqrt{R^2 + X_L^2}$$

$$= 2\sqrt{200^2 + 50^2}$$

$$= 2\sqrt{42{,}500}$$

$$= (2)(206.16)$$

$$V = 412.3 \text{ volts}$$

Note: Inverse trigonometric functions such as arctan are discussed in the Mathematics Fundamentals Manual, Module 4, Trigonometry, pages 6 and 7 should the student require review.

3. $P = EI \cos\theta$

$= (412.3)(2)(0.97)$

$P = 799.86$ watts

4. $Q = EI \sin\theta$

$= (412.3)(2)(0.242)$

$Q = 199.6$ VAR

5. $S = EI$

$= (412.3)(2)$

$S = 824.6$ VA

Power in Parallel R-L Circuit

Example: A 600 Ω resistor and 200 Ω X_L are in parallel with a 440V source, as shown in Figure 5.

Find:
1. I_T
2. pf
3. P
4. Q
5. S

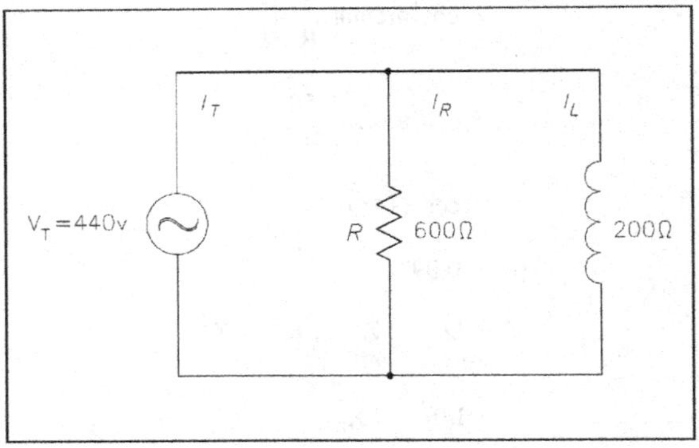

Figure 5 Parallel R-L Circuit

Solution:

1. $I_T = \sqrt{I_R^2 + I_L^2} \qquad I_R = \frac{V_T}{R} \qquad I_L = \frac{V_T}{X_L}$

$$= \sqrt{\left(\frac{V_T}{R}\right)^2 + \left(\frac{V_T}{X_L}\right)^2}$$

$$= \sqrt{\left(\frac{440}{600}\right)^2 + \left(\frac{440}{200}\right)^2}$$

$$= \sqrt{(0.73)^2 + (2.2)^2}$$

$$I_T = 2.3 \text{ amps}$$

2. $pf = \cos\theta \qquad \theta = \arctan\left(-\frac{I_L}{I_R}\right)$

$$= \cos\left(\arctan\left(-\frac{I_L}{I_R}\right)\right)$$

$$= \cos\left(\arctan\left(-\frac{2.2}{0.73}\right)\right)$$

$$= \cos(\arctan(-3))$$

$$= \cos(-71.5°)$$

$$pf = 0.32$$

3. $P = EI\cos\theta$

$$= (440)(2.3)(0.32)$$

$$P = 323.84 \text{ watts}$$

4. $Q = EI\sin\theta$

$$= (440)(2.3)(0.948)$$

$$Q = 959.4 \text{ VAR}$$

5. $S = EI$

$= (440)(2.3)$

$S = 1012 \text{ VA}$

Power in Series R-C Circuit

Example: An 80 Ω X_c and a 60 Ω resistance are in series with a 120V source, as shown in Figure 6.

Find:
1. Z
2. I_T
3. pf
4. P
5. Q
6. S

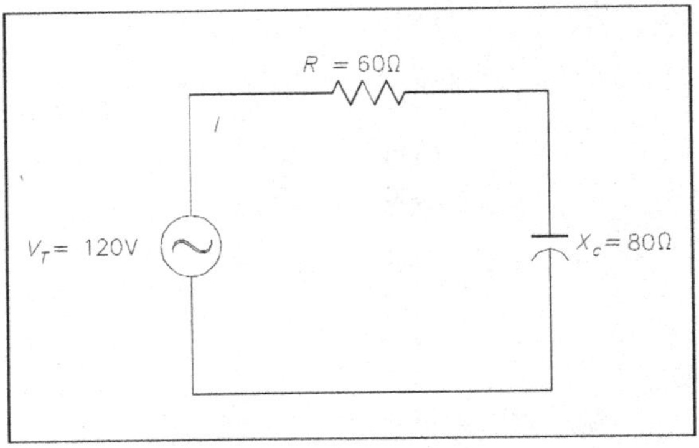

Figure 6 Series R-C Circuit

Solution:

1. $Z = \sqrt{R^2 + X_C^2}$

$= \sqrt{60^2 + 80^2}$

$= \sqrt{3600 + 6400}$

$Z = 100\Omega$

2. $I_T = \frac{V_T}{Z}$

$= \frac{120}{100}$

$I_T = 1.2 \text{ amps}$

3. $pf = \cos\theta \qquad \theta = \arctan\left(-\frac{X_C}{R}\right)$

$$= \cos\left(\arctan\left(-\frac{X_C}{R}\right)\right)$$

$$= \cos\left(\arctan\left(-\frac{80}{60}\right)\right)$$

$$= \cos(\arctan(-1.33))$$

$$= \cos(-53°)$$

$$pf = 0.60$$

4. $P = EI\cos\theta$

$$= (120)(1.2)(0.60)$$

$$P = 86.4 \text{ watts}$$

5. $Q = EI\sin\theta$

$$= (120)(1.2)(0.798)$$

$$Q = 114.9 \text{ VAR}$$

6. $S = EI$

$$= (120)(1.2)$$

$$S = 144 \text{ VA}$$

Power in Parallel R-C Circuit

Example: A 30 Ω resistance and a 40 Ω X_C are in parallel with a 120V power source, as shown in Figure 7.

Find:
1. I_T
2. Z
3. pf
4. P
5. Q
6. S

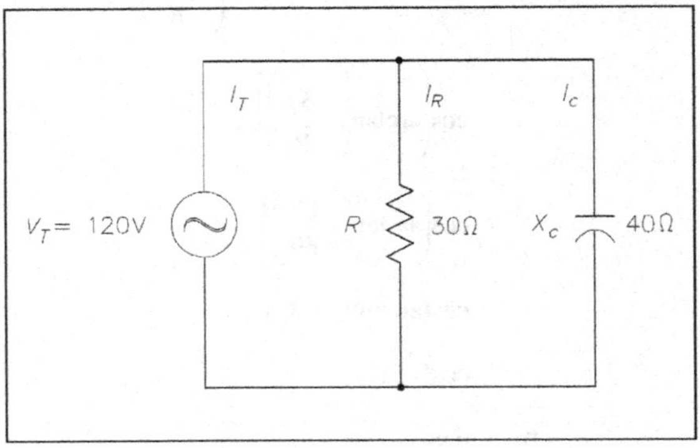

Figure 7 Parallel R-C Circuit

Solution:

1. $I_T = \sqrt{I_R^2 + I_C^2} \qquad I_R = \frac{V_T}{R} \qquad I_C = \frac{V_T}{X_C}$

$$= \sqrt{\left(\frac{V_T}{R}\right)^2 + \left(\frac{V_T}{X_C}\right)^2}$$

$$= \sqrt{\left(\frac{120}{30}\right)^2 + \left(\frac{120}{40}\right)^2}$$

$$= \sqrt{4^2 + 3^2}$$

$$I_T = 5 \text{ amps}$$

2. $Z = \frac{V_T}{I_T}$

$$= \frac{120}{5}$$

$$Z = 24\Omega$$

3. $pf = \cos\theta \qquad \theta = \arctan\left(\frac{I_C}{I_R}\right)$

$$= \cos\left(\arctan\left(\frac{I_C}{I_R}\right)\right)$$

$$= \cos\left(\arctan\left(\frac{3}{4}\right)\right)$$

$$= \cos(\arctan(36.9°))$$

$$pf = 0.80$$

4. $P = EI\cos\theta$

$$= (120)(5)(0.80)$$

$$P = 480 \text{ watts}$$

5. $Q = EI\sin\theta$

$$= (120)(5)(0.6)$$

$$Q = 360 \text{ VAR}$$

6. $S = EI$

$$= (120)(5)$$

$$S = 600 \text{ VA}$$

Power in Series R-C-L Circuit

Example: An 8 Ω resistance, a 40 Ω X_L, and a 24 Ω X_C are in series with a 60 Hz source with a current flow of 4 amps, as shown in Figure 8.

Find: 1. Z
2. V_T
3. pf
4. P
5. Q
6. S

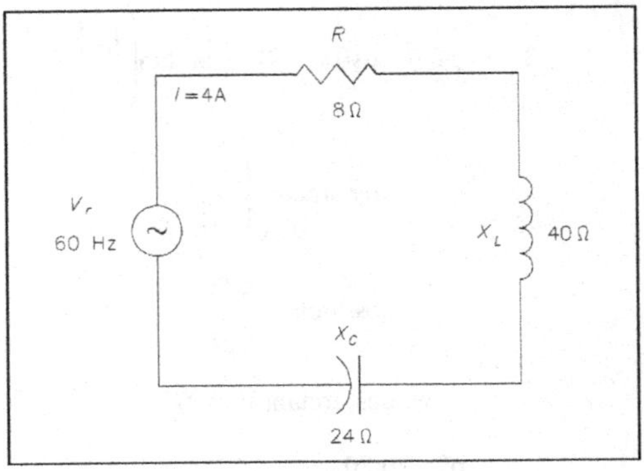

Figure 8 Series R-C-L Circuit

1. $$Z = \sqrt{R^2 + (X_L - X_C)^2}$$

$$= \sqrt{8^2 + (40 - 24)^2}$$

$$= \sqrt{8^2 + 16^2}$$

$$Z = 17.9\ \Omega$$

2. $$V_T = IZ$$

$$= (4)(17.9)$$

$$V_T = 71.6 \text{ volts}$$

3. $pf = \cos\theta \qquad \theta = \arctan\left(\frac{X}{R}\right)$

$$= \cos\left(\arctan\left(\frac{X}{R}\right)\right)$$

$$= \cos\left(\arctan\left(\frac{16}{8}\right)\right)$$

$$= \cos(\arctan(2))$$

$$= \cos(63.4°)$$

$$pf = 0.45$$

4. $P = EI\cos\theta$

$$= (71.6)(4)(0.45)$$

$$P = 128.9 \text{ watts}$$

5. $Q = EI\sin\theta$

$$= (71.6)(4)(0.89)$$

$$Q = 254.9 \text{ VAR}$$

6. $S = EI$

$$= (71.6)(4)$$

$$S = 286.4 \text{ VA}$$

Power in Parallel R-C-L Circuits

Example: An 800 Ω resistance, 100 Ω X_L, and an 80 Ω X_C are in parallel with a 120V, 60Hz source, as shown in Figure 9.

Find:
1. I_T
2. pf
3. P
4. Q
5. S

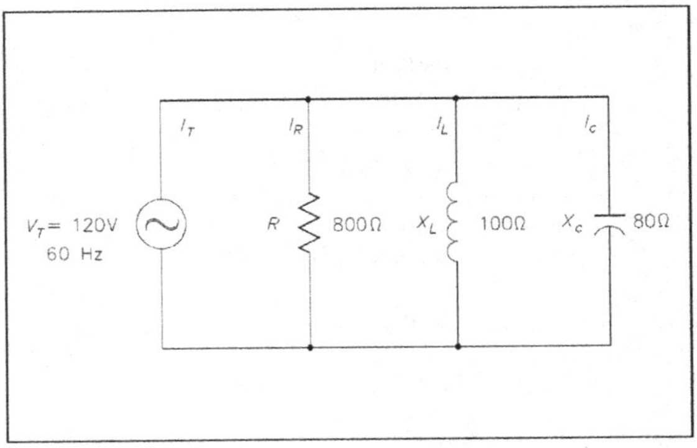

Figure 9　Parallel R-C-L Circuit

Solution:

1. $I_T = \sqrt{I_R^2 + (I_C - I_L)^2}$　　$I_R = \frac{V_T}{R}$　　$I_L = \frac{V_T}{X_L}$　　$I_C = \frac{V_T}{X_C}$

$$= \sqrt{\left(\frac{V_T}{R}\right)^2 + \left(\frac{V_T}{X_L} - \frac{V_T}{X_C}\right)^2}$$

$$= \sqrt{\left(\frac{120}{800}\right)^2 + \left(\frac{120}{100} - \frac{120}{80}\right)^2}$$

$$= \sqrt{0.15^2 + (1.2 - 1.5)^2}$$

$$= \sqrt{0.15^2 + (-0.3)^2}$$

$$I_T = 0.34 \text{ amps}$$

2. $pf = \cos\theta \qquad \theta = \arctan\left(\frac{I_C - I_L}{I_R}\right)$

$$= \cos\left(\arctan\left(\frac{I_C - I_L}{I_R}\right)\right)$$

$$= \cos\left(\arctan\left(\frac{1.5 - 1.2}{0.15}\right)\right)$$

$$= \cos(\arctan(2))$$

$$= \cos(63.4°)$$

$$pf = 0.45$$

3. $P = EI\cos\theta$

$$= (120)(0.34)(0.45)$$

$$P = 18.36 \text{ watts}$$

4. $Q = EI\sin\theta$

$$= (120)(0.34)(0.89)$$

$$Q = 36.4 \text{ VAR}$$

5. $S = EI$

$$= (120)(0.34)$$

$$S = 40.8 \text{ VA}$$

Summary

AC power relationships are summarized below.

AC Power Relationships Summary

- Observe the equations for apparent, true, and reactive power, and power factor:
  - Apparent power (S) = $I^2Z = I_TE$
  - True power (P) = $I^2R = EI \cos\theta$
  - Reactive power (Q) = $I^2X = EI \sin\theta$
  - Power factor (pf) = $\frac{P}{S} = \cos\theta$
- From observation, you can see that three power equations have the angle θ in common. θ is the angle between voltage and current. From this relationship, a power triangle, as shown in Figure 1, is formed.
- ELI the ICE man is a mnemonic device that describes the reactive characteristics of an AC circuit.
  - Current (I) lags voltage (E) in an inductive circuit (L)
  - Current (I) leads voltage (E) in a capacitive circuit (C)

THREE-PHASE CIRCUITS

The design of three-phase AC circuits lends itself to a more efficient method of producing and utilizing an AC voltage.

EO 1.5 **STATE the reasons that three-phase power systems are used in the industry.**

EO 1.6 **Given values for current, voltage, and power factor in a three-phase system, CALCULATE the following:**
- **a. Real power**
- **b. Reactive power**
- **c. Apparent power**

EO 1.7 **Given a diagram of a wye- or delta-connected three-phase system, DESCRIBE the voltage/current relationships of the circuit.**

EO 1.8 **STATE the indications of an unbalanced load in a three-phase power system.**

Three-Phase Systems

A three-phase (3ϕ) system is a combination of three single-phase systems. In a 3ϕ balanced system, power comes from a 3ϕ AC generator that produces three separate and equal voltages, each of which is 120° out of phase with the other voltages (Figure 10).

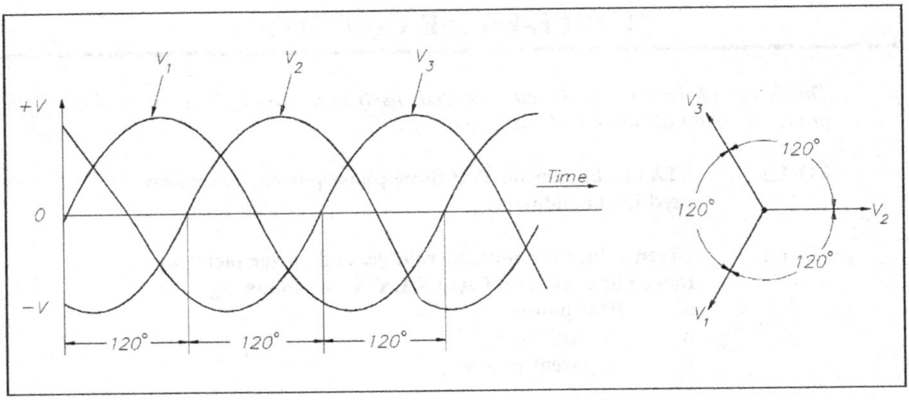

Figure 10 Three-Phase AC

Three-phase equipment (motors, transformers, etc.) weighs less than single-phase equipment of the same power rating. They have a wide range of voltages and can be used for single-phase loads. Three-phase equipment is smaller in size, weighs less, and is more efficient than single-phase equipment.

Three-phase systems can be connected in two different ways. If the three common ends of each phase are connected at a common point and the other three ends are connected to a 3ϕ line, it is called a wye, or Y-, connection (Figure 11). If the three phases are connected in series to form a closed loop, it is called a delta, or Δ-, connection.

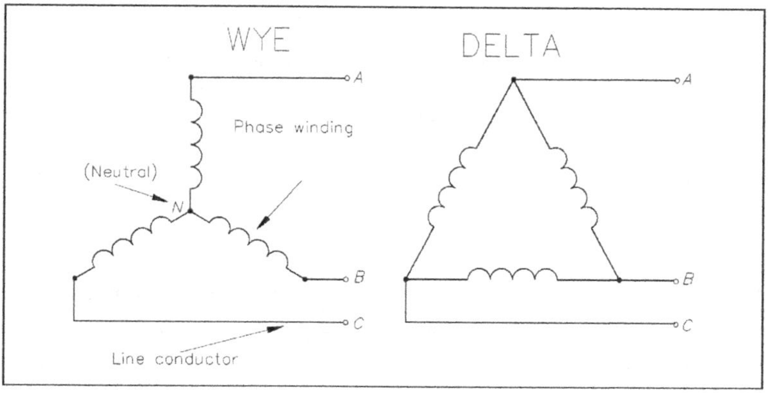

Figure 11 3ϕ AC Power Connections

Power in Balanced 3ϕ Loads

Balanced loads, in a 3ϕ system, have identical impedance in each secondary winding (Figure 12). The impedance of each winding in a delta load is shown as Z_Δ (Figure 12a), and the impedence in a wye load is shown as Z_y (Figure 12b). For either the delta or wye connection, the lines A, B, and C supply a 3ϕ system of voltages.

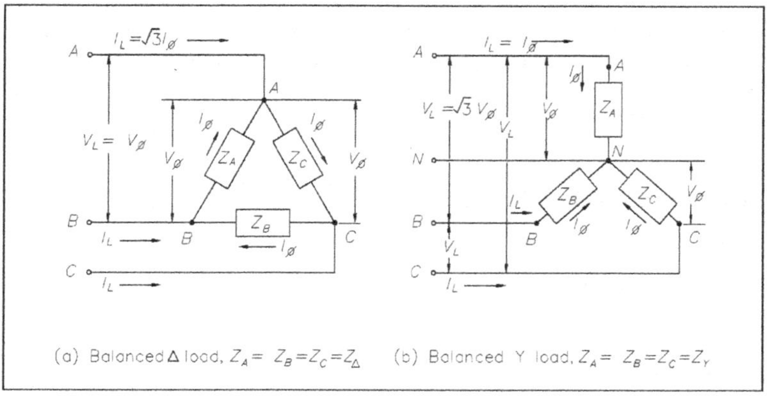

Figure 12 3ϕ Balanced Loads

In a balanced delta load, the line voltage (V_L) is equal to the phase voltage (V_ϕ), and the line current (I_L) is equal to the square root of three times the phase current ($\sqrt{3}\, I_\phi$). Equation (9-5) is a mathematical representation of V_L in a balanced delta load. Equation (9-6) is a mathematical representation of I_L in a balanced delta load.

$$V_L = V_\phi \tag{9-5}$$

$$I_L = \sqrt{3}\, I_\phi \tag{9-6}$$

In a balanced wye load, the line voltage (V_L) is equal to the square root of three times phase voltage ($\sqrt{3}\, V_\phi$), and line current (I_L) is equal to the phase current (I_ϕ). Equation (9-7) is a mathematical representation of V_L in a balanced wye load. Equation (9-8) is a mathematical representation of I_L in a balanced wye load.

$$V_L = \sqrt{3}\, V_\phi \tag{9-7}$$

$$I_L = I_\phi \tag{9-8}$$

Because the impedance of each phase of a balanced delta or wye load has equal current, phase power is one third of the total power. Equation (9-10) is the mathematical representation for phase power (P_ϕ) in a balanced delta or wye load.

$$P_\phi = V_\phi\, I_\phi \cos\theta \tag{9-10}$$

Total power (P_T) is equal to three times the single-phase power. Equation (9-11) is the mathematical representation for total power in a balanced delta or wye load.

$$P_T = 3V_\phi\, I_\phi \cos\theta \tag{9-11}$$

In a delta-connected load, $V_L = V_\phi$ and $I_\phi = \frac{\sqrt{3}\, I_L}{3}$ so:

$$P_T = \sqrt{3}\, V_L I_L \cos\theta$$

In a wye-connected load, $I_L = I_\phi$ and $V_\phi = \frac{\sqrt{3}\, V_L}{3}$ so:

$$P_T = \sqrt{3}\, V_L I_L \cos\theta$$

As you can see, the total power formulas for delta- and wye-connected loads are identical.

Total apparent power (S_T) in volt-amperes and total reactive power (Q_T) in volt-amperes-reactive are related to total real power (P_T) in watts (Figure 13).

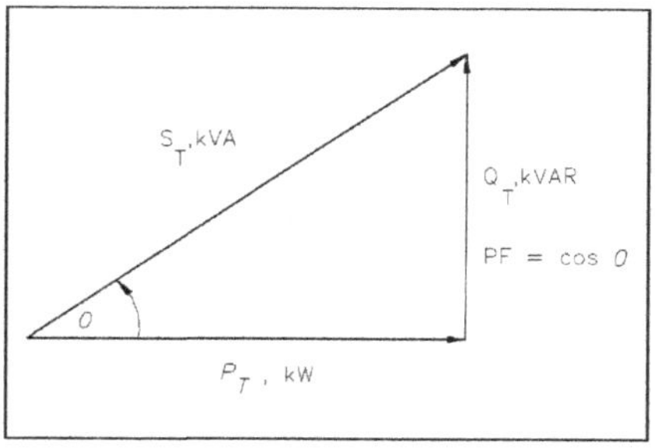

Figure 13 3ϕ Power Triangle

A balanced three-phase load has the real, apparent, and reactive powers given by:

$$P_T = \sqrt{3}\, V_T I_L \cos\theta$$

$$S_T = \sqrt{3}\, V_T I_L$$

$$Q_T = \sqrt{3}\, V_T I_L \sin\theta$$

Example 1: Each phase of a delta-connected 3ϕ AC generator supplies a full load current of 200 A at 440 volts with a 0.6 lagging power factor, as shown in Figure 14.

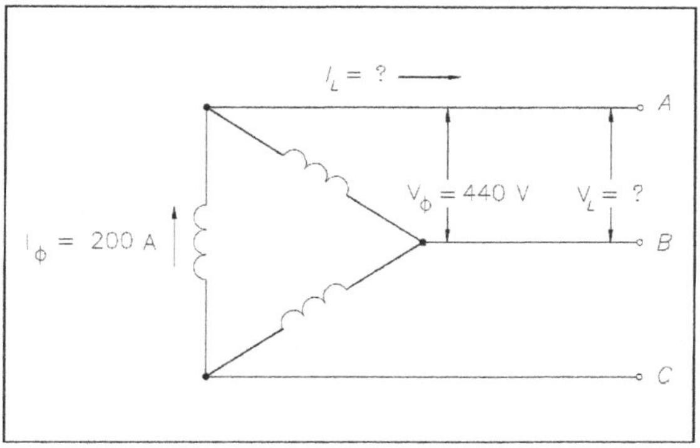

Figure 14 Three-Phase Delta Generator

Find:
1. V_L
2. I_L
3. P_T
4. Q_T
5. S_T

Solution:

1. $V_L = V_\phi$

V_L = 440 volts

2. $I_L = \sqrt{3}\, I_\phi$

$= (1.73)(200)$

$I_L = 346$ amps

3. $P_T = \sqrt{3}\, V_L I_L \cos\theta$

$= (1.73)(440)(346)(0.6)$

$P_T = 158.2$ kW

4. $Q_T = \sqrt{3}\, V_L I_L \sin\theta$

$= (1.73)(440)(346)(0.8)$

$Q_T = 210.7$ kVAR

5. $S_T = \sqrt{3}\, V_L I_L$

$= (1.73)(440)(346)$

$S_T = 263.4$ kVA

Example 2: Each phase of a wye-connected 3ϕ AC generator supplies a 100 A current at a phase voltage of 240V and a power factor of 0.9 lagging, as shown in Figure 15.

Find:
1. V_L
2. P_T
3. Q_T
4. S_T

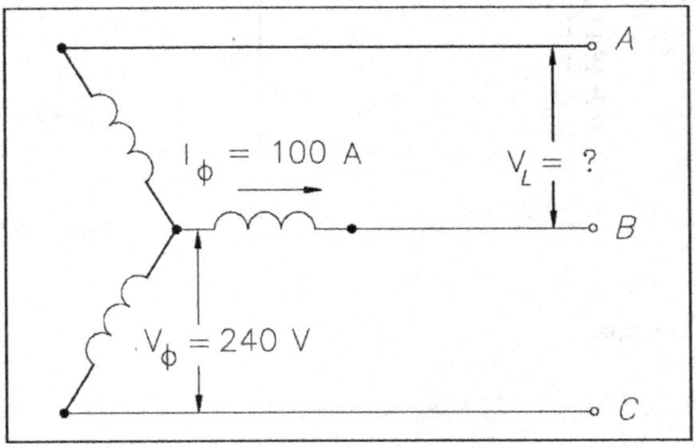

Figure 15 Three-Phase Wye Generator

Solution:

1. $V_L = \sqrt{3}\, V_\phi$

   $= (1.73)(240)$

   $V_L = 415.2 \text{ volts}$

2. $P_T = \sqrt{3}\, V_L I_L \cos\theta$

   $= (1.73)(415.2)(100)(0.9)$

   $P_T = 64.6 \text{ kW}$

3. $Q_T = \sqrt{3}\, V_L I_L \sin\theta$

   $= (1.73)(415.2)(100)(0.436)$

   $Q_T = 31.3 \text{ kVAR}$

4. $S_T = \sqrt{3}\, V_L I_L$

   $= (1.73)(415.2)(100)$

   $S_T = 71.8 \text{ kVA}$

Unbalanced 3ϕ Loads

An important property of a three-phase balanced system is that the phasor sum of the three line or phase voltages is zero, and the phasor sum of the three line or phase currents is zero. When the three load impedances are not equal to one another, the phasor sums and the neutral current (I_n) are not zero, and the load is, therefore, unbalanced. The imbalance occurs when an open or short circuit appears at the load.

If a three-phase system has an unbalanced load and an unbalanced power source, the methods of fixing the system are complex. Therefore, we will only consider an unbalanced load with a balanced power source.

Example: A 3ϕ balanced system, as shown in Figure 16a, contains a wye load. The line-to- line voltage is 240V, and the resistance is 40 Ω in each branch.

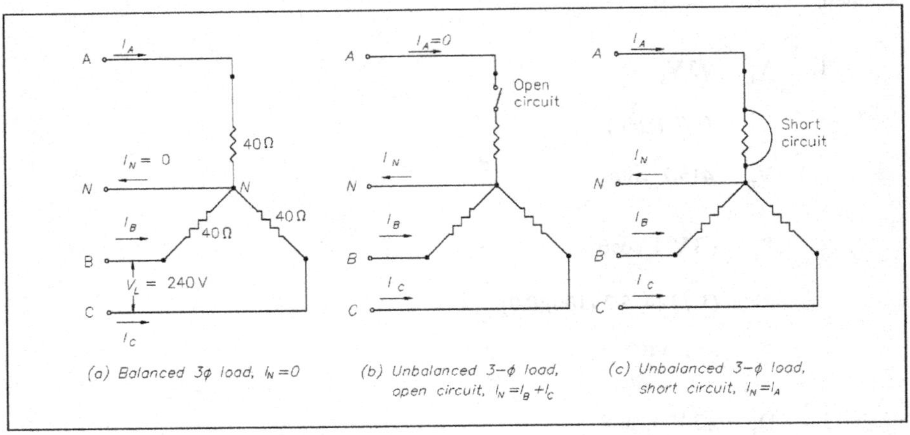

Figure 16 3φ Unbalanced Load

Find line current and neutral current for the following load conditions.

1. balanced load
2. open circuit phase A (Figure 16b)
3. short circuit in phase A (Figure 16c)

1. $I_L = I_\phi \qquad I_\phi = \frac{V_\phi}{R_\phi} \qquad V_\phi = \frac{V_L}{\sqrt{3}}$

$$I_L = \frac{\left(\frac{V_L}{\sqrt{3}}\right)}{R_\phi}$$

$$= \frac{\left(\frac{240}{1.73}\right)}{40}$$

$$= \frac{138.7}{40}$$

$I_L = 3.5$ amps $\qquad I_N = 0$

2. Current flow in lines B and C becomes the resultant of the loads in B and C connected in series.

$$I_B = \frac{V_L}{R_B + R_C} \qquad I_C = I_B$$

$$= \frac{240}{40 + 40}$$

$$I_B = 3 \text{ amps} \qquad I_C = 3 \text{ amps}$$

$$I_N = I_B + I_C$$

$$= 3 + 3$$

$$I_N = 6 \text{ amps}$$

3. $$I_B = \frac{V_L}{R_B} \qquad I_C = I_B$$

$$= \frac{240}{40}$$

$$I_B = 6 \text{ amps} \qquad I_C = 6 \text{ amps}$$

The current in Phase A is equal to the neutral line current, $I_A = I_N$. Therefore, I_N is the phasor sum of I_B and I_C.

$$I_N = \sqrt{3}\, I_B$$

$$= (1.73)(6)$$

$$I_N = 10.4 \text{ amps}$$

In a fault condition, the neutral connection in a wye-connected load will carry more current than the phase under a balanced load. Unbalanced three-phase circuits are indicated by abnormally high currents in one or more of the phases. This may cause damage to equipment if the imbalance is allowed to continue.

Summary

Three-phase circuits are summarized below.

Three-Phase Circuits Summary

- Three-phase power systems are used in the industry because:

  - Three-phase circuits weigh less than single-phase circuits of the same power rating.
  - They have a wide range of voltages and can be used for single-phase loads.
  - Three-phase equipment is smaller in size, weighs less, and is more efficient than single-phase equipment.

- Unbalanced three-phase circuits are indicated by abnormally high currents in one or more of the phases.

Department of Energy
Fundamentals Handbook

ELECTRICAL SCIENCE
Module 10
AC Generators

www.KnowledgePublications.com

TABLE OF CONTENTS

LIST OF FIGURES ii

LIST OF TABLES iii

REFERENCES iv

OBJECTIVES v

AC GENERATOR COMPONENTS 1

Field 1
Armature 1
Prime Mover 1
Rotor 2
Stator 2
Slip Rings 3
Summary 4

AC GENERATOR THEORY 5

Theory of Operation 5
Losses in an AC Generator 6
Hysteresis Losses 7
Mechanical Losses 7
Efficiency 7
Summary 8

AC GENERATOR OPERATION 9

Ratings 9
Paralleling AC Generators 10
Types of AC Generators 10
Three-Phase AC Generators 11
AC Generator Connections 12
Summary 14

LIST OF FIGURES

Figure 1	Basic AC Generator	2
Figure 2	Comparison of DC and AC Generator Outputs	3
Figure 3	Simple AC Generator	5
Figure 4	AC Generator Nameplate Ratings	9
Figure 5	Stationary Field, Rotating Armature AC Generator	10
Figure 6	Simple AC Generator - Rotating Field, Stationary Armature	11
Figure 7	Stationary Armature 3ϕ Generator	11
Figure 8	Delta Connection	12
Figure 9	Characteristics of a Delta-Connected Generator	12
Figure 10	Wye Connection	13
Figure 11	Characteristics of a Wye-Connected AC Generator	13

LIST OF TABLES

NONE

REFERENCES

- Gussow, Milton, Schaum's Outline Series, Basic Electricity, McGraw-Hill.
- Academic Program for Nuclear Power Plant Personnel, Volume IV, Columbia, MD: General Physics Corporation, Library of Congress Card #A 326517, 1982.
- Academic Program for Nuclear Power Plant Personnel, Volume II, Columbia, MD: General Physics Corporation, Library of Congress Card #A 326517, 1982.
- Nasar and Unnewehr, Electromechanics and Electric Machines, John Wiley and Sons.
- Van Valkenburgh, Nooger, and Neville, Basic Electricity, Vol. 5, Hayden Book Company.
- Lister, Eugene C., Electric Circuits and Machines, 5th Edition, McGraw-Hill.
- Croft, Carr, Watt, and Summers, American Electricians Handbook, 10th Edition, McGraw-Hill.
- Mason, C. Russel, The Art and Science of Protective Relaying, John Wiley and Sons.
- Mileaf, Harry, Electricity One - Seven, Revised 2nd Edition, Hayden Book Company.
- Buban and Schmitt, Understanding Electricity and Electronics, 3rd Edition, McGraw-Hill.
- Kidwell, Walter, Electrical Instruments and Measurements, McGraw-Hill.

TERMINAL OBJECTIVE

1.0 Given the type and application of an AC generator, **DESCRIBE** the operating characteristics of that generator including methods of voltage production, advantages of each type, and methods for paralleling.

ENABLING OBJECTIVES

1.1 **STATE** the purpose of the following components of an AC generator:
   a. Field
   b. Armature
   c. Prime mover
   d. Rotor
   e. Stator
   f. Slip rings

1.2 Given the speed of rotation and number of poles, **CALCULATE** the frequency output of an AC generator.

1.3 **LIST** the three losses found in an AC generator.

1.4 Given the prime mover input and generator output, **DETERMINE** the efficiency of an AC generator.

1.5 **DESCRIBE** the bases behind the kW and current ratings of an AC generator.

1.6 **DESCRIBE** the conditions that must be met prior to paralleling two AC generators including consequences of not meeting these conditions.

1.7 **DESCRIBE** the difference between a stationary field, rotating armature AC generator and a rotating field, stationary armature AC generator.

1.8 **EXPLAIN** the differences between a wye-connected and delta-connected AC generator including advantages and disadvantages of each type.

Intentionally Left Blank

AC GENERATOR COMPONENTS

AC generators are widely used to produce AC voltage. To understand how these generators operate, the function of each component of the generator must first be understood.

EO 1.1 STATE the purpose of the following components of an AC generator:

a. Field
b. Armature
c. Prime mover
d. Rotor
e. Stator
f. Slip rings

Field

The *field* in an AC generator consists of coils of conductors within the generator that receive a voltage from a source (called excitation) and produce a magnetic flux. The magnetic flux in the field cuts the armature to produce a voltage. This voltage is ultimately the output voltage of the AC generator.

Armature

The *armature* is the part of an AC generator in which voltage is produced. This component consists of many coils of wire that are large enough to carry the full-load current of the generator.

Prime Mover

The *prime mover* is the component that is used to drive the AC generator. The prime mover may be any type of rotating machine, such as a diesel engine, a steam turbine, or a motor.

Rotor

The *rotor* of an AC generator is the rotating component of the generator, as shown in Figure 1. The rotor is driven by the generator's prime mover, which may be a steam turbine, gas turbine, or diesel engine. Depending on the type of generator, this component may be the armature or the field. The rotor will be the armature if the voltage output is generated there; the rotor will be the field if the field excitation is applied there.

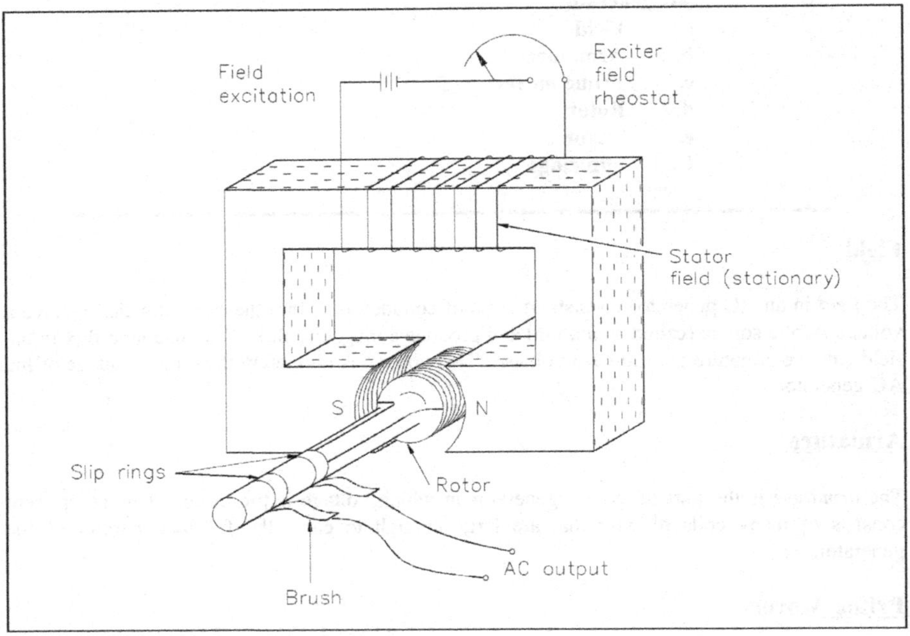

Figure 1 Basic AC Generator

Stator

The *stator* of an AC generator is the part that is stationary (refer to Figure 1). Like the rotor, this component may be the armature or the field, depending on the type of generator. The stator will be the armature if the voltage output is generated there; the stator will be the field if the field excitation is applied there.

Slip Rings

Slip rings are electrical connections that are used to transfer power to and from the rotor of an AC generator (refer to Figure 1). The slip ring consists of a circular conducting material that is connected to the rotor windings and insulated from the shaft. Brushes ride on the slip ring as the rotor rotates. The electrical connection to the rotor is made by connections to the brushes.

Slip rings are used in AC generators because the desired output of the generator is a sine wave. In a DC generator, a commutator was used to provide an output whose current always flowed in the positive direction, as shown in Figure 2. This is not necessary for an AC generator. Therefore, an AC generator may use slip rings, which will allow the output current and voltage to oscillate through positive and negative values. This oscillation of voltage and current takes the shape of a sine wave.

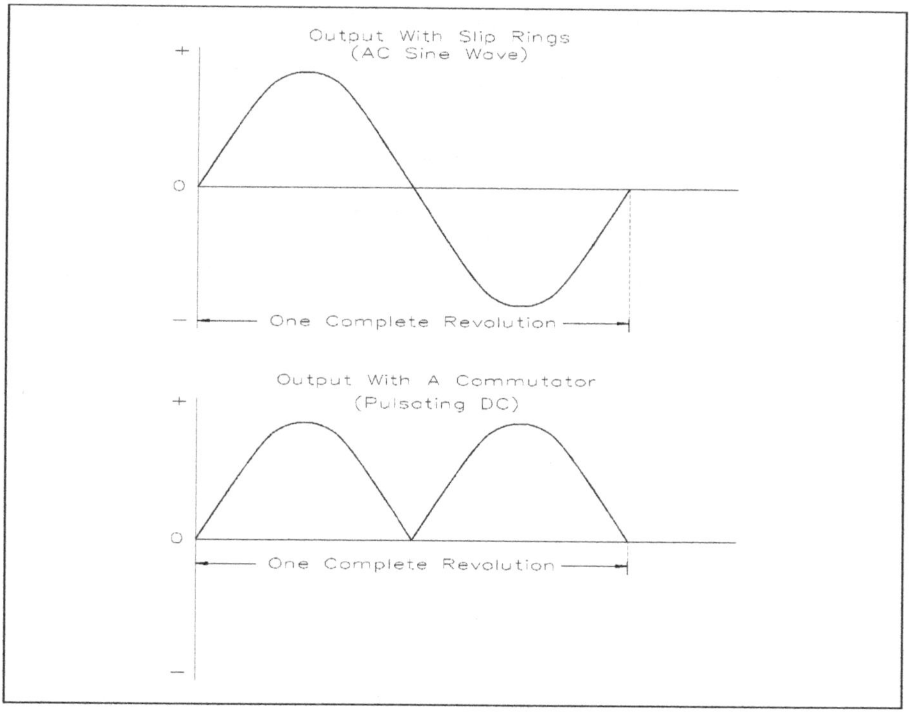

Figure 2 - Comparison of DC and AC Generator Outputs

Summary

The important information in this chapter is summarized below.

AC Generator Components Summary

- The field in an AC generator consists of coils of conductors within the generator that receive a voltage from a source (called excitation) and produce a magnetic flux.
- The armature is the part of an AC generator in which output voltage is produced.
- The prime mover is the component that is used to drive the AC generator.
- The rotor of an AC generator is the part that is driven by the prime mover and that rotates.
- The stator of an AC generator is the part that is stationary.
- Slip rings are electrical connections that are used to transfer power to and from the rotor of an AC generator.

AC GENERATOR THEORY

AC generators are widely used to produce AC voltage. To understand how these generators operate, the basic theory of operation must first be understood.

EO 1.2 **Given the speed of rotation and number of poles, CALCULATE the frequency output of an AC generator.**

EO 1.3 **LIST the three losses found in an AC generator.**

EO 1.4 **Given the prime mover input and generator output, DETERMINE the efficiency of an AC generator.**

Theory of Operation

A simple AC generator consists of: (a) a strong magnetic field, (b) conductors that rotate through that magnetic field, and (c) a means by which a continuous connection is provided to the conductors as they are rotating (Figure 3). The strong magnetic field is produced by a current flow through the field coil of the rotor. The field coil in the rotor receives excitation through the use of slip rings and brushes. Two brushes are spring-held in contact with the slip rings to provide the continuous connection between the field coil and the external excitation circuit. The armature is contained within the windings of the stator and is connected to the output. Each time the rotor makes one complete revolution, one complete cycle of AC is developed. A generator has many turns of wire wound into the slots of the rotor.

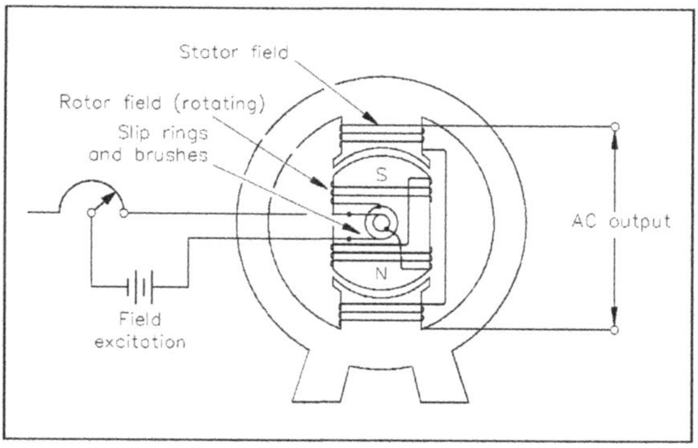

Figure 3 Simple AC Generator

The magnitude of AC voltage generated by an AC generator is dependent on the field strength and speed of the rotor. Most generators are operated at a constant speed; therefore, the generated voltage depends on field excitation, or strength.

The frequency of the generated voltage is dependent on the number of field poles and the speed at which the generator is operated, as indicated in Equation (10-1).

$$f = \frac{NP}{120} \tag{10-1}$$

where

f = frequency (Hz)
P = total number of poles
N = rotor speed (rpm)
120 = conversion from minutes to seconds and from poles to pole pairs

The 120 in Equation (10-1) is derived by multiplying the following conversion factors.

$$\frac{60 \text{ seconds}}{1 \text{ minute}} \times \frac{2 \text{ poles}}{\text{pole pair}}$$

In this manner, the units of frequency (hertz or cycles/sec.) are derived.

Losses in an AC Generator

The load current flows through the armature in all AC generators. Like any coil, the armature has some amount of resistance and inductive reactance. The combination of these make up what is known as the *internal resistance*, which causes a loss in an AC generator. When the load current flows, a voltage drop is developed across the internal resistance. This voltage drop subtracts from the output voltage and, therefore, represents generated voltage and power that is lost and not available to the load. The voltage drop in an AC generator can be found using Equation (10-2).

$$\text{Voltage drop} = I_a R_a + I_a X_{La} \tag{10-2}$$

where

I_a = armature current
R_a = armature resistance
X_{La} = armature inductive reactance

Hysteresis Losses

Hysteresis losses occur when iron cores in an AC generator are subject to effects from a magnetic field. The magnetic domains of the cores are held in alignment with the field in varying numbers, dependent upon field strength. The magnetic domains rotate, with respect to the domains not held in alignment, one complete turn during each rotation of the rotor. This rotation of magnetic domains in the iron causes friction and heat. The heat produced by this friction is called magnetic hysteresis loss.

To reduce hysteresis losses, most AC armatures are constructed of heat-treated silicon steel, which has an inherently low hysteresis loss. After the heat-treated silicon steel is formed to the desired shape, the laminations are heated to a dull red and then allowed to cool. This process, known as annealing, reduces hysteresis losses to a very low value.

Mechanical Losses

Rotational or *mechanical losses* can be caused by bearing friction, brush friction on the commutator, and air friction (called windage), which is caused by the air turbulence due to armature rotation. Careful maintenance can be instrumental in keeping bearing friction to a minimum. Clean bearings and proper lubrication are essential to the reduction of bearing friction. Brush friction is reduced by ensuring: proper brush seating, proper brush use, and maintenance of proper brush tension. A smooth and clean commutator also aids in the reduction of brush friction. In very large generators, hydrogen is used within the generator for cooling; hydrogen, being less dense than air, causes less windage losses than air.

Efficiency

Efficiency of an AC generator is the ratio of the useful power output to the total power input. Because any mechanical process experiences some losses, no AC generators can be 100 percent efficient. Efficiency of an AC generator can be calculated using Equation (10-3).

$$\text{Efficiency} = \frac{\text{Output}}{\text{Input}} \times 100 \qquad (10\text{-}3)$$

Example: Given a 5 hp motor acting as the prime mover of a generator that has a load demand of 2 kW, what is the efficiency of the generator?

Solution:

In order to calculate efficiency, the input and output power must be in the same units. As described in Thermodynamics, the horsepower and the watt are equivalent units of power.

Therefore, the equivalence of these units is expressed with a conversion factor as follows.

$$\left(\frac{550\ \frac{\text{ft-lbf}}{\text{sec}}}{1\text{hp}}\right)\left(\frac{1\ \text{kW}}{737.6\ \frac{\text{ft-lbf}}{\text{sec}}}\right)\left(\frac{1000\ \text{w}}{1\ \text{kW}}\right) = 746\ \frac{\text{W}}{\text{hp}}$$

$$\text{Input Power} = 5\ \text{hp} \times 746\ \frac{\text{W}}{\text{hp}} = 3730\ \text{W}$$

$$\text{Output Power} = 2\ \text{kW} = 2000\ \text{W}$$

$$\text{Efficiency} = \frac{\text{Output}}{\text{Input}} = \frac{2000\ \text{W}}{3730\ \text{W}} = 0.54 \times 100 = 54\%$$

Summary

The important information covered in this chapter is summarized below.

AC Generator Theory Summary

- The frequency of the generated voltage in an AC generator can be calculated by multiplying the number of poles by the speed of the generator and dividing by a factor of 120.

- The three losses found in an AC generator are:
  - Internal voltage drops due to the internal resistance and impedance of the generator
  - Hysteresis losses
  - Mechanical losses

- Efficiency of an AC generator can be calculated by dividing the output by the input and multiplying by 100.

AC GENERATOR OPERATION

Because of the nature of AC voltage and current, the operation of an AC generator requires that rules and procedures be followed. In addition, there are various types of AC generators available, each type having advantages and disadvantages.

EO 1.5 **DESCRIBE the bases behind the kW and current ratings of an AC generator.**

EO 1.6 **DESCRIBE the conditions that must be met prior to paralleling two AC generators including consequences of not meeting these conditions.**

EO 1.7 **DESCRIBE the difference between a stationary field, rotating armature AC generator and a rotating field, stationary armature AC generator.**

EO 1.8 **EXPLAIN the differences between a wye-connected and delta-connected AC generator including advantages and disadvantages of each type.**

Ratings

Typical name plate data for an AC generator (Figure 4) includes: (1) manufacturer; (2) serial number and type number; (3) speed (rpm), number of poles, frequency of output, number of phases, and maximum supply voltage; (4) capacity rating in KVA and kW at a specified power factor and maximum output voltage; (5) armature and field current per phase; and (6) maximum temperature rise.

Power (kW) ratings of an AC generator are based on the ability of the prime mover to overcome generator losses and the ability of the machine to dissipate the internally generated heat. The current rating of an AC generator is based on the insulation rating of the machine.

Westinghouse
AC generator air cooled NO. 6750616 Type ATB 3600 RPM
2 poles 60 hertz 3-phase wye-connected for 13800 volts
Rating 15625 KVA 12500 kW 0.80 PF exciter 250 volts
Armature 654 amp field 183 amp
Guaranteed temp. rise not to exceed 60° C on armature by detector 80° C on field by resistance

Figure 4 AC Generator Nameplate Ratings

Paralleling AC Generators

Most electrical power grids and distribution systems have more than one AC generator operating at one time. Normally, two or more generators are operated in parallel in order to increase the available power. Three conditions must be met prior to paralleling (or synchronizing) AC generators.

- Their terminal voltages must be equal. If the voltages of the two AC generators are not equal, one of the AC generators could be picked up as a reactive load to the other AC generator. This causes high currents to be exchanged between the two machines, possibly causing generator or distribution system damage.
- Their frequencies must be equal. A mismatch in frequencies of the two AC generators will cause the generator with the lower frequency to be picked up as a load on the other generator (a condition referred to as "motoring"). This can cause an overload in the generators and the distribution system.
- Their output voltages must be in phase. A mismatch in the phases will cause large opposing voltages to be developed. The worst case mismatch would be 180° out of phase, resulting in an opposing voltage between the two generators of twice the output voltage. This high voltage can cause damage to the generators and distribution system due to high currents.

During paralleling operations, voltages of the two generators that are to be paralleled are indicated through the use of voltmeters. Frequency matching is accomplished through the use of output frequency meters. Phase matching is accomplished through the use of a synchroscope, a device that senses the two frequencies and gives an indication of phase differences and a relative comparison of frequency differences.

Types of AC Generators

As previously discussed, there are two types of AC generators: the stationary field, rotating armature; and the rotating field, stationary armature.

Small AC generators usually have a stationary field and a rotating armature (Figure 5). One important disadvantage to this arrangement is that the slip-ring and brush assembly is in series with the load circuits and, because of worn or dirty components, may interrupt the flow of current.

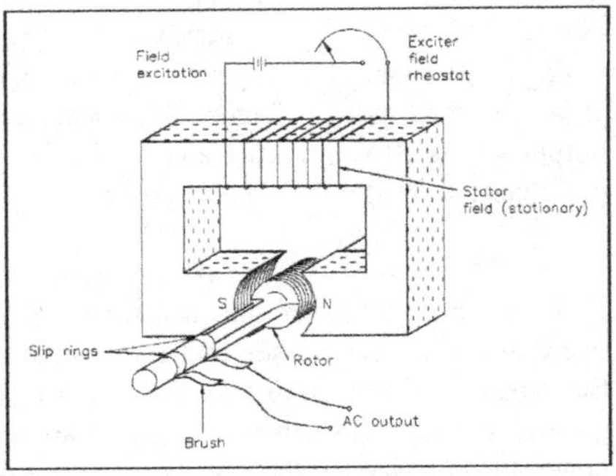

Figure 5 Stationary Field, Rotating Armature AC Generator

If DC field excitation is connected to the rotor, the stationary coils will have AC induced into them (Figure 6). This arrangement is called a rotating field, stationary armature AC generator.

The rotating field, stationary armature type AC generator is used when large power generation is involved. In this type of generator, a DC source is supplied to the rotating field coils, which produces a magnetic field around the rotating element. As the rotor is turned by the prime mover, the field will cut the conductors of the stationary armature, and an EMF will be induced into the armature windings.

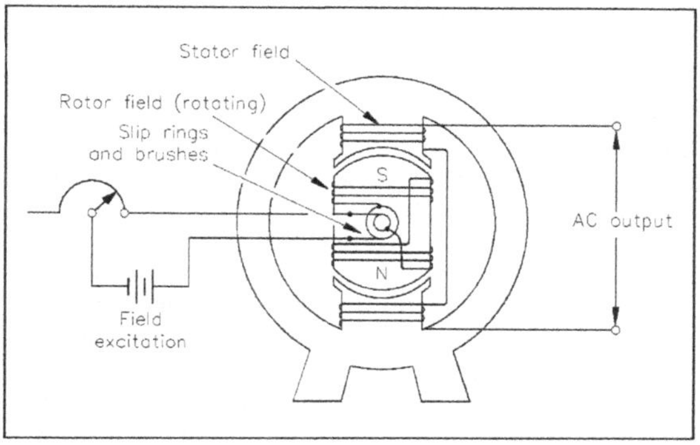

Figure 6 Simple AC Generator - Rotating Field, Stationary Armature

This type of AC generator has several advantages over the stationary field, rotating armature AC generator: (1) a load can be connected to the armature without moving contacts in the circuit; (2) it is much easier to insulate stator fields than rotating fields; and (3) much higher voltages and currents can be generated.

Three-Phase AC Generators

The principles of a three-phase generator are basically the same as that of a single-phase generator, except that there are three equally-spaced windings and three output voltages that are all 120° out of phase with one another. Physically adjacent loops (Figure 7) are separated by 60° of rotation; however, the loops are connected to the slip rings in such a manner that there are 120 electrical degrees between phases.

The individual coils of each winding are combined and represented as a single coil. The significance of Figure 7 is that it shows that the three-phase generator has three separate armature windings that are 120 electrical degrees out of phase.

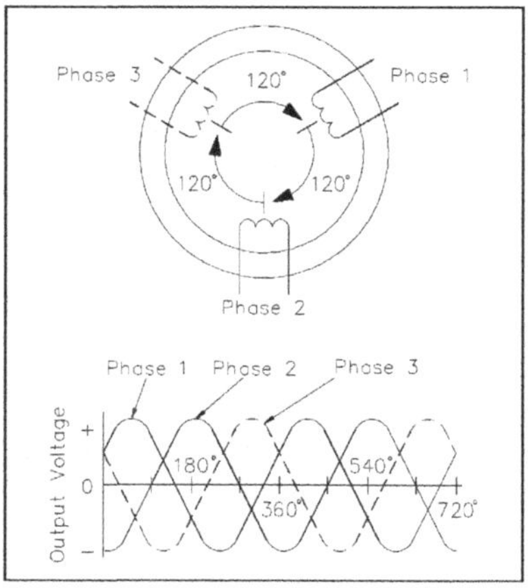

Figure 7 Stationary Armature 3ϕ Generator

AC Generator Connections

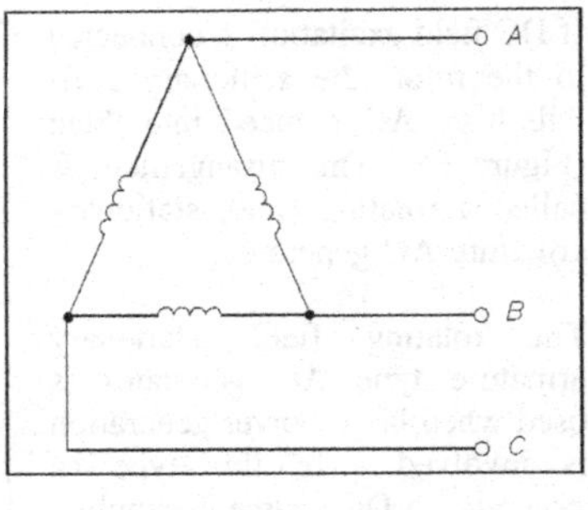

Figure 8 Delta Connection

As shown in Figure 7, there are six leads from the armature of a three-phase generator, and the output is connected to an external load. In actual practice, the windings are connected together, and only three leads are brought out and connected to the external load.

Two means are available to connect the three armature windings. In one type of connection, the windings are connected in series, or delta-connected (Δ) (Figure 8).

In a delta-connected generator, the voltage between any two of the phases, called line voltage, is the same as the voltage generated in any one phase. As shown in Figure 9, the three phase voltages are equal, as are the three line voltages. The current in any line is $\sqrt{3}$ times the phase current. You can see that a delta-connected generator provides an increase in current, but no increase in voltage.

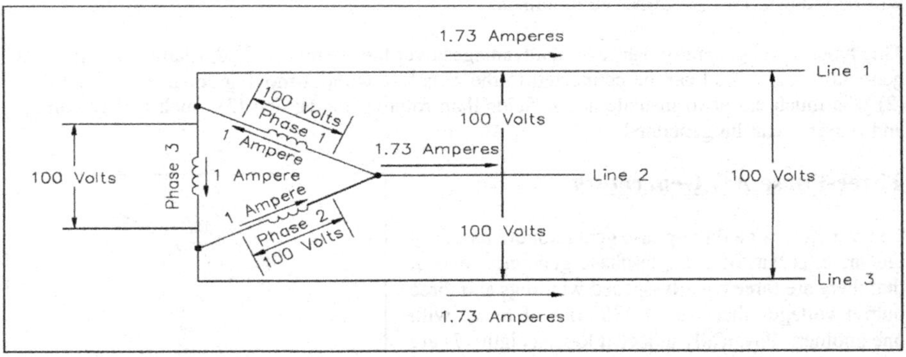

Figure 9 Characteristics of a Delta-Connected Generator

An advantage of the delta-connected AC generator is that if one phase becomes damaged or open, the remaining two phases can still deliver three-phase power. The capacity of the generator is reduced to 57.7% of what it was with all three phases in operation.

In the other type of connection, one of the leads of each winding is connected, and the remaining three leads are connected to an external load. This is called a wye connection (Y) (Figure 10).

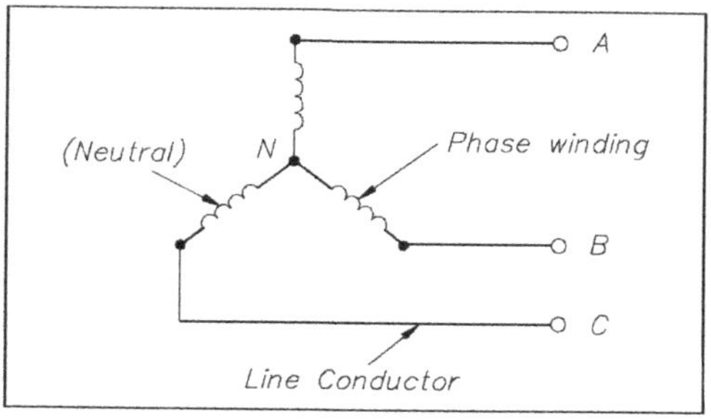

Figure 10 Wye Connection

The voltage and current characteristics of the wye-connected AC generator are opposite to that of the delta connection. Voltage between any two lines in a wye-connected AC generator is 1.73 (or $\sqrt{3}$) times any one phase voltage, while line currents are equal to phase currents. The wye-connected AC generator provides an increase in voltage, but no increase in current (Figure 11).

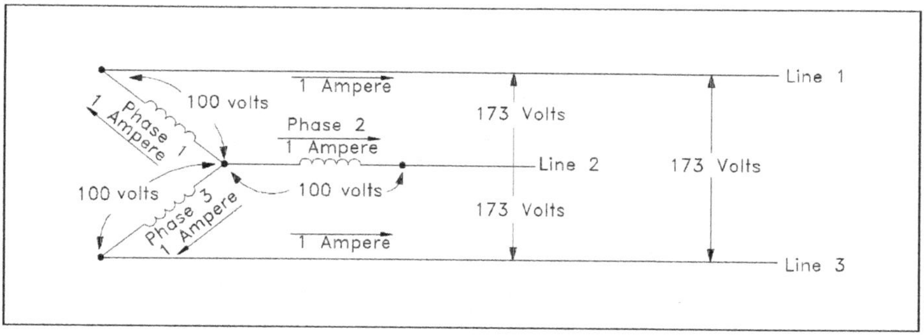

Figure 11 Characteristics of a Wye-Connected AC Generator

An advantage of a wye-connected AC generator is that each phase only has to carry 57.7% of line voltage and, therefore, can be used for high voltage generation.

Summary

The important information covered in this chapter is summarized below.

AC Generator Operation Summary

- Power (kW) ratings of an AC generator are based on the ability of the prime mover to overcome generation losses and the ability of the machine to dissipate the heat generated internally. The current rating of an AC generator is based on the insulation rating of the machine.

- There are three requirements that must be met to parallel AC generators:

    1) Their terminal voltages must be equal. A mismatch may cause high currents and generator or distribution system damage.

    2) Their frequencies must be equal. A mismatch in frequencies can cause one generator to "motor," causing an overload in the generators and the distribution system.

    3) Their output voltages must be in phase. A mismatch in the phases will cause large opposing voltages to be developed, resulting in damage to the generators and distribution system due to high currents.

- The disadvantage of a stationary field, rotating armature is that the slip-ring and brush assembly is in series with the load circuits and, because of worn or dirty components, may interrupt the flow of current.

- A stationary armature, rotating field generator has several advantages: (1) a load can be connected to the armature without moving contacts in the circuit; (2) it is much easier to insulate stator fields than rotating fields; and (3) much higher voltages and currents can be generated.

- The advantage of the delta-connected AC generator is that if one phase becomes damaged or open, the remaining two phases can still deliver three-phase power at a reduced capacity of 57.7%.

- The advantage of a wye-connected AC generator is that each phase only has to carry 57.7% of line voltage and, therefore, can be used for high voltage generation.

Department of Energy
Fundamentals Handbook

ELECTRICAL SCIENCE
Module 11
Voltage Regulators

TABLE OF CONTENTS

LIST OF FIGURES ii

LIST OF TABLES iii

REFERENCES iv

OBJECTIVES v

VOLTAGE REGULATORS 1

- Purpose 1
- Block Diagram Description 1
- Sensing Circuit 2
- Reference Circuit 2
- Comparison Circuit 2
- Amplification Circuit 2
- Signal Output Circuit 3
- Feedback Circuit 3
- Changing Output Voltage 3
- Summary 4

LIST OF FIGURES

Figure 1 Voltage Regulator Block Diagram 2

LIST OF TABLES

NONE

REFERENCES

- Gussow, Milton, Schaum's Outline Series, Basic Electricity, McGraw-Hill.
- Academic Program for Nuclear Power Plant Personnel, Volume IV, Columbia, MD: General Physics Corporation, Library of Congress Card #A 326517, 1982.
- Academic Program for Nuclear Power Plant Personnel, Volume II, Columbia, MD: General Physics Corporation, Library of Congress Card #A 326517, 1982.
- Nasar and Unnewehr, Electromechanics and Electric Machines, John Wiley and Sons.
- Van Valkenburgh, Nooger, and Neville, Basic Electricity, Vol. 5, Hayden Book Company.
- Lister, Eugene C., Electric Circuits and Machines, 5th Edition, McGraw-Hill.
- Croft, Carr, Watt, and Summers, American Electricians Handbook, 10th Edition, McGraw-Hill.
- Mileaf, Harry, Electricity One - Seven, Revised 2nd Edition, Hayden Book Company.
- Buban and Schmitt, Understanding Electricity and Electronics, 3rd Edition, McGraw-Hill.
- Kidwell, Walter, Electrical Instruments and Measurements, McGraw-Hill.

TERMINAL OBJECTIVE

1.0 Given a block diagram, **DESCRIBE** the operation of a typical voltage regulator.

ENABLING OBJECTIVES

1.1 **STATE** the purpose for voltage regulation equipment.

1.2 Given a block diagram of a typical voltage regulator, **DESCRIBE** the function of each of the following components:

a. Sensing circuit
b. Reference circuit
c. Comparison circuit
d. Amplification circuit(s)
e. Signal output circuit
f. Feedback circuit

Intentionally Left Blank

VOLTAGE REGULATORS

Because the voltage from an AC generator varies as the output load and power factor change, a voltage regulator circuit is necessary to permit continuity of the desired output voltage.

EO 1.1 **STATE the purpose for voltage regulation equipment.**

EO 1.2 **Given a block diagram of a typical voltage regulator, DESCRIBE the function of each of the following components:**
- **a. Sensing circuit**
- **b. Reference circuit**
- **c. Comparison circuit**
- **d. Amplification circuit(s)**
- **e. Signal output circuit**
- **f. Feedback circuit**

Purpose

The purpose of a *voltage regulator* is to maintain the output voltage of a generator at a desired value. As load on an AC generator changes, the voltage will also tend to change. The main reason for this change in voltage is the change in the voltage drop across the armature winding caused by a change in load current. In an AC generator, there is an IR drop and an IX_L drop caused by the AC current flowing through the resistance and inductance of the windings. The IR drop is dependent on the amount of the load change only. The IX_L drop is dependent on not only the load change, but also the power factor of the circuit. Therefore, the output voltage of an AC generator varies with both changes in load (i.e., current) and changes in power factor. Because of changes in voltage, due to changes in load and changes in power factor, AC generators require some auxiliary means of regulating output voltage.

Block Diagram Description

Figure 1 shows a typical block diagram of an AC generator voltage regulator. This regulator consists of six basic circuits that together regulate the output voltage of an AC generator from no-load to full-load.

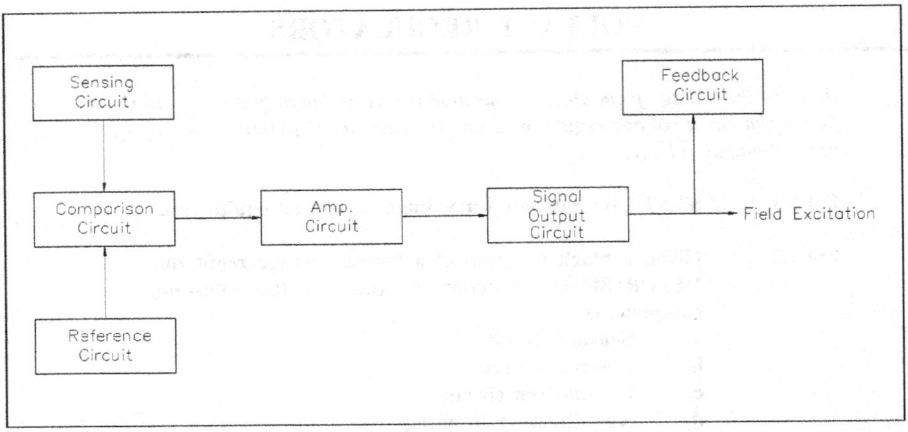

Figure 1 Voltage Regulator Block Diagram

Sensing Circuit

The *sensing circuit* senses output voltage of the AC generator. As the generator is loaded or unloaded, the output voltage changes, and the sensing circuit provides a signal of these voltage changes. This signal is proportional to output voltage and is sent to the comparison circuit.

Reference Circuit

The *reference circuit* maintains a constant output for reference. This reference is the desired voltage output of the AC generator.

Comparison Circuit

The *comparison circuit* electrically compares the reference voltage to the sensed voltage and provides an error signal. This error signal represents an increase or decrease in output voltage. The signal is sent to the amplification circuit.

Amplification Circuit

The *amplification circuit*, which can be a magnetic amplifier or transistor amplifier, takes the signal from the comparison circuit and amplifies the milliamp input to an amp output, which is then sent to the signal output, or field, circuit.

Signal Output Circuit

The *signal output circuit*, which controls field excitation of the AC generator, increases or decreases field excitation to either raise or lower the AC output voltage.

Feedback Circuit

The *feedback circuit* takes some of the output of the signal output circuit and feeds it back to the amplification circuit. It does this to prevent overshooting or undershooting of the desired voltage by slowing down the circuit response.

Changing Output Voltage

Let us consider an increase in generator load and, thereby, a drop in output voltage. First, the sensing circuit senses the decrease in output voltage as compared to the reference and lowers its input to the comparison circuit. Since the reference circuit is always a constant, the comparison circuit will develop an error signal due to the difference between the sensed voltage and the reference voltage. The error signal developed will be of a positive value with the magnitude of the signal dependent on the difference between the sensed voltage and the reference voltage. This output from the comparison circuit will then be amplified by the amplifier circuit and sent to the signal output circuit. The signal output circuit then increases field excitation to the AC generator. This increase in field excitation causes generated voltage to increase to the desired output.

If the load on the generator were decreased, the voltage output of the machine would rise. The actions of the voltage regulator would then be the opposite of that for a lowering output voltage. In this case, the comparison circuit will develop a negative error signal whose magnitude is again dependent on the difference between the sensed voltage and the reference voltage. As a result, the signal output circuit will decrease field excitation to the AC generator, causing the generated voltage to decrease to the desired output.

Summary

Voltage regulators are summarized below.

Voltage Regulators Summary

- Purpose - to maintain the output voltage of a generator at a desired value
- Sensing circuit - senses output voltage of the AC generator
- Reference circuit - maintains a constant output for reference, or desired, voltage output of the AC generator
- Comparison circuit - compares reference voltage to output voltage and provides an error signal to the amplification circuit
- Amplification circuit(s) - takes the signal from the comparison circuit and amplifies the milliamp input to an amp output
- Signal output circuit - controls field excitation of the AC generator
- Feedback circuit - prevents overshooting or undershooting of the desired voltage by slowing down the circuit response

DOE-HDBK-1011/4-92
JUNE 1992

DOE FUNDAMENTALS HANDBOOK

ELECTRICAL SCIENCE

Volume 4 of 4

U.S. Department of Energy
Washington, D.C. 20585

FSC-6910

This document has been reproduced directly from the best available copy.

Available to DOE and DOE contractors from the Office of Scientific and Technical Information. P. O. Box 62, Oak Ridge, TN 37831; (615) 576-8401.

Available to the public from the National Technical Information Service, U.S. Department of Commerce, 5285 Port Royal Rd., Springfield, VA 22161.

Order No. DE92019788

ABSTRACT

The *Electrical Science Fundamentals Handbook* was developed to assist nuclear facility operating contractors provide operators, maintenance personnel, and the technical staff with the necessary fundamentals training to ensure a basic understanding of electrical theory, terminology, and application. The handbook includes information on alternating current (AC) and direct current (DC) theory, circuits, motors, and generators; AC power and reactive components; batteries; AC and DC voltage regulators; transformers; and electrical test instruments and measuring devices. This information will provide personnel with a foundation for understanding the basic operation of various types of DOE nuclear facility electrical equipment.

Key Words: Training Material, Magnetism, DC Theory, DC Circuits, Batteries, DC Generators, DC Motors, AC Theory, AC Power, AC Generators, Voltage Regulators, AC Motors, Transformers, Test Instruments, Electrical Distribution

FOREWORD

The *Department of Energy (DOE) Fundamentals Handbooks* consist of ten academic subjects, which include Mathematics; Classical Physics; Thermodynamics, Heat Transfer, and Fluid Flow; Instrumentation and Control; Electrical Science; Material Science; Mechanical Science; Chemistry; Engineering Symbology, Prints, and Drawings; and Nuclear Physics and Reactor Theory. The handbooks are provided as an aid to DOE nuclear facility contractors.

These handbooks were first published as Reactor Operator Fundamentals Manuals in 1985 for use by DOE category A reactors. The subject areas, subject matter content, and level of detail of the Reactor Operator Fundamentals Manuals were determined from several sources. DOE Category A reactor training managers determined which materials should be included, and served as a primary reference in the initial development phase. Training guidelines from the commercial nuclear power industry, results of job and task analyses, and independent input from contractors and operations-oriented personnel were all considered and included to some degree in developing the text material and learning objectives.

The *DOE Fundamentals Handbooks* represent the needs of various DOE nuclear facilities' fundamental training requirements. To increase their applicability to nonreactor nuclear facilities, the Reactor Operator Fundamentals Manual learning objectives were distributed to the Nuclear Facility Training Coordination Program Steering Committee for review and comment. To update their reactor-specific content, DOE Category A reactor training managers also reviewed and commented on the content. On the basis of feedback from these sources, information that applied to two or more DOE nuclear facilities was considered generic and was included. The final draft of each of the handbooks was then reviewed by these two groups. This approach has resulted in revised modular handbooks that contain sufficient detail such that each facility may adjust the content to fit their specific needs.

Each handbook contains an abstract, a foreword, an overview, learning objectives, and text material, and is divided into modules so that content and order may be modified by individual DOE contractors to suit their specific training needs. Each subject area is supported by a separate examination bank with an answer key.

The *DOE Fundamentals Handbooks* have been prepared for the Assistant Secretary for Nuclear Energy, Office of Nuclear Safety Policy and Standards, by the DOE Training Coordination Program. This program is managed by EG&G Idaho, Inc.

OVERVIEW

The *Department of Energy Fundamentals Handbook* entitled *Electrical Science* was prepared as an information resource for personnel who are responsible for the operation of the Department's nuclear facilities. A basic understanding of electricity and electrical systems is necessary for DOE nuclear facility operators, maintenance personnel, and the technical staff to safely operate and maintain the facility and facility support systems. The information in the handbook is presented to provide a foundation for applying engineering concepts to the job. This knowledge will help personnel more fully understand the impact that their actions may have on the safe and reliable operation of facility components and systems.

The *Electrical Science* handbook consists of fifteen modules that are contained in four volumes. The following is a brief description of the information presented in each module of the handbook.

Volume 1 of 4

Module 1 - Basic Electrical Theory

This module describes basic electrical concepts and introduces electrical terminology.

Module 2 - Basic DC Theory

This module describes the basic concepts of direct current (DC) electrical circuits and discusses the associated terminology.

Volume 2 of 4

Module 3 - DC Circuits

This module introduces the rules associated with the reactive components of inductance and capacitance and how they affect DC circuits.

Module 4 - Batteries

This module introduces batteries and describes the types of cells used, circuit arrangements, and associated hazards.

Module 5 - DC Generators

This module describes the types of DC generators and their application in terms of voltage production and load characteristics.

Module 6 - DC Motors

This module describes the types of DC motors and includes discussions of speed control, applications, and load characteristics.

Volume 3 of 4

Module 7 - Basic AC Theory

This module describes the basic concepts of alternating current (AC) electrical circuits and discusses the associated terminology.

Module 8 - AC Reactive Components

This module describes inductance and capacitance and their effects on AC circuits.

Module 9 - AC Power

This module presents power calculations for single-phase and three-phase AC circuits and includes the power triangle concept.

Module 10 - AC Generators

This module describes the operating characteristics of AC generators and includes terminology, methods of voltage production, and methods of paralleling AC generation sources.

Module 11 - Voltage Regulators

This module describes the basic operation and application of voltage regulators.

Volume 4 of 4

Module 12 - AC Motors

This module explains the theory of operation of AC motors and discusses the various types of AC motors and their application.

Module 13 - Transformers

This module introduces transformer theory and includes the types of transformers, voltage/current relationships, and application.

Module 14 - Test Instruments and Measuring Devices

This module describes electrical measuring and test equipment and includes the parameters measured and the principles of operation of common instruments.

Module 15 - Electrical Distribution Systems

This module describes basic electrical distribution systems and includes characteristics of system design to ensure personnel and equipment safety.

The information contained in this handbook is by no means all encompassing. An attempt to present the entire subject of electrical science would be impractical. However, the *Electrical Science* handbook does present enough information to provide the reader with a fundamental knowledge level sufficient to understand the advanced theoretical concepts presented in other subject areas, and to better understand basic system and equipment operations.

Module 13 - Transformers

This module introduces transformer theory and includes the types of transformers, voltage/current relationships, and application.

Module 14 - Test Instruments and Measuring Devices

This module describes electrical measuring and test equipment and includes the parameters measured and the principles of operation of common instruments.

Module 15 - Electrical Distribution Systems

This module describes basic electrical distribution systems and includes characteristics of system design to ensure personnel and equipment safety.

The information contained in this handbook is by no means all encompassing. An attempt to present the entire subject of electrical science would be impractical. However, the *Electrical Science* handbook does present enough information to provide the reader with a fundamental knowledge level sufficient to understand the advanced theoretical concepts presented in other subject areas, and to better understand basic system and equipment operations.

Department of Energy
Fundamentals Handbook

ELECTRICAL SCIENCE
Module 12
AC Motors

TABLE OF CONTENTS

LIST OF FIGURES ii

LIST OF TABLES iii

REFERENCES iv

OBJECTIVES v

AC MOTOR THEORY 1

- Principles of Operation 1
- Rotating Field 1
- Torque Production 5
- Slip 5
- Torque 7
- Summary 8

AC MOTOR TYPES 9

- Induction Motor 9
- Single-Phase AC Induction Motors 11
- Synchronous Motors 12
- Starting a Synchronous Motor 12
- Field Excitation 14
- Summary 15

LIST OF FIGURES

Figure 1 Three-Phase Stator 2

Figure 2 Rotating Magnetic Field 3

Figure 3 Induction Motor 5

Figure 4 Torque vs Slip 7

Figure 5 Squirrel-Cage Induction Rotor 10

Figure 6 Split-Phase Motor 11

Figure 7 Wound Rotor 12

Figure 8 Torque Angle 13

Figure 9 Synchronous Motor Field Excitation 14

LIST OF TABLES

NONE

REFERENCES

- Gussow, Milton, Schaum's Outline Series, Basic Electricity, McGraw-Hill.
- Academic Program for Nuclear Power Plant Personnel, Volume IV, Columbia, MD: General Physics Corporation, Library of Congress Card #A 326517, 1982.
- Academic Program for Nuclear Power Plant Personnel, Volume II, Columbia, MD: General Physics Corporation, Library of Congress Card #A 326517, 1982.
- Nasar and Unnewehr, Electromechanics and Electric Machines, John Wiley and Sons.
- Van Valkenburgh, Nooger, and Neville, Basic Electricity, Vol. 5, Hayden Book Company.
- Lister, Eugene C., Electric Circuits and Machines, 5th Edition, McGraw-Hill.
- Croft, Carr, Watt, and Summers, American Electricians Handbook, 10th Edition, McGraw-Hill.
- Mason, C. Russel, The Art and Science of Protective Relaying, John Wiley and Sons.
- Mileaf, Harry, Electricity One - Seven, Revised 2nd Edition, Hayden Book Company.
- Buban and Schmitt, Understanding Electricity and Electronics, 3rd Edition, McGraw-Hill.
- Kidwell, Walter, Electrical Instruments and Measurements, McGraw-Hill.

TERMINAL OBJECTIVE

1.0 Given the type and application of an AC motor, **DESCRIBE** the operating characteristics of that motor including methods of torque production and advantages of that type.

ENABLING OBJECTIVES

1.1 **DESCRIBE** how a rotating magnetic field is produced in an AC motor.

1.2 **DESCRIBE** how torque is produced in an AC motor.

1.3 Given field speed and rotor speed, **CALCULATE** percent slip in an AC motor.

1.4 **EXPLAIN** the relationship between speed and torque in an AC induction motor.

1.5 **DESCRIBE** how torque is produced in a single-phase AC motor.

1.6 **EXPLAIN** why an AC synchronous motor does not have starting torque.

1.7 **DESCRIBE** how an AC synchronous motor is started.

1.8 **DESCRIBE** the effects of over and under-exciting an AC synchronous motor.

1.9 **STATE** the applications of the following types of AC motors:
   a. Induction
   b. Single-phase
   c. Synchronous

Intentionally Left Blank

AC MOTOR THEORY

AC motors are widely used to drive machinery for a wide variety of applications. To understand how these motors operate, a knowledge of the basic theory of operation of AC motors is necessary.

EO 1.1 **DESCRIBE how a rotating magnetic field is produced in an AC motor.**

EO 1.2 **DESCRIBE how torque is produced in an AC motor.**

EO 1.3 **Given field speed and rotor speed, CALCULATE percent slip in an AC motor.**

EO 1.4 **EXPLAIN the relationship between slip and torque in an AC induction motor.**

Principles of Operation

The principle of operation for all AC motors relies on the interaction of a revolving magnetic field created in the stator by AC current, with an opposing magnetic field either induced on the rotor or provided by a separate DC current source. The resulting interaction produces usable torque, which can be coupled to desired loads throughout the facility in a convenient manner. Prior to the discussion of specific types of AC motors, some common terms and principles must be introduced.

Rotating Field

Before discussing how a rotating magnetic field will cause a motor rotor to turn, we must first find out how a rotating magnetic field is produced. Figure 1 illustrates a three-phase stator to which a three-phase AC current is supplied.

The windings are connected in wye. The two windings in each phase are wound in the same direction. At any instant in time, the magnetic field generated by one particular phase will depend on the current through that phase. If the current through that phase is zero, the resulting magnetic field is zero. If the current is at a maximum value, the resulting field is at a maximum value. Since the currents in the three windings are 120° out of phase, the magnetic fields produced will also be 120° out of phase. The three magnetic fields will combine to produce one field, which will act upon the rotor. In an AC induction motor, a magnetic field is induced in the rotor opposite in polarity of the magnetic field in the stator. Therefore, as the magnetic field rotates in the stator, the rotor also rotates to maintain its alignment with the stator's magnetic field. The remainder of this chapter's discussion deals with AC induction motors.

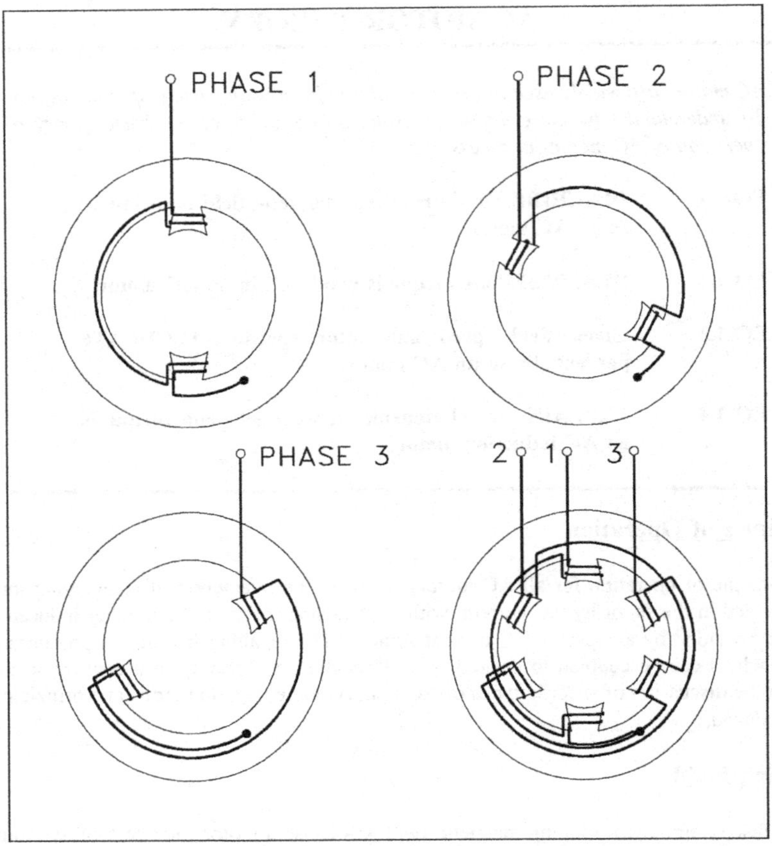

Figure 1 Three-Phase Stator

From one instant to the next, the magnetic fields of each phase combine to produce a magnetic field whose position shifts through a certain angle. At the end of one cycle of alternating current, the magnetic field will have shifted through 360°, or one revolution (Figure 2). Since the rotor has an opposing magnetic field induced upon it, it will also rotate through one revolution.

For purpose of explanation, rotation of the magnetic field is developed in Figure 2 by "stopping" the field at six selected positions, or instances. These instances are marked off at 60° intervals on the sine waves representing the current flowing in the three phases, A, B, and C. For the following discussion, when the current flow in a phase is positive, the magnetic field will develop a north pole at the poles labeled A, B, and C. When the current flow in a phase is negative, the magnetic field will develop a north pole at the poles labeled A', B', and C'.

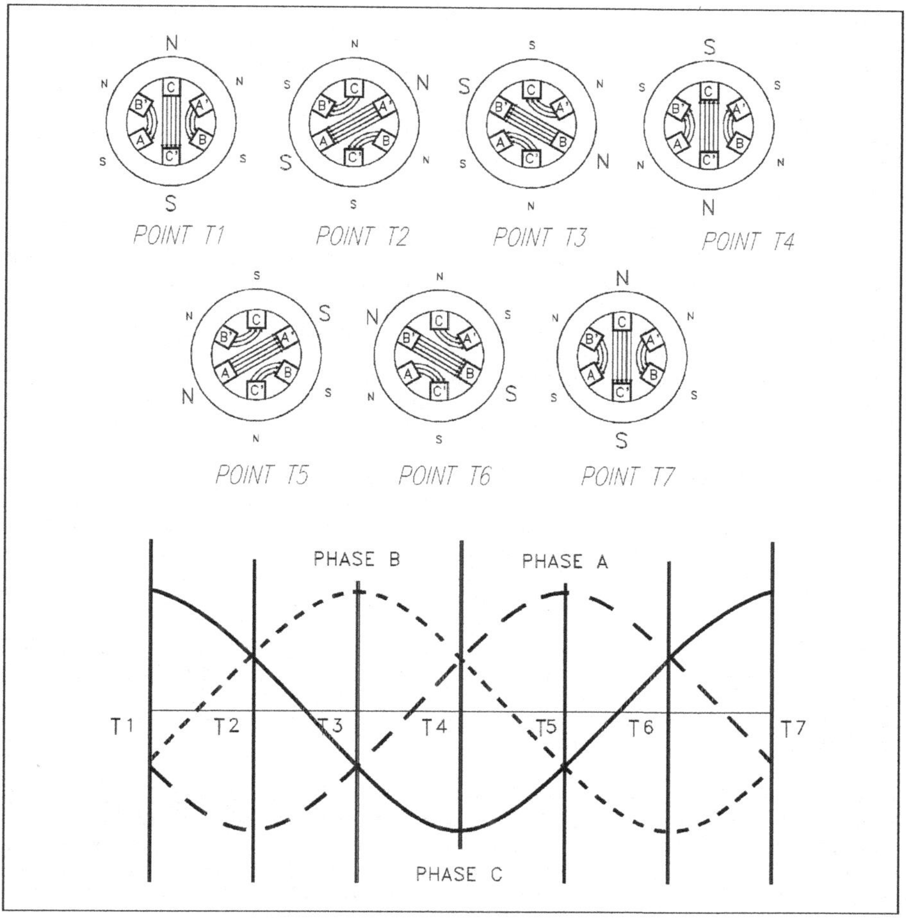

Figure 2 Rotating Magnetic Field

At point T1, the current in phase C is at its maximum positive value. At the same instance, the currents in phases A and B are at half of the maximum negative value. The resulting magnetic field is established vertically downward, with the maximum field strength developed across the C phase, between pole C (north) and pole C' (south). This magnetic field is aided by the weaker fields developed across phases A and B, with poles A' and B' being north poles and poles A and B being south poles.

At Point T2, the current sine waves have rotated through 60 electrical degrees. At this point, the current in phase A has increased to its maximum negative value. The current in phase B has reversed direction and is at half of the maximum positive value. Likewise, the current in phase C has decreased to half of the maximum positive value. The resulting magnetic field is established downward to the left, with the maximum field strength developed across the A phase, between poles A' (north) and A (south). This magnetic field is aided by the weaker fields developed across phases B and C, with poles B and C being north poles and poles B' and C' being south poles. Thus, it can be seen that the magnetic field within the stator of the motor has physically rotated 60°.

At Point T3, the current sine waves have again rotated 60 electrical degrees from the previous point for a total rotation of 120 electrical degrees. At this point, the current in phase B has increased to its maximum positive value. The current in phase A has decreased to half of its maximum negative value, while the current in phase C has reversed direction and is at half of its maximum negative value also. The resulting magnetic field is established upward to the left, with the maximum field strength developed across phase B, between poles B (north) and B' (south). This magnetic field is aided by the weaker fields developed across phases A and C, with poles A' and C' being north poles and poles A and C being south poles. Thus, it can be seen that the magnetic field on the stator has rotated another 60° for a total rotation of 120°.

At Point T4, the current sine waves have rotated 180 electrical degrees from Point T1 so that the relationship of the phase currents is identical to Point T1 except that the polarity has reversed. Since phase C is again at a maximum value, the resulting magnetic field developed across phase C will be of maximum field strength. However, with current flow reversed in phase C the magnetic field is established vertically upward between poles C' (north) and C (south). As can be seen, the magnetic field has now physically rotated a total of 180° from the start.

At Point T5, phase A is at its maximum positive value, which establishes a magnetic field upward to the right. Again, the magnetic field has physically rotated 60° from the previous point for a total rotation of 240°. At Point T6, phase B is at its maximum negative value, which will establish a magnetic field downward to the right. The magnetic field has again rotated 60° from Point T5 for a total rotation of 300°.

Finally, at Point T7, the current is returned to the same polarity and values as that of Point T1. Therefore, the magnetic field established at this instance will be identical to that established at Point T1. From this discussion it can be seen that for one complete revolution of the electrical sine wave (360°), the magnetic field developed in the stator of a motor has also rotated one complete revolution (360°). Thus, you can see that by applying three-phase AC to three windings symmetrically spaced around a stator, a rotating magnetic field is generated.

Torque Production

When alternating current is applied to the stator windings of an AC induction motor, a rotating magnetic field is developed. The rotating magnetic field cuts the bars of the rotor and induces a current in them due to generator action. The direction of this current flow can be found using the left-hand rule for generators. This induced current will produce a magnetic field, opposite in polarity of the stator field, around the conductors of the rotor, which will try to line up with the magnetic field of the stator. Since the stator field is rotating continuously, the rotor cannot line up with, or lock onto, the stator field and, therefore, must follow behind it (Figure 3).

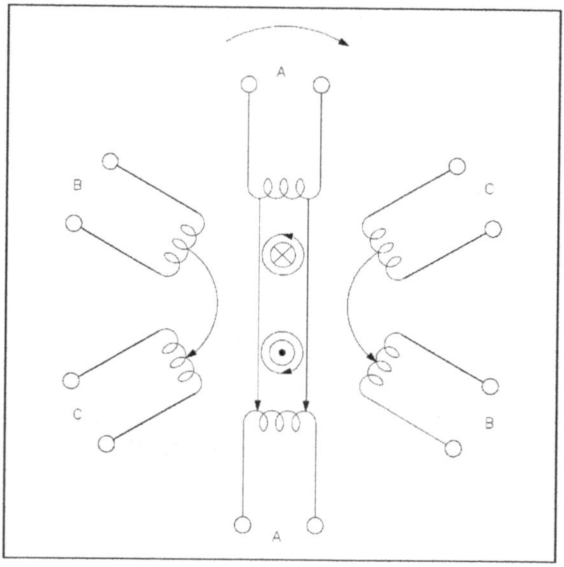

Figure 3 Induction Motor

Slip

It is virtually impossible for the rotor of an AC induction motor to turn at the same speed as that of the rotating magnetic field. If the speed of the rotor were the same as that of the stator, no relative motion between them would exist, and there would be no induced EMF in the rotor. (Recall from earlier modules that relative motion between a conductor and a magnetic field is needed to induce a current.) Without this induced EMF, there would be no interaction of fields to produce motion. The rotor must, therefore, rotate at some speed less than that of the stator if relative motion is to exist between the two.

The percentage difference between the speed of the rotor and the speed of the rotating magnetic field is called *slip*. The smaller the percentage, the closer the rotor speed is to the rotating magnetic field speed. Percent slip can be found by using Equation (12-1).

$$\text{SLIP} = \frac{N_S - N_R}{N_S} \times 100\% \qquad (12\text{-}1)$$

where

N_S = synchronous speed (rpm)
N_R = rotor speed (rpm)

The speed of the rotating magnetic field or synchronous speed of a motor can be found by using Equation (12-2).

$$N_S = \frac{120\ f}{P} \tag{12-2}$$

where

N_s = speed of rotating field (rpm)
f = frequency of rotor current (Hz)
P = total number of poles

Example: A two pole, 60 Hz AC induction motor has a full load speed of 3554 rpm. What is the percent slip at full load?

Solution:

Synchronous speed:

$$N_S = \frac{120\ f}{P}$$

$$N_S = \frac{120\ (60\ \text{Hz})}{2}$$

$$N_S = 3600\ \text{rpm}$$

Slip:

$$\text{SLIP} = \frac{N_S - N_R}{N_S} \times 100\%$$

$$\text{SLIP} = \frac{3600 - 3554\ \text{rpm}}{3600\ \text{rpm}} \times 100\% = 1.3\%$$

Torque

The torque of an AC induction motor is dependent upon the strength of the interacting rotor and stator fields and the phase relationship between them. Torque can be calculated by using Equation (12-3).

$$T = K \Phi I_R \cos \theta_R \qquad (12\text{-}3)$$

where

T = torque (lb-ft)
K = constant
Φ = stator magnetic flux
I_R = rotor current (A)
$\cos \theta_R$ = power factor of rotor

During normal operation, K, Φ, and $\cos \theta_R$ are, for all intents and purposes, constant, so that torque is directly proportional to the rotor current. Rotor current increases in almost direct proportion to slip. The change in torque with respect to slip (Figure 4) shows that, as slip increases from zero to ~10%, the torque increases linearly. As the load and slip are increased beyond full-load torque, the torque will reach a maximum value at about 25% slip. The maximum value of torque is called the *breakdown torque* of the motor. If load is increased beyond this point, the motor will stall and come to a rapid stop. The typical induction motor breakdown torque varies from 200 to 300% of full load torque. Starting torque is the value of torque at 100% slip and is normally 150 to 200% of full-load torque. As the rotor accelerates, torque will increase to breakdown torque and then decrease to the value required to carry the load on the motor at a constant speed, usually between 0-10%.

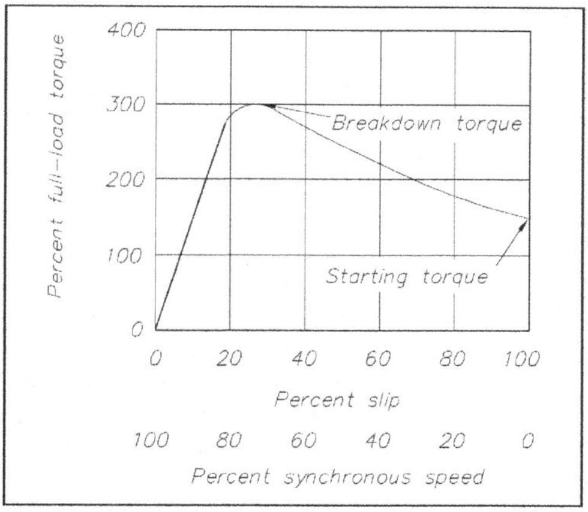

Figure 4 Torque vs Slip

Summary

The important information covered in this chapter is summarized below.

AC Motor Theory Summary

- A magnetic field is produced in an AC motor through the action of the three-phase voltage that is applied. Each of the three phases is 120° from the other phases. From one instant to the next, the magnetic fields combine to produce a magnetic field whose position shifts through a certain angle. At the end of one cycle of alternating current, the magnetic field will have shifted through 360°, or one revolution.

- Torque in an AC motor is developed through interactions with the rotor and the rotating magnetic field. The rotating magnetic field cuts the bars of the rotor and induces a current in them due to generator action. This induced current will produce a magnetic field around the conductors of the rotor, which will try to line up with the magnetic field of the stator.

- Slip is the percentage difference between the speed of the rotor and the speed of the rotating magnetic field.

- In an AC induction motor, as slip increases from zero to ~10%, the torque increases linearly. As the load and slip are increased beyond full-load torque, the torque will reach a maximum value at about 25% slip. If load is increased beyond this point, the motor will stall and come to a rapid stop. The typical induction motor breakdown torque varies from 200 to 300% of full-load torque. Starting torque is the value of torque at 100% slip and is normally 150 to 200% of full-load torque.

AC MOTOR TYPES

Various types of AC motors are used for specific applications. By matching the type of motor to the appropriate application, increased equipment performance can be obtained.

EO 1.5 **DESCRIBE how torque is produced in a single-phase AC motor.**

EO 1.6 **EXPLAIN why an AC synchronous motor does not have starting torque.**

EO 1.7 **DESCRIBE how an AC synchronous motor is started.**

EO 1.8 **DESCRIBE the effects of over and under-exciting an AC synchronous motor.**

EO 1.9 **STATE the applications of the following types of AC motors:**

a. Induction
b. Single-phase
c. Synchronous

Induction Motor

Previous explanations of the operation of an AC motor dealt with induction motors. The induction motor is the most commonly used AC motor in industrial applications because of its simplicity, rugged construction, and relatively low manufacturing costs. The reason that the induction motor has these characteristics is because the rotor is a self-contained unit, with no external connections. This type of motor derives its name from the fact that AC currents are induced into the rotor by a rotating magnetic field.

The induction motor rotor (Figure 5) is made of a laminated cylinder with slots in its surface. The windings in the slots are one of two types. The most commonly used is the "squirrel-cage" rotor. This rotor is made of heavy copper bars that are connected at each end by a metal ring made of copper or brass. No insulation is required between the core and the bars because of the low voltages induced into the rotor bars. The size of the air gap between the rotor bars and stator windings necessary to obtain the maximum field strength is small.

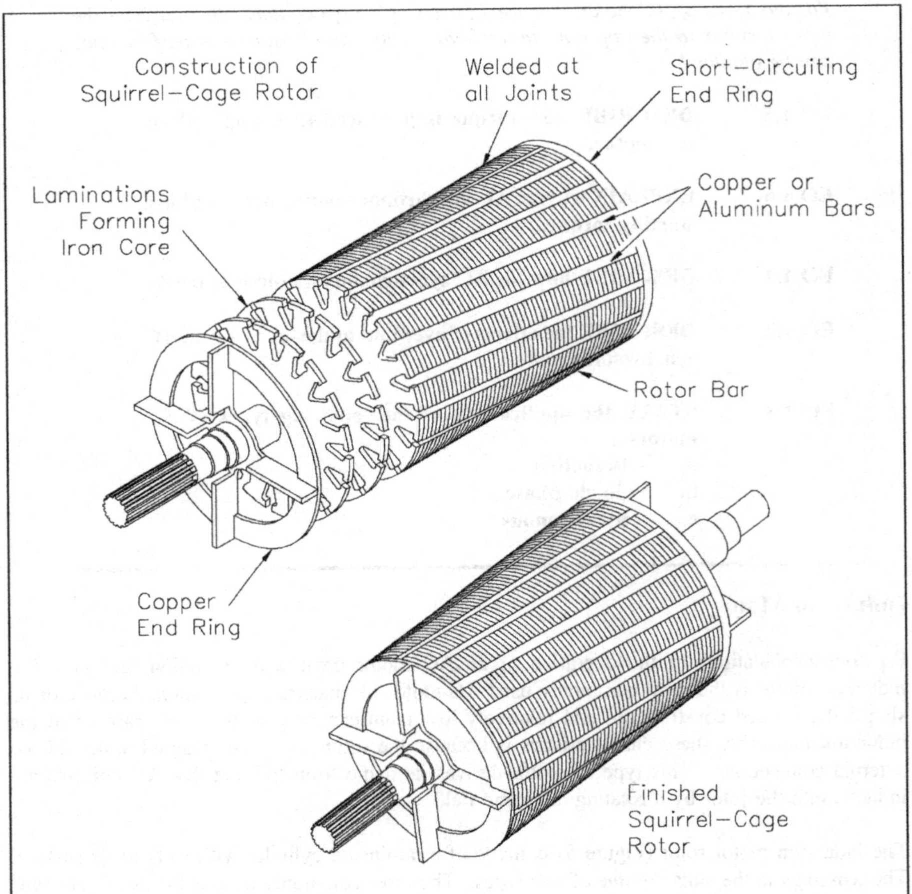

Figure 5 Squirrel-Cage Induction Rotor

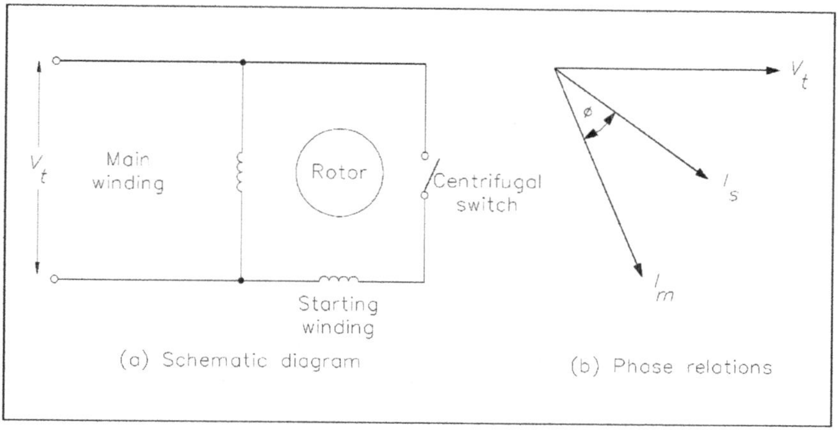

Figure 6 Split-Phase Motor

Single-Phase AC Induction Motors

If two stator windings of unequal impedance are spaced 90 electrical degrees apart and connected in parallel to a single-phase source, the field produced will appear to rotate. This is called phase splitting.

In a split-phase motor, a starting winding is utilized. This winding has a higher resistance and lower reactance than the main winding (Figure 6). When the same voltage V_T is applied to the starting and main windings, the current in the main winding (I_M) lags behind the current of the starting winding I_S (Figure 6). The angle between the two windings is enough phase difference to provide a rotating magnetic field to produce a starting torque. When the motor reaches 70 to 80% of synchronous speed, a centrifugal switch on the motor shaft opens and disconnects the starting winding.

Single-phase motors are used for very small commercial applications such as household appliances and buffers.

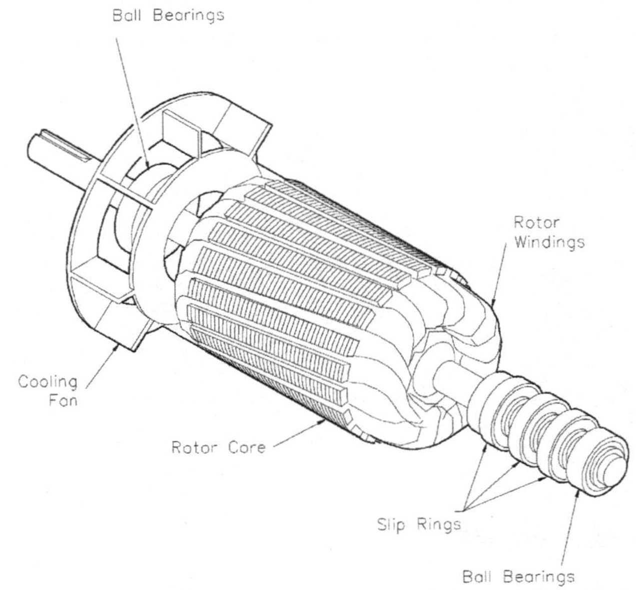

Figure 7 Wound Rotor

Synchronous Motors

Synchronous motors are like induction motors in that they both have stator windings that produce a rotating magnetic field. Unlike an induction motor, the synchronous motor is excited by an external DC source and, therefore, requires slip rings and brushes to provide current to the rotor. In the synchronous motor, the rotor locks into step with the rotating magnetic field and rotates at synchronous speed. If the synchronous motor is loaded to the point where the rotor is pulled out of step with the rotating magnetic field, no torque is developed, and the motor will stop. A synchronous motor is not a self-starting motor because torque is only developed when running at synchronous speed; therefore, the motor needs some type of device to bring the rotor to synchronous speed.

Synchronous motors use a wound rotor. This type of rotor contains coils of wire placed in the rotor slots. Slip rings and brushes are used to supply current to the rotor. (Figure 7).

Starting a Synchronous Motor

A synchronous motor may be started by a DC motor on a common shaft. When the motor is brought to synchronous speed, AC current is applied to the stator windings. The DC motor now acts as a DC generator and supplies DC field excitation to the rotor of the synchronous motor. The load may now be placed on the synchronous motor. Synchronous motors are more often started by means of a squirrel-cage winding embedded in the face of the rotor poles. The motor is then started as an induction motor and brought to ~95% of synchronous speed, at which time direct current is applied, and the motor begins to pull into synchronism. The torque required to pull the motor into synchronism is called the pull-in torque.

As we already know, the synchronous motor rotor is locked into step with the rotating magnetic field and must continue to operate at synchronous speed for all loads. During no-load conditions, the center lines of a pole of the rotating magnetic field and the DC field pole coincide (Figure 8a). As load is applied to the motor, there is a backward shift of the rotor pole, relative to the stator pole (Figure 8b). There is no change in speed. The angle between the rotor and stator poles is called the *torque angle* (α).

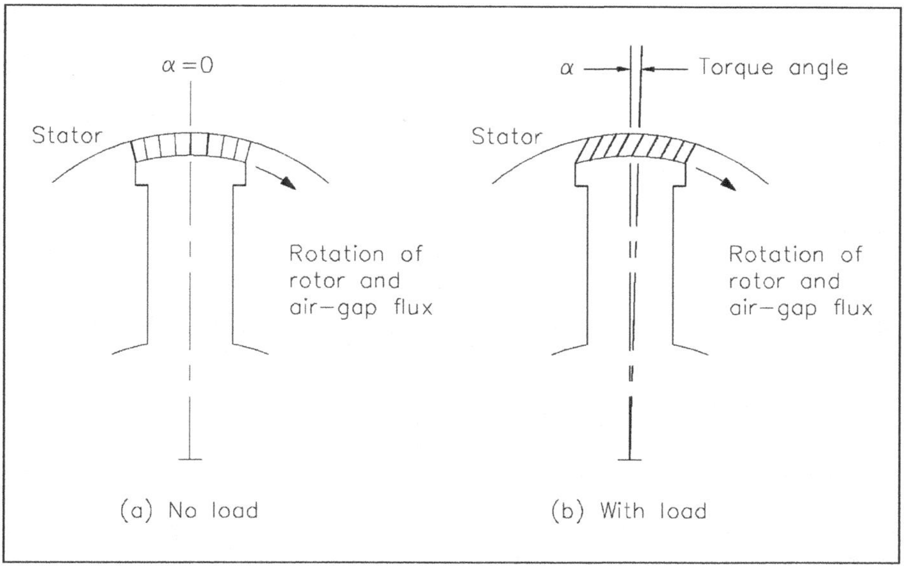

Figure 8 Torque Angle

If the mechanical load on the motor is increased to the point where the rotor is pulled out of synchronism ($\alpha \cong 90°$), the motor will stop. The maximum value of torque that a motor can develop without losing synchronism is called its pull-out torque.

Field Excitation

For a constant load, the power factor of a synchronous motor can be varied from a leading value to a lagging value by adjusting the DC field excitation (Figure 9). Field excitation can be adjusted so that PF = 1 (Figure 9a). With a constant load on the motor, when the field excitation is increased, the counter EMF (V_G) increases. The result is a change in phase between stator current (I) and terminal voltage (V_t), so that the motor operates at a leading power factor (Figure 9b). V_p in Figure 9 is the voltage drop in the stator winding's due to the impedance of the windings and is 90° out of phase with the stator current. If we reduce field excitation, the motor will operate at a lagging power factor (Figure 9c). Note that torque angle, α, also varies as field excitation is adjusted to change power factor.

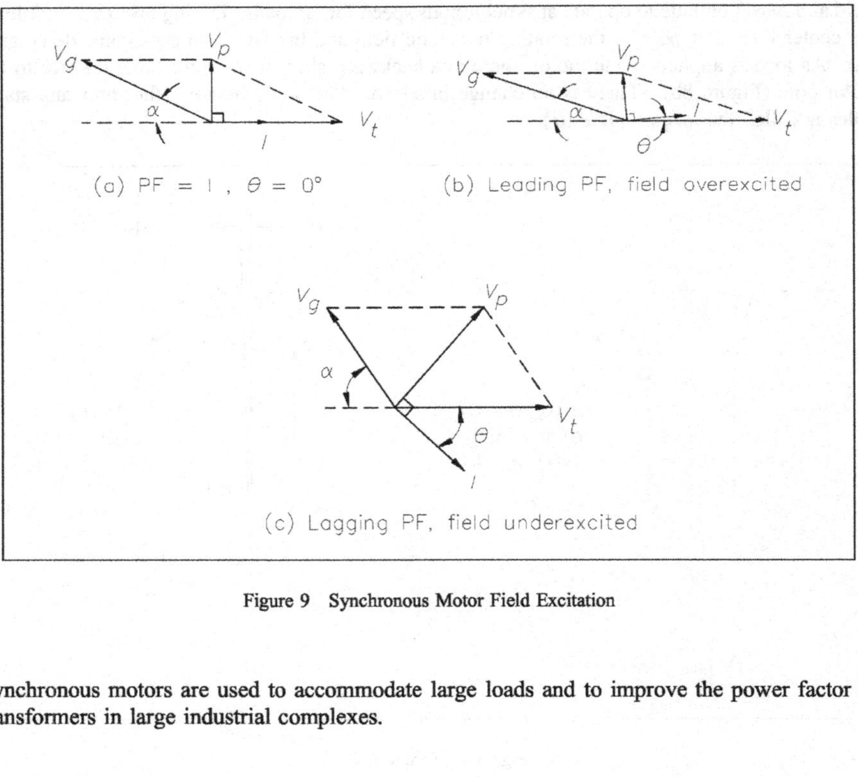

Figure 9 Synchronous Motor Field Excitation

Synchronous motors are used to accommodate large loads and to improve the power factor of transformers in large industrial complexes.

Summary

The important information in this chapter is summarized below.

AC Motor Types Summary

- In a split-phase motor, a starting winding is utilized. This winding has a higher resistance and lower reactance than the main winding. When the same voltage (V_T) is applied to the starting and main windings, the current in the main winding lags behind the current of the starting winding. The angle between the two windings is enough phase difference to provide a rotating magnetic field to produce a starting torque.
- A synchronous motor is not a self-starting motor because torque is only developed when running at synchronous speed.
- A synchronous motor may be started by a DC motor on a common shaft or by a squirrel-cage winding imbedded in the face of the rotor poles.
- Keeping the same load, when the field excitation is increased on a synchronous motor, the motor operates at a leading power factor. If we reduce field excitation, the motor will operate at a lagging power factor.
- The induction motor is the most commonly used AC motor in industrial applications because of its simplicity, rugged construction, and relatively low manufacturing costs.
- Single-phase motors are used for very small commercial applications such as household appliances and buffers.
- Synchronous motors are used to accommodate large loads and to improve the power factor of transformers in large industrial complexes.

Intentionally Left Blank

Department of Energy
Fundamentals Handbook

ELECTRICAL SCIENCE
Module 13
Transformers

TABLE OF CONTENTS

LIST OF FIGURES ii

LIST OF TABLES iii

REFERENCES iv

OBJECTIVES v

TRANSFORMER THEORY 1

- Mutual Induction 1
- Turns Ratio 2
- Impedance Ratio 2
- Efficiency 3
- Theory of Operation 3
- Voltage Ratio 5
- Current Ratio 7
- Three-Phase Transformer Connections 8
- Delta Connection 8
- Wye Connection 8
- Combinations of Delta and Wye Transformer Connections 9
- Transformer Losses and Efficiency 11
- Transformer Operation Under No-Load 13
- Coil Polarity 14
- Summary 16

TRANSFORMER TYPES 17

- Types of Transformers 17
- Distribution Transformer 17
- Power Transformer 17
- Control Transformer 18
- Auto Transformer 18
- Isolation Transformer 18
- Instrument Potential Transformer 19
- Instrument Current Transformer 19
- Summary 20

LIST OF FIGURES

Figure 1 Core-Type Transformer 4

Figure 2 Example 1 Transformer 5

Figure 3 Delta Connection 8

Figure 4 Wye Connection 9

Figure 5 3ϕ Transformer Connections 9

Figure 6 Open Circuit Secondary 13

Figure 7 Polarity of Transformer Coils 15

Figure 8 Auto Transformer Schematic 18

LIST OF TABLES

Table 1 Voltage and Current Ratings of Transformers 10

REFERENCES

- Gussow, Milton, Schaum's Outline Series, Basic Electricity, McGraw-Hill.
- Academic Program for Nuclear Power Plant Personnel, Volume IV, Columbia, MD: General Physics Corporation, Library of Congress Card #A 326517, 1982.
- Nasar and Unnewehr, Electromechanics and Electric Machines, John Wiley and Sons.
- Van Valkenburgh, Nooger, and Neville, Basic Electricity, Vol. 5, Hayden Book Company.
- Croft, Carr, Watt, and Summers, American Electricians Handbook, 10th Edition, McGraw-Hill.
- Mileaf, Harry, Electricity One - Seven, Revised 2nd Edition, Hayden Book Company.
- Buban and Schmitt, Understanding Electricity and Electronics, 3rd Edition, McGraw-Hill.

TERMINAL OBJECTIVE

1.0 Given the type of a transformer, **DESCRIBE** the operating characteristics and applications for that transformer type.

ENABLING OBJECTIVES

1.1 **DEFINE** the following terms as they pertain to transformers:
a. Mutual induction
b. Turns ratio
c. Impedance ratio
d. Efficiency

1.2 **DESCRIBE** the differences between a wye-connected and delta-connected transformer.

1.3 Given the type of connection and turns ratios for the primary and secondary of a transformer, **CALCULATE** voltage, current, and power for each of the following types:
a. Δ - Δ
b. Δ - Y
c. Y - Δ
d. Y - Y

1.4 **STATE** the applications of each of the following types of transformers:
a. Distribution
b. Power
c. Control
d. Auto
e. Isolation
f. Instrument potential
g. Instrument current

Intentionally Left Blank

TRANSFORMER THEORY

Transformers are used extensively for AC power transmissions and for various control and indication circuits. Knowledge of the basic theory of how these components operate is necessary to understand the role transformers play in today's nuclear facilities.

EO 1.1 DEFINE the following terms as they pertain to transformers:
- **a. Mutual induction**
- **b. Turns ratio**
- **c. Impedance ratio**
- **d. Efficiency**

EO 1.2 DESCRIBE the differences between a wye-connected and delta-connected transformer.

EO 1.3 Given the type of connection and turns ratios for the primary and secondary of a transformer, CALCULATE voltage, current, and power for each of the following types:
- **a. Δ - Δ**
- **b. Δ - Y**
- **c. Y - Δ**
- **d. Y - Y**

Mutual Induction

If flux lines from the expanding and contracting magnetic field of one coil cut the windings of another nearby coil, a voltage will be induced in that coil. The inducing of an EMF in a coil by magnetic flux lines generated in another coil is called *mutual induction*. The amount of electromotive force (EMF) that is induced depends on the relative positions of the two coils.

Turns Ratio

Each winding of a transformer contains a certain number of turns of wire. The *turns ratio* is defined as the ratio of turns of wire in the primary winding to the number of turns of wire in the secondary winding. Turns ratio can be expressed using Equation (13-1).

$$\text{Turns ratio} = \frac{N_P}{N_S} \qquad (13\text{-}1)$$

where

N_P = number of turns on the primary coil
N_S = number of turns on the secondary coil

The coil of a transformer that is energized from an AC source is called the primary winding (coil), and the coil that delivers this AC to the load is called the secondary winding (coil) (Figure 1).

Impedance Ratio

Maximum power is transferred from one circuit to another through a transformer when the impedances are equal, or matched. A transformer winding constructed with a definite turns ratio can perform an impedance matching function. The turns ratio will establish the proper relationship between the primary and secondary winding impedances. The ratio between the two impedances is referred to as the *impedance ratio* and is expressed by using Equation (13-2).

$$\left(\frac{N_P}{N_S}\right)^2 = \frac{Z_P}{Z_S} \qquad (13\text{-}2)$$

Another way to express the impedance ratio is to take the square root of both sides of Equation (13-2). This puts the ratio in terms of the turns ratio, which is always given for a transformer.

$$0\frac{N_P}{N_S} = \sqrt{\frac{Z_P}{Z_S}}$$

where

N_P = number of turns in the primary
N_S = number of turns in the secondary
Z_P = impedance of primary
Z_S = impedance of secondary

Efficiency

Efficiency of a transformer is the ratio of the power output to the power input, as illustrated by Equation (13-3).

$$\text{Efficiency} = \frac{\text{Power Output}}{\text{Power Input}} = \frac{P_S}{P_P} \times 100 \tag{13-3}$$

where

P_S = power of secondary
P_P = power of primary

Theory of Operation

A transformer works on the principle that energy can be transferred by magnetic induction from one set of coils to another set by means of a varying magnetic flux. The magnetic flux is produced by an AC source.

The coil of a transformer that is energized from an AC source is called the primary winding (coil), and the coil that delivers this AC to the load is called the secondary winding (coil) (Figure 1).

In Figure 1, the primary and secondary coils are shown on separate legs of the magnetic circuit so that we can easily understand how the transformer works. Actually, half of the primary and secondary coils are wound on each of the two legs, with sufficient insulation between the two coils and the core to properly insulate the windings from one another and the core. A transformer wound, such as in Figure 1, will operate at a greatly reduced efficiency due to the magnetic leakage. Magnetic leakage is the part of the magnetic flux that passes through either one of the coils, but not through both. The larger the distance between the primary and secondary windings, the longer the magnetic circuit and the greater the leakage.

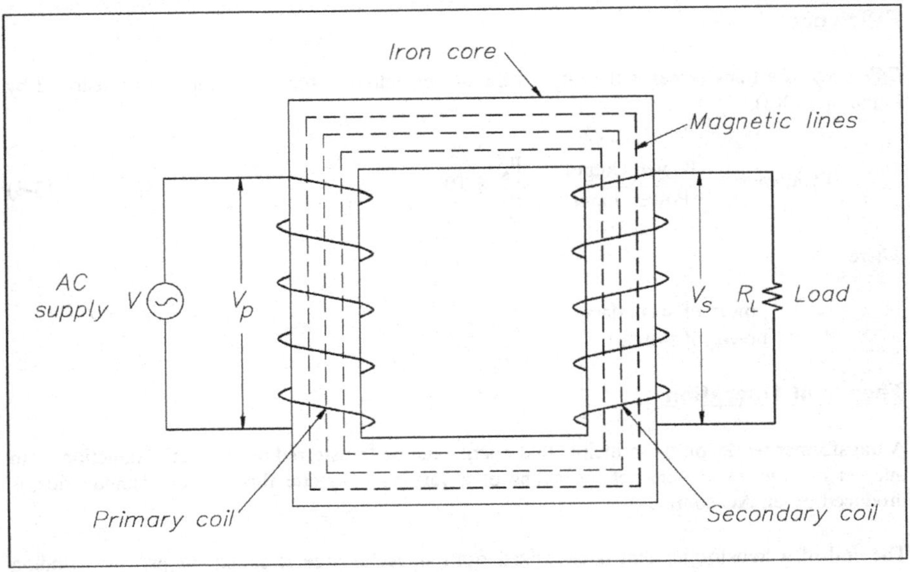

Figure 1 Core-Type Transformer

When alternating voltage is applied to the primary winding, an alternating current will flow that will magnetize the magnetic core, first in one direction and then in the other direction. This alternating flux flowing around the entire length of the magnetic circuit induces a voltage in both the primary and secondary windings. Since both windings are linked by the same flux, the voltage induced per turn of the primary and secondary windings must be the same value and same direction. This voltage opposes the voltage applied to the primary winding and is called counter-electromotive force (CEMF).

Voltage Ratio

The voltage of the windings in a transformer is directly proportional to the number of turns on the coils. This relationship is expressed in Equation (13-4).

$$\frac{V_P}{V_S} = \frac{N_P}{N_S} \tag{13-4}$$

where

V_P = voltage on primary coil
V_S = voltage on secondary coil
N_P = number of turns on the primary coil
N_S = number of turns on the secondary coil

The ratio of primary voltage to secondary voltage is known as the *voltage ratio* (VR). As mentioned previously, the ratio of primary turns of wire to secondary turns of wire is known as the turns ratio (TR). By substituting into the Equation (13-4), we find that the voltage ratio is equal to the turns ratio.

$$VR = TR$$

A voltage ratio of 1:5 means that for each volt on the primary, there will be 5 volts on the secondary. If the secondary voltage of a transformer is greater than the primary voltage, the transformer is referred to as a "step-up" transformer. A ratio of 5:1 means that for every 5 volts on the primary, there will only be 1 volt on the secondary. When secondary voltage is less than primary voltage, the transformer is referred to as a "step-down" transformer.

Example 1: A transformer (Figure 2) reduces voltage from 120 volts in the primary to 6 volts in the secondary. If the primary winding has 300 turns and the secondary has 15 turns, find the voltage and turns ratio.

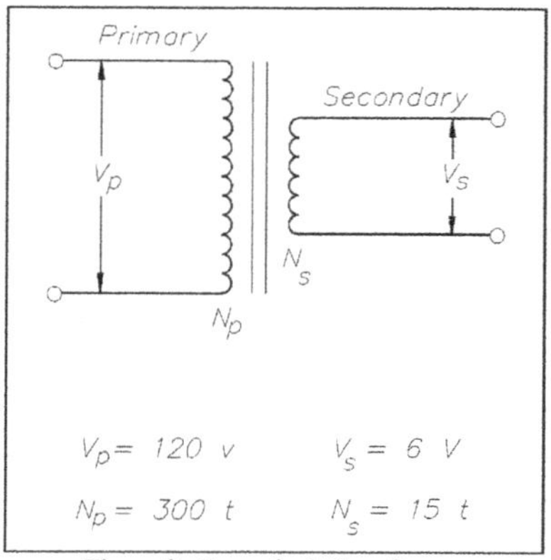

Figure 2 Example 1 Transformer

Solution:

$$VR = \frac{V_P}{V_S} = \frac{120}{60} = \frac{20}{1} = 20{:}1$$

$$TR = \frac{N_P}{N_S} = \frac{300}{15} = \frac{20}{1} = 20{:}1$$

Example 2: An iron core transformer with a primary voltage of 240 volts has 250 turns in the primary and 50 turns in the secondary. Find the secondary voltage.

Solution:

$$\frac{V_P}{V_S} = \frac{N_P}{N_S}$$

Next, solve for V_S.

$$V_S = \frac{N_S}{N_P} V_P$$

$$V_S = \frac{50}{250} \, 240 \text{ volts}$$

$$V_S = 48 \text{ volts}$$

Example 3: A power transformer has a turns ratio of 1:4. If the secondary coil has 5000 turns and secondary voltage is 60 volts, find the voltage ratio, V_P, and N_P.

Solution:

$$VR = TR$$

$$VR = 1{:}4$$

$$\frac{V_P}{V_S} = VR = 1{:}4 = \frac{1}{4}$$

$$V_P = \frac{1}{4} V_S = \frac{60}{4} = 15 \text{ volts}$$

$$TR = \frac{N_P}{N_S} = \frac{1}{4}$$

$$N_P = \frac{1}{4} N_S = \frac{5000}{4} = 1250 \text{ turns}$$

Current Ratio

The current in the windings of a transformer is inversely proportional to the voltage in the windings. This relationship is expressed in Equation (13-5).

$$\frac{V_P}{V_S} = \frac{I_S}{I_P} \tag{13-5}$$

where

I_P = primary coil current
I_S = secondary coil current

Since the voltage ratio is equal to the turns ratio, we can express the current ratio in terms of the turns ratio, as in Equation (13-6).

$$\frac{N_P}{N_S} = \frac{I_S}{I_P} \tag{13-6}$$

Example 1: When operated at 120 V in the primary of an iron core transformer, the current in the primary is 4 amps. Find the current in the secondary if the voltage is stepped up to 500 V.

Solution:

$$\frac{V_P}{V_S} = \frac{I_S}{I_P}$$

Next, we solve for I_S.

$$I_S = \frac{V_P}{V_S} I_P$$

$$I_S = \frac{120}{500} 4 \text{ amps}$$

$$I_S = 0.96 \text{ amps}$$

Example 2: A transformer with 480 turns on the primary and 60 turns on the secondary draws 0.6 amps from a 120 V line. Find I_S.

Solution:

$$\frac{N_P}{N_S} = \frac{I_S}{I_P}$$

Next, we solve for I_S.

$$I_S = \frac{N_P}{N_S} I_P$$

$$I_S = \frac{480}{60} 0.6 \text{ amps}$$

$$I_S = 4.8 \text{ amps}$$

The student should note from the previous examples that a transformer that "steps-up" voltage, "steps-down" the current proportionally.

Three-Phase Transformer Connections

So far, our discussion has dealt with the operation of single-phase transformers. Three-phase transformer operation is identical except that three single-phase windings are used. These windings may be connected in wye, delta, or any combination of the two.

Delta Connection

In the delta connection, all three phases are connected in series to form a closed loop (Figure 3).

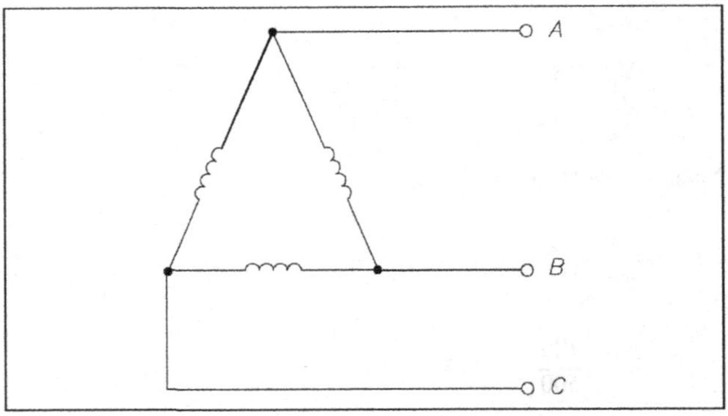

Figure 3 Delta Connection

Wye Connection

In the wye connection, three common ends of each phase are connected together at a common terminal (marked "N" for neutral), and the other three ends are connected to a three-phase line (Figure 4).

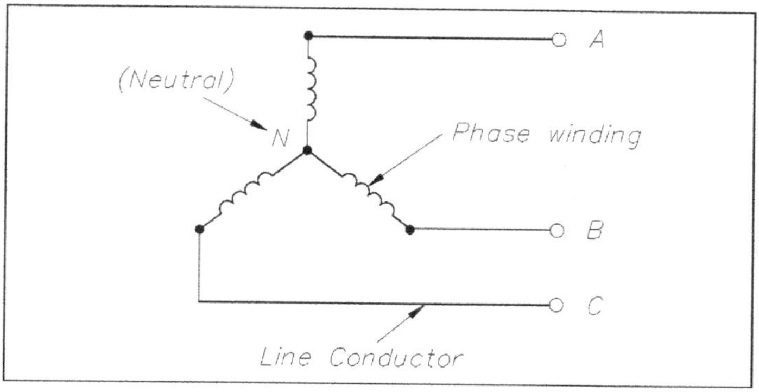

Figure 4 Wye Connection

Combinations of Delta and Wye Transformer Connections

A three-phase transformer may have three separate but identical single-phase (1ϕ) transformers or a single 3ϕ unit containing three-phase windings. The transformer windings may be connected to form a 3ϕ bank in any of four different ways (Figure 5).

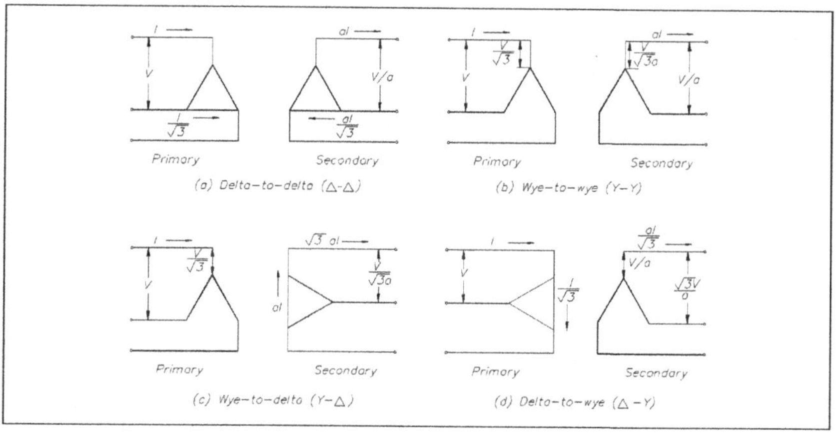

Figure 5 3ϕ Transformer Connections

Figure 5 shows the voltages and currents in terms of applied line voltage (V) and line current (I), where the turns ratio (a) is equal to one. Voltage and current ratings of the individual transformers depend on the connections (Figure 5) and are indicated by Table 1 for convenience of calculations.

TABLE 1: Voltage and Current Ratings of Transformers

Transformer Connection (Primary to Secondary)	Primary				Secondary			
	Line		Phase		Line		Phase	
	Volt.	Current	Volt.	Current	Volt. *	Current	Volt.	Current
Δ-Δ	V	I	V	$\frac{I}{\sqrt{3}}$	$\frac{V}{a}$	aI	$\frac{V}{a}$	$\frac{aI}{\sqrt{3}}$
Y-Y	V	I	$\frac{V}{\sqrt{3}}$	I	$\frac{V}{a}$	aI	$\frac{V}{\sqrt{3}\,a}$	aI
Y-Δ	V	I	$\frac{V}{\sqrt{3}}$	I	$\frac{V}{\sqrt{3}\,a}$	$\sqrt{3}\,aI$	$\frac{V}{\sqrt{3}\,a}$	aI
Δ-Y	V	I	V	$\frac{I}{\sqrt{3}}$	$\frac{\sqrt{3}\,V}{a}$	$\frac{aI}{\sqrt{3}}$	$\frac{V}{a}$	$\frac{aI}{\sqrt{3}}$

*$a = N_1/N_2$; $\sqrt{3} = 1.73$

Example 1: If line voltage is 440 V to a 3ϕ transformer bank, find the voltage across each primary winding for all four types of transformer connections.

Δ-Δ: primary voltage = V = 440 volts

$$\text{Y-Y: primary voltage} = \frac{V}{\sqrt{3}} = \frac{440}{1.73} = 254.3 \text{ volts}$$

$$\text{Y-}\Delta\text{: primary voltage} = \frac{V}{\sqrt{3}} = \frac{440}{1.73} = 254.3 \text{ volts}$$

Δ-Y: primary voltage = V = 440 volts

Example 2: If line current is 10.4 A in a 3ϕ transformer connection, find the primary phase current.

$$\Delta\text{-}\Delta\text{: primary phase current} = \frac{I}{\sqrt{3}} = \frac{10.4}{1.73} = 6 \text{ amps}$$

Y-Y: primary phase current = I = 10.4 amps

Y-Δ: primary phase current = I = 10.4 amps

Δ-Y: primary phase current = $\frac{I}{\sqrt{3}} = \frac{10.4}{1.73} = 6$ amps

Example 3: Find the secondary line current and phase current for each type of transformer connection, if primary line current is 20 amps, and the turns ratio is 4:1.

Δ-Δ: secondary line current = 4(20) = 80 amps

secondary phase current = $\frac{aI}{\sqrt{3}} = \frac{4\ (20)}{1.73} = 46.2$ amps

Y-Y: second line current = aI = 4(20) = 80 amps

second phase current = aI = 4(20) = 80 amps

Y-Δ: secondary line current = $\sqrt{3}$ aI = (1.73)(4)(20) = 138.4 amps

secondary phase current = aI = 4(20) = 80 amps

Δ-Y: secondary line current = $\frac{aI}{\sqrt{3}} = \frac{4\ (20)}{1.73} = 46.2$ amps

secondary phase current = $\frac{aI}{\sqrt{3}} = \frac{4\ (20)}{1.73} = 46.2$ amps

Transformer Losses and Efficiency

All transformers have copper and core losses. Copper loss is power lost in the primary and secondary windings of a transformer due to the ohmic resistance of the windings. Copper loss, in watts, can be found using Equation (13-7).

$$\text{Copper Loss} = I_P^2 R_P + I_S^2 R_S \tag{13-7}$$

where

I_P = primary current
I_S = secondary current
R_P = primary winding resistance
R_S = secondary winding resistance

Core losses are caused by two factors: hysteresis and eddy current losses. Hysteresis loss is that energy lost by reversing the magnetic field in the core as the magnetizing AC rises and falls and reverses direction. Eddy current loss is a result of induced currents circulating in the core.

The efficiency of a transformer can be calculated using Equations (13-8), (13-9), and (13-10).

$$\text{Efficiency} = \frac{\text{Power Output}}{\text{Power Input}} = \frac{P_S}{P_P} \text{ x } 100 \tag{13-8}$$

$$\text{Efficiency} = \frac{\text{Power Output}}{\text{Power Output} + \text{Copper Loss} + \text{Core Loss}} \text{ x } 100 \tag{13-9}$$

$$\text{Efficiency} = \frac{V_S\, I_S \text{ x PF}}{(V_S\, I_S \text{ x PF}) + \text{Copper Loss} + \text{Core Loss}} \text{ x } 100 \tag{13-10}$$

where

PF = power factor of the load

Example 1: A 5:1 step-down transformer has a full-load secondary current of 20 amps. A short circuit test for copper loss at full load gives a wattmeter reading of 100 W. If $R_P = 0.3\Omega$, find R_S and power loss in the secondary.

Solution:

$$\text{Copper Loss} = I_P^2\, R_P + I_S^2\, R_S = 100 \text{ W}$$

To find I_P:

$$\frac{N_P}{N_S} = \frac{I_S}{I_P}$$

$$I_P = \frac{N_S}{N_P}\, I_S = \frac{1}{5}\, 20 = 4 \text{ amps}$$

To find R_S:

$$I_S^2\, R_S = 100 - I_P^2\, R_P$$

$$R_S = \frac{100 - I_P^2\, R_P}{I_S^2} = \frac{100 - 0.3(4)^2}{20^2} = 0.24$$

Power loss in secondary = $I_S^2 R_S = (20)^2 (0.24) = 96$ W

Example 2: An open circuit test for core losses in a 10 kVA transformer [Example (1)] gives a reading of 70 W. If the PF of the load is 90%, find efficiency at full load.

Solution:

$$\text{Eff.} = \frac{V_S I_S \text{ x PF}}{(V_S I_S \text{ x PF}) + \text{Copper Loss} + \text{Core Loss}} \text{ x } 100$$

$V_S I_S$ = transformer rating = 10 kVA = 10,000 VA

PF = 0.90; Copper loss = 100 W; Core loss = 70 W

$$\text{Eff} = \frac{10{,}000\ (0.90)}{10{,}000\ (0.90) + 100 + 70} \text{ x } 100 = \frac{9000}{9170} \text{ x } 100 = 98.2\%$$

Transformer Operation Under No-Load

If the secondary of a transformer is left open-circuited (Figure 6), primary current is very low and is called the *no-load current*. No-load current produces the magnetic flux and supplies the hysteresis and eddy current losses in the core. The no-load current (I_E) consists of two components: the magnetizing current (I_m) and the core loss (I_H). Magnetizing current lags applied voltage by 90°, while core loss is in phase with the applied voltage (Figure 6b). V_P and V_S are shown 180° out of phase. I_H is very small in comparison with I_m, and I_m is nearly equal to I_E. No-load current, I_E, is also referred to as exciting current.

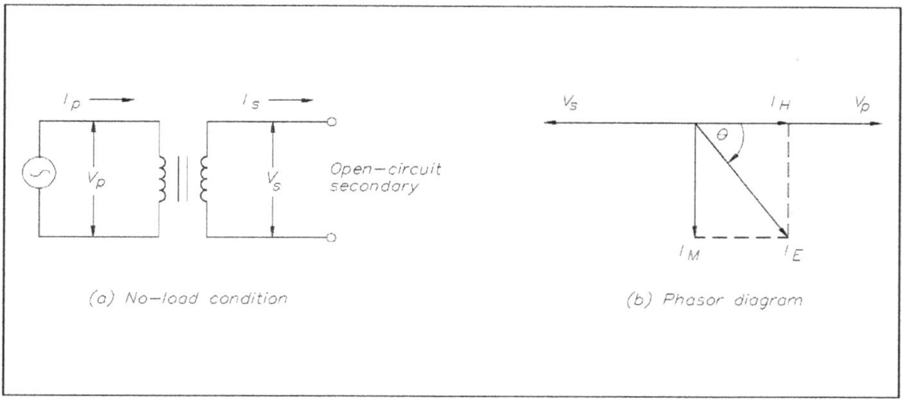

Figure 6 Open-Circuit Secondary

Example: When the secondary of a 120/440 V transformer is open, primary current is 0.2 amps at a PF of .3. The transformer is a 5 kVA transformer. Find: (a) I_P, (b) I_E, (c) I_H, and (d) I_m.

(a) Full load current $= \dfrac{\text{kVA Rating}}{V_P}$

(b) I_P at no load is equal to I_E

$I_E = 0.2$ amp

(c) $I_H = I_E \cos\theta = I_E \times PF$

$= 0.2\ (0.3)$

$I_H = 0.06$ amps

(d) $I_M = I_E \sin\theta$

$\theta = \arccos 0.3 = 72.5°$

$= (0.2) \sin 72.5° = (0.2)\ (0.95)$

$I_M = 0.19$ amps

Coil Polarity

The symbol for a transformer gives no indication of the phase of the voltage across the secondary. The phase of that voltage depends on the direction of the windings around the core. In order to solve this problem, polarity dots are used to show the phase of primary and secondary signals. The voltages are either in phase (Figure 7a) or 180° out of phase with respect to primary voltage (Figure 7b).

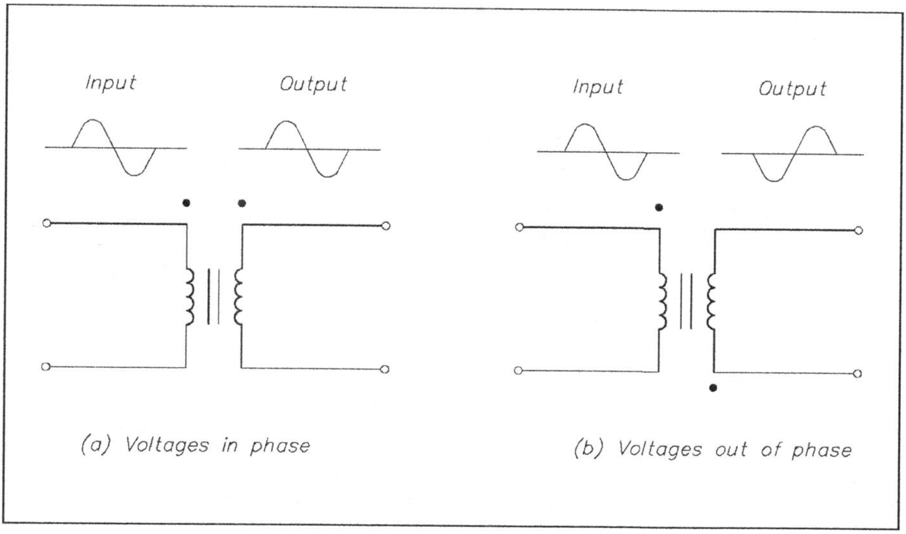

Figure 7 Polarity of Transformer Coils

Summary

The important information covered in this chapter is summarized below.

Transformer Theory Summary

- The induction of an EMF in a coil by magnetic flux lines generated in another coil is called mutual induction.
- The turns ratio is defined as the ratio of turns of wire in the primary winding to the number of turns of wire in the secondary winding.
- The ratio between the primary and secondary impedances is referred to as the impedance ratio.
- Efficiency of a transformer is the ratio of the power output to the power input.
- In a delta connection, all three phases are connected in series to form a closed loop.
- In a wye connection, three common ends of each phase are connected together at a common terminal, and the other three ends are connected to a three-phase line.
- In a Δ connected transformer:

$$V_L = V\phi$$

$$I_L = \sqrt{3}\ I\phi$$

- In a Y connected transformer:

$$I_L = \sqrt{3}\ V\phi$$

$$I_L = I\phi$$

TRANSFORMER TYPES

Transformers can be constructed so that they are designed to perform a specific function. A basic understanding of the various types of transformers is necessary to understand the role transformers play in today's nuclear facilities.

EO 1.4 STATE the applications of each of the following types of transformers:

a. Distribution
b. Power
c. Control
d. Auto
e. Isolation
f. Instrument potential
g. Instrument current

Types of Transformers

Transformers are constructed so that their characteristics match the application for which they are intended. The differences in construction may involve the size of the windings or the relationship between the primary and secondary windings. Transformer types are also designated by the function the transformer serves in a circuit, such as an isolation transformer.

Distribution Transformer

Distribution transformers are generally used in electrical power distribution and transmission systems. This class of transformer has the highest power, or volt-ampere ratings, and the highest continuous voltage rating. The power rating is normally determined by the type of cooling methods the transformer may use. Some commonly-used methods of cooling are by using oil or some other heat-conducting material. Ampere rating is increased in a distribution transformer by increasing the size of the primary and secondary windings; voltage ratings are increased by increasing the voltage rating of the insulation used in making the transformer.

Power Transformer

Power transformers are used in electronic circuits and come in many different types and applications. Electronics or power transformers are sometimes considered to be those with ratings of 300 volt-amperes and below. These transformers normally provide power to the power supply of an electronic device, such as in power amplifiers in audio receivers.

Control Transformer

Control transformers are generally used in electronic circuits that require constant voltage or constant current with a low power or volt-amp rating. Various filtering devices, such as capacitors, are used to minimize the variations in the output. This results in a more constant voltage or current.

Auto Transformer

The auto transformer is generally used in low power applications where a variable voltage is required. The auto transformer is a special type of power transformer. It consists of only one winding. By tapping or connecting at certain points along the winding, different voltages can be obtained (Figure 8).

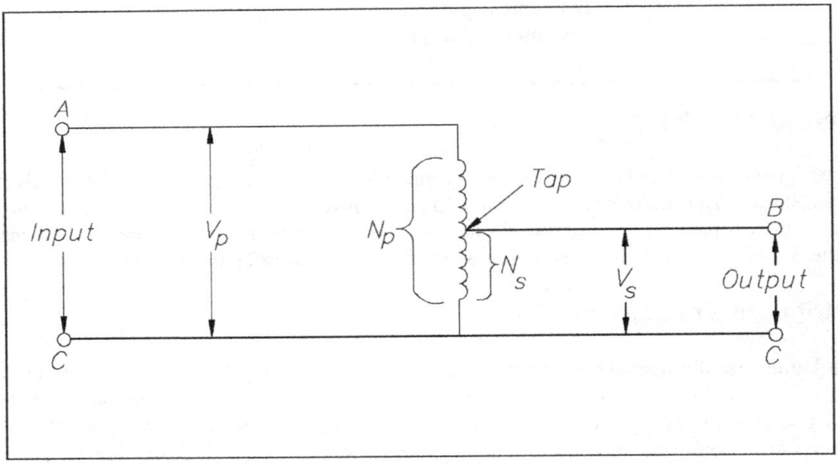

Figure 8 Auto Transformer Schematic

Isolation Transformer

Isolation transformers are normally low power transformers used to isolate noise from or to ground electronic circuits. Since a transformer cannot pass DC voltage from primary to secondary, any DC voltage (such as noise) cannot be passed, and the transformer acts to isolate this noise.

Instrument Potential Transformer

The instrument potential transformer (PT) steps down voltage of a circuit to a low value that can be effectively and safely used for operation of instruments such as ammeters, voltmeters, watt meters, and relays used for various protective purposes.

Instrument Current Transformer

The instrument current transformer (CT) steps down the current of a circuit to a lower value and is used in the same types of equipment as a potential transformer. This is done by constructing the secondary coil consisting of many turns of wire, around the primary coil, which contains only a few turns of wire. In this manner, measurements of high values of current can be obtained.

A current transformer should always be short-circuited when not connected to an external load. Because the magnetic circuit of a current transformer is designed for low magnetizing current when under load, this large increase in magnetizing current will build up a large flux in the magnetic circuit and cause the transformer to act as a step-up transformer, inducing an excessively high voltage in the secondary when under no load.

Summary

The important information covered in this chapter is summarized below.

Transformer Types Summary

- Distribution transformers are generally used in power distribution and transmission systems.
- Power transformers are used in electronic circuits and come in many different types and applications.
- Control transformers are generally used in circuits that require constant voltage or constant current with a low power or volt-amp rating.
- Auto transformers are generally used in low power applications where a variable voltage is required.
- Isolation transformers are normally low power transformers used to isolate noise from or to ground electronic circuits.
- Instrument potential and instrument current transformers are used for operation of instruments such as ammeters, voltmeters, watt meters, and relays used for various protective purposes.

Department of Energy
Fundamentals Handbook

ELECTRICAL SCIENCE
Module 14
Test Instruments & Measuring Devices

TABLE OF CONTENTS

LIST OF FIGURES iii

LIST OF TABLES iv

REFERENCES v

OBJECTIVES vi

METER MOVEMENTS 1

- D'Arsonval Movement 1
- Electrodynamometer Movement 2
- Moving Iron Vane Movement 3
- Summary 4

VOLTMETERS 5

- Voltmeters 5
- Summary 9

AMMETERS 10

- Ammeters 10
- Summary 14

OHM METERS 15

- Ohm Meters 15
- Summary 19

WATTMETERS 20

- Wattmeters 20
- Three-Phase Wattmeter 21
- Summary 22

TABLE OF CONTENTS (Cont.)

OTHER ELECTRICAL MEASURING DEVICES 23

Ampere-Hour Meter 23
Power Factor Meter 24
Ground Detector 24
Synchroscope 26
Summary 27

TEST EQUIPMENT 28

Multimeter 28
Megger 28
Summary 30

LIST OF FIGURES

Figure 1 D'Arsonval Meter Movement 1

Figure 2 Electrodynamometer Movement 2

Figure 3 Moving Iron Vane Meter Movement 3

Figure 4 Simple DC Voltmeter 5

Figure 5 Measuring Circuit Voltage 8

Figure 6 Ammeter 10

Figure 7 Ammeter Accuracy 11

Figure 8 Ammeter with Shunt 13

Figure 9 Simple Ohm Meter Circuit 15

Figure 10 Ohm Meter Scale 19

Figure 11 Wattmeter Schematic 20

Figure 12 Wattmeters in Each Phase 21

Figure 13 Two Wattmeters to Measure 3ϕ Power 22

Figure 14 3ϕ Power Factor Meter Schematic 24

Figure 15 Simple Ohm Meter Ground Detector 25

Figure 16 Ground Detector Lamp Circuit 26

Figure 17 Simple Megger Circuit Diagram 29

LIST OF TABLES

NONE

REFERENCES

- Gussow, Milton, Schaum's Outline Series, Basic Electricity, McGraw-Hill.
- Academic Program for Nuclear Power Plant Personnel, Volume IV, Columbia, MD: General Physics Corporation, Library of Congress Card #A 326517, 1982.
- Academic Program for Nuclear Power Plant Personnel, Volume II, Columbia, MD: General Physics Corporation, Library of Congress Card #A 326517, 1982.
- Nasar and Unnewehr, Electromechanics and Electric Machines, John Wiley and Sons.
- Van Valkenburgh, Nooger, and Neville, Basic Electricity, Vol. 5, Hayden Book Company.
- Lister, Eugene C., Electric Circuits and Machines, 5th Edition, McGraw-Hill.
- Croft, Carr, Watt, and Summers, American Electricians Handbook, 10th Edition, McGraw-Hill.
- Mason, C. Russel, The Art and Science of Protective Relaying, John Wiley and Sons.
- Mileaf, Harry, Electricity One - Seven, Revised 2nd Edition, Hayden Book Company.
- Buban and Schmitt, Understanding Electricity and Electronics, 3rd Edition, McGraw-Hill.
- Kidwell, Walter, Electrical Instruments and Measurements, McGraw-Hill.

TERMINAL OBJECTIVE

1.0 Given a piece of test equipment or measuring device, **DESCRIBE** the use of that piece of electrical equipment, to include the meter movement, electrical parameter measurement, and connection of the device to a circuit.

ENABLING OBJECTIVES

1.1 **EXPLAIN** the following meter movements:
- a. D'Arsonval
- b. Electrodynamometer
- c. Moving iron vane

1.2 **STATE** the electrical parameters measured by each of the following in-place measuring devices:
- a. Voltmeter
- b. Ammeter
- c. Ohm meter
- d. Wattmeter
- e. Ampere-hour meter
- f. Power factor meter
- g. Ground detector
- h. Synchroscope

1.3 **EXPLAIN** how the following electrical test equipment and measuring devices are connected to a circuit:
- a. Voltmeter
- b. Ammeter
- c. Ohm meter
- d. Wattmeter
- e. Ampere-hour meter
- f. Power factor meter
- g. Ground detector
- h. Synchroscope
- i. Megger

1.4 **STATE** the electrical parameters measured by each of the following test instruments:
- a. Multimeter
- b. Megger

METER MOVEMENTS

There are three basic meter movements utilized in electrical meters: D'Arsonval, electrodynamometer, and the moving iron vane. Some meter movements can be used for both AC or DC measurements, but in general, each meter movement is best suited for a particular type.

EO 1.1 EXPLAIN the following meter movements:
- **a. D'Arsonval**
- **b. Electrodynamometer**
- **c. Moving iron vane**

D'Arsonval Movement

The most commonly used sensing mechanism used in DC ammeters, voltmeters, and ohm meters is a current-sensing device called a D'Arsonval meter movement (Figure 1). The D'Arsonval movement is a DC moving coil-type movement in which an electromagnetic core is suspended between the poles of a permanent magnet.

The current measured is directed through the coils of the electromagnet so that the magnetic field produced by the current opposes the field of the permanent magnet and causes rotation of the core. The core is restrained by springs so that the needle will deflect or move in proportion to the current intensity. The more current applied to the core, the stronger the opposing field, and the larger the deflection, up to the limit of the current capacity of the coil. When the current is interrupted, the opposing field collapses, and the needle is returned to zero by the restraining springs. The limit of the current that can be applied to this type movement is usually less than one milliampere.

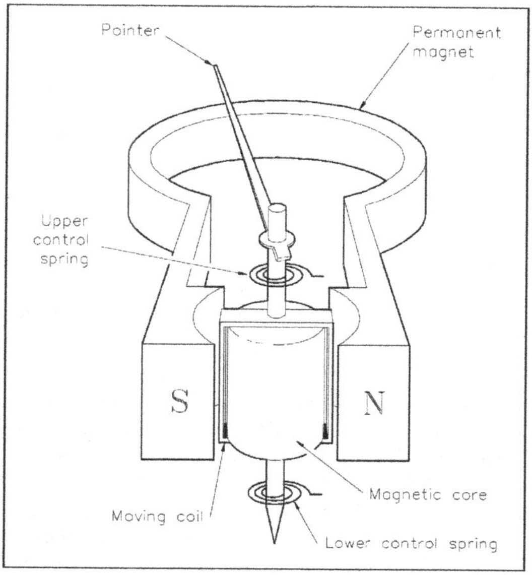

Figure 1 D'Arsonval Meter Movement

A common variation of the D'Arsonval movement is the Weston movement, which uses essentially the same principle built to a more rugged construction by employing jeweled supports for the core and employing a heavier winding in the electromagnet. Remember that the D'Arsonval movement is a DC device and can only measure DC current or AC current rectified to DC.

Electrodynamometer Movement

The electrodynamometer movement (Figure 2) has the same basic operating principle as the D'Arsonval meter movement, except that the permanent magnet is replaced by fixed coils. The moving coil and pointer, which are attached to the coil, are suspended between and connected in series with the two field coils. The two field coils and moving coil are connected in series such that the same current flows through each coil.

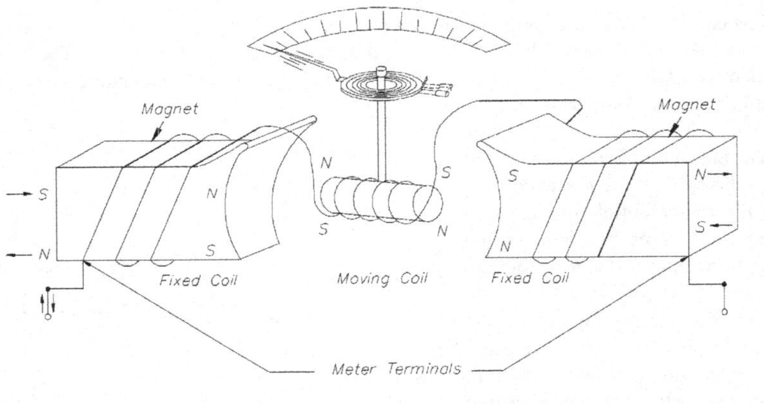

Figure 2 Electrodynamometer Movement

Current flow through the three coils in either direction causes a magnetic field to be produced between the field coils. The same current flow through the moving coil causes it to act as a magnet exerting a force against the spring. If the current is reversed, the field polarity and the polarity of the moving coil reverse, and the force continues in the same direction. Due to this characteristic of the electrodynamometer movement, it can be used in both AC and DC systems to measure current. Some voltmeters and ammeters use the electrodynamometer. However, its most important use is in the wattmeter, which will be discussed later in this module.

Moving Iron Vane Movement

The moving iron vane movement (Figure 3) can be used to measure both AC current and voltage. By changing the meter scale calibration, the movement can be used to measure DC current and voltage. The moving iron vane meter operates on the principle of magnetic repulsion between like poles. The measured current flows through a field coil which produces a magnetic field proportional to the magnitude of current. Suspended in this field are two iron vanes attached to a pointer. The two iron vanes consist of one fixed and one moveable vane. The magnetic field produced by the current flow magnetizes the two iron vanes with the same polarity regardless of the direction of current through the coil. Since like poles repel one another, the moving iron vane pulls away from the fixed vane and moves the meter pointer. This motion exerts a force against a spring. The distance the moving iron vane will travel against the spring depends on the strength of the magnetic field. The strength of the magnetic field depends on the magnitude of current flow.

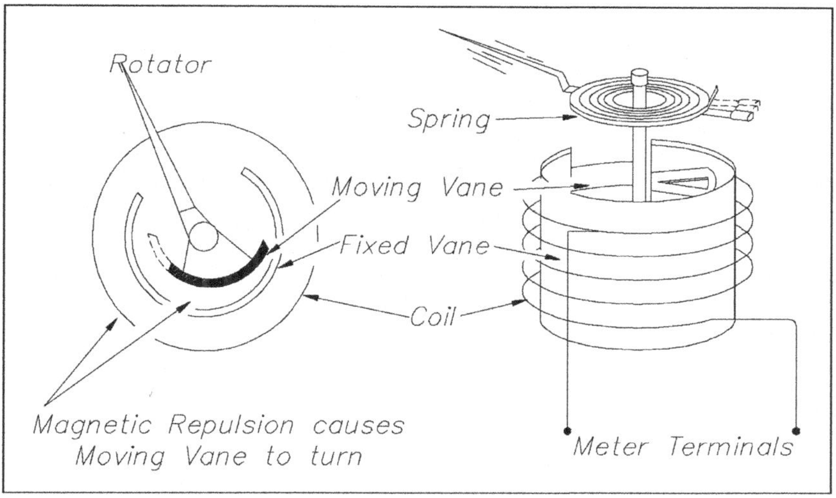

Figure 3 Moving Iron Vane Meter Movement

As stated previously, this type of meter movement may also be used to measure voltage. When this type of movement is used to measure voltage, the field coil consists of many turns of fine wire used to generate a strong magnetic field with only a small current flow.

Summary

Meter movements are summarized below.

Meter Movement Summary

- D'Arsonval - A DC moving coil movement where the moving coil is suspended between the poles of a permanent magnet restrained by helical springs, and the measured current flowing through the moving coil produces a torque on the attached pointer proportional to the current.
- Electrodynamometer - The moving coil and attached pointer are suspended between and connected in series with the two stationary field coils so that the same current flows through each. A measured current flowing through the three coils in either direction causes a magnetic repulsion between the field coils and the moving coil. The magnetic repulsion exerts a force against the spring and provides a measurement of either DC or AC current.
- Moving iron vane - The moving iron vane meter operates on the principle of magnetic repulsion between like poles. The measured current flows through a field coil which induces a like magnetic field into a fixed and moving vane causing the moving vane to deflect a pointer in proportion to the current or voltage applied to the coil.

VOLTMETERS

Voltmeters are used extensively in industry where the surveillance of input and/or output voltages is vital for plant operation.

EO 1.2 **STATE the electrical parameters measured by each of the following in-place measuring devices:**
a. Voltmeter

EO 1.3 **EXPLAIN how the following electrical test equipment and measuring devices are connected to a circuit:**
a. Voltmeter

Voltmeter

A simple DC voltmeter can be constructed by placing a resistor (R_S), called a multiplier, in series with the ammeter meter movement, and marking the meter face to read voltage (Figure 4). Voltmeters are connected in parallel with the load (R_L) being measured.

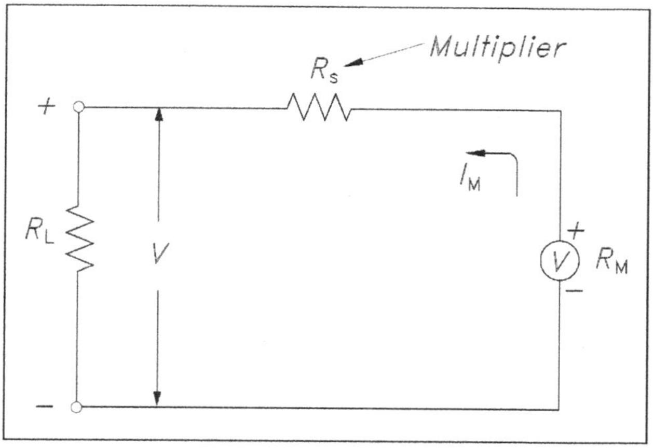

Figure 4 Simple DC Voltmeter

When constructing a voltmeter, the resistance of the multiplier must be determined to measure the desired voltage. Equation (14-1) is a mathematical representation of the voltmeter's multiplier resistance.

$$V = I_m R_s + I_m R_m$$

$$I_m R_s = V - I_m R_m$$

$$R_s = \frac{V}{I_m} - R_m \qquad (14\text{-}1)$$

where

V = voltage range desired
I_m = meter current
R_m = meter resistance
R_s = multiplier resistance or series resistance

Example: A 2 mA meter movement with internal resistance of 25 ohms is to be constructed as a voltmeter.

What value must the series resistance be to measure full scale voltage of 100 volts?

Solution:

$$R_s = \frac{V}{I_m} - R_m$$

Since R_m is negligibly low, then:

$$R_s = \frac{V}{I_m}$$

$$= \frac{100}{2 \times 10^{-3}}$$

$$R_s = 50 k\Omega$$

When a voltmeter is connected in a circuit, the voltmeter will draw current from that circuit. This current causes a voltage drop across the resistance of the meter, which is subtracted from the voltage being measured by the meter. This reduction in voltage is known as the loading effect and can have a serious effect on measurement accuracy, especially for low current circuits.

The accuracy of a voltmeter (K_v) is defined as the ratio of measured voltage when the meter is in the circuit (V_w) to the voltage measured with the meter out of the circuit. Equation (14-2) is a mathematical representation of the accuracy of a voltmeter, or true voltage (V_o).

$$K_v = \frac{V_w}{V_o} \tag{14-2}$$

Meter accuracy can also be determined by comparing the relationship between the input and circuit resistances using Ohm's Law as described below.

$$K_v = \frac{V_w}{V_o} \qquad V_w = I_m R_{in}$$

$$= \frac{I_m R_{in}}{V_o} \qquad I_m = \frac{V_o}{R_o + R_{in}}$$

$$= \frac{\left(\dfrac{V_o \cdot R_{in}}{R_o + R_{in}}\right)}{V_o}$$

$$K_v = \frac{R_{in}}{R_o + R_{in}}$$

where

I_m = meter current
V_o = true voltage
R_o = circuit resistance
R_{in} = input resistance of the voltmeter
K_w = indicated voltage
K_v = meter accuracy

Example: A voltmeter in the 100 volt range with a sensitivity of 40 KΩ/V is to measure the voltage across terminals ab (Figure 5).

Find:
1. V_o
2. V_w
3. K_v

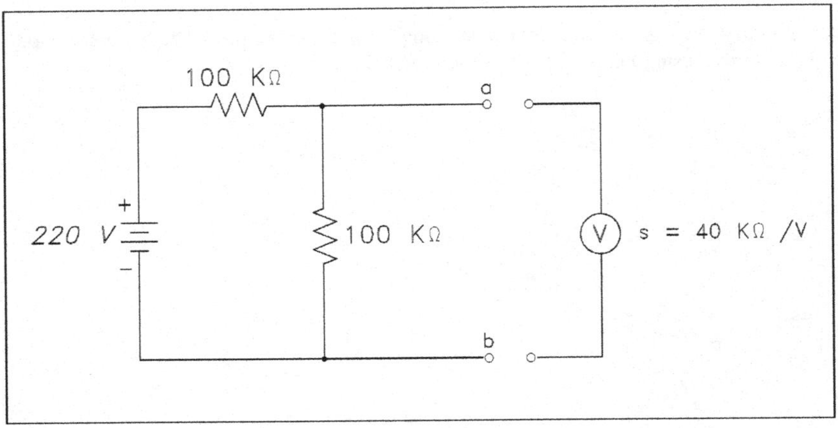

Figure 5 Measuring Circuit Voltage

Solution:

1. $$V_o = \frac{100\,K\Omega}{100\,K\Omega + 100\,K\Omega} \times 220V$$

$$V_o = 110 \text{ volts}$$

2. $R_o = \dfrac{(100)(100)}{100 + 100} = 50K\Omega \qquad R_{in} = SV = (40K\Omega/V)(100V) = 4.4M\Omega$

$$V_w = \frac{R_{in}}{R_o + R_{in}} V_o$$

$$= \left(\frac{4.4 \times 10^6}{50 \times 10^3 + 4.4 \times 10^6}\right)(110)$$

$$= (0.99)(110)$$

$$V_w = 108.9 \text{ volts}$$

3. $K_v = \dfrac{V_w}{V_o}$

$$= \frac{108.9}{110}$$

$$K_v = 0.99 \text{ or } 99\%$$

Summary

Voltmeters are summarized below.

Voltmeter Summary

- Measures voltage
- Connected in parallel with the load being measured

AMMETERS

Measurement of current being supplied to or from a component is measured by an ammeter.

EO 1.2 **STATE the electrical parameters measured by each of the following in-place measuring devices:**
b. Ammeter

EO 1.3 **EXPLAIN how the following electrical test equipment and measuring devices are connected to a circuit:**
b. Ammeter

Ammeter

The ammeter measures electric current. It may be calibrated in amperes, milliamperes, or microamperes. In order to measure current, the ammeter must be placed in series with the circuit to be tested (Figure 6).

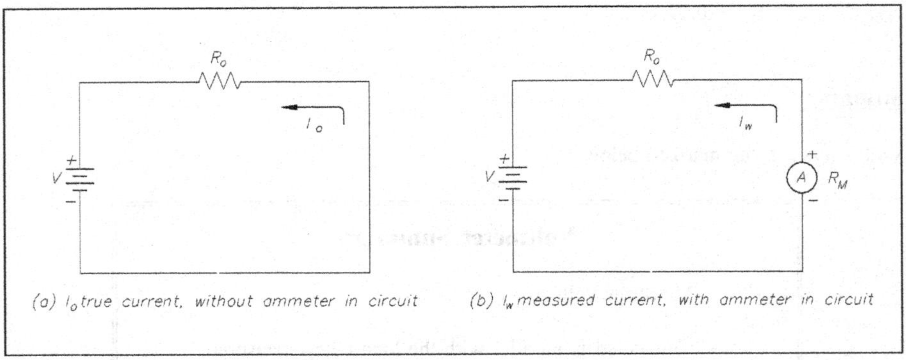

Figure 6 Ammeter

When an ammeter is placed in series with a circuit, it will increase the resistance of that circuit by an amount equal to the internal resistance of the meter R_m. Equation (14-3) is the mathematical representation of the current without the meter installed.

$$I_o = \frac{V}{R_o} \tag{14-3}$$

Equation (14-4) is the mathematical representation of the current with the meter installed in the circuit.

$$I_w = \frac{V}{R_o + R_m} \tag{14-4}$$

The accuracy of the ammeter K_A is the ratio of the current when the meter is in the circuit, I_w, to the current with the meter out of the circuit, I_o. Equation (14-5) is the mathematical representation for solving for the accuracy of the ammeter (K_A).

$$K_A = \frac{I_w}{I_o} \tag{14-5}$$

By substitution laws, Equation (14-6) is a mathematical representation of the accuracy using circuit resistance.

$$K_A = \left(\frac{V}{R_o + R_m}\right)\left(\frac{R_o}{V}\right) = \frac{R_o}{R_o + R_m} \tag{14-6}$$

The percent loading error is that percent of error due to loading effects that result from the added resistance of the meter. Equation (14-7) is a mathematical representation of the percent loading error.

$$\% \text{ loading error} = (1 - K_A)(100\ \%) \tag{14-7}$$

A second error which occurs in an ammeter is calibration error. Calibration error is an error that occurs due to inaccurately marked meter faces. Typical values of calibration error in terms of full scale current are about 3 percent.

Example: An ammeter, with a 10 mA full scale deflection and an internal resistance of 400 Ω, is placed in a circuit with a 20 V power source and a 2 KΩ resistor (Figure 7).

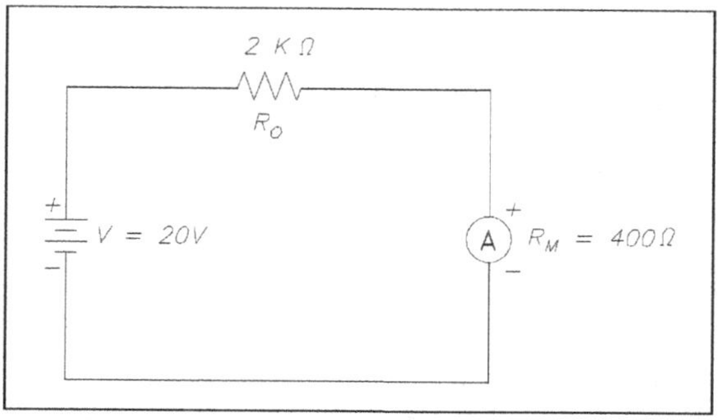

Figure 7 Ammeter Accuracy

Find: 1. accuracy
2. %loading error
3. true current
4. measured current

1. $$K_A = \frac{R}{R_o + R_m}$$

$$K_A = \frac{2000}{2000 + 400}$$

$$K_A = 0.833 \text{ or } 83.3\%$$

2. $$\%\text{ loading error} = (1 - K_A)(100\%)$$

$$\%\text{ loading error} = (1 - 0.833)(100\%)$$

$$\%\text{ loading error} = 16.7\%$$

3. $$I_o = \frac{V}{R_o}$$

$$= \frac{20}{2000}$$

$$I_o = 0.01\text{A or } 10\text{mA}$$

4. $$I_w = \frac{V}{R_o + R_m}$$

$$= \frac{20}{2000 + 400}$$

$$I_w = 8.33 \times 10^{-3}\text{A or } 8.33\text{mA}$$

An ammeter with a full scale I_m can be shunted with a resistor R_{SH} in order to measure currents in excess of I_m (Figure 8). The reason for shunting an ammeter is to extend the range of the ammeter and, thereby, measure currents higher than the original full scale value.

By Kirchhoff's current law,

$$I_{SH} = I_T - I_m$$

Since the voltage across the shunt must be equal to the voltage across the ammeter, shunt resistance is calculated as follows:

$$I_{SH}R_{SH} = I_m R_m$$

$$R_{SH} = \frac{I_m R_m}{I_{SH}}$$

$$R_{SH} = \frac{I_m R_m}{I_T - I_m}$$

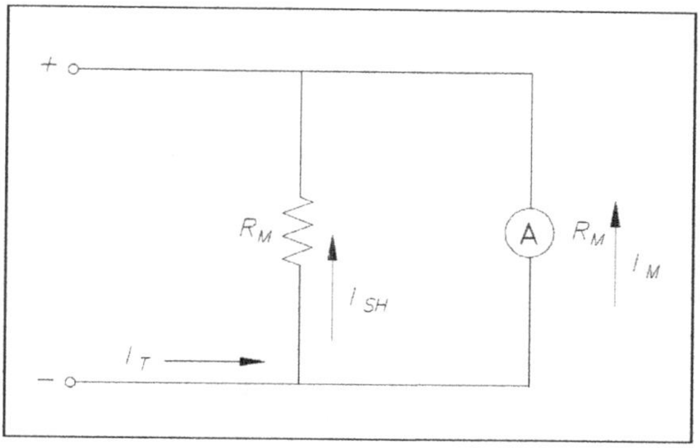

Figure 8 Ammeter with Shunt

Therefore, the input resistance of a shunted ammeter is related to the meter and shunt resistance. Equation (14-8) is a mathematical representation of this relationship.

NOTE: When computing accuracy for a shunted ammeter, use R^1_m in place of R_m.

$$R^1_m = \frac{R_m R_{SH}}{R_m + R_{SH}} \qquad (14\text{-}8)$$

Equation (14-9) is a mathematical representation of the relationship between input voltage and current to the ammeter and the value of input resistance.

$$R^1_m = \frac{V_{in}}{I_{in}} = \frac{I_m R_m}{I_T} \qquad (14\text{-}9)$$

Example: An ammeter, with a 100 Ω meter resistance and a full scale deflection current of 4 mA, is to be shunted to measure currents from 1 to 20 mA.

Find: 1. R_{SH}
2. R^1_m

Solution:

1. $$R_{SH} = \frac{I_m R_m}{I_T - I_m}$$

$$= \frac{(4)(100)}{20 - 4}$$

$$R_{SH} = 25\Omega$$

2. $$R_m^1 = \frac{I_m R_m}{I_T}$$

$$= \frac{(4)(100)}{20}$$

$$R_m^1 = 20\Omega$$

Summary

Ammeters are summarized below.

Ammeter Summary

- Measure circuit current flow
- Connected in series with the circuit

OHM METERS

The resistance of a wire or a circuit is measured by an ohm meter. An ohm meter aids the troubleshooter in determining if a ground or a short exists in a circuit.

EO 1.2 **STATE the electrical parameters measured by each of the following in-place measuring devices:**
c. Ohm meter

EO 1.3 **EXPLAIN how the following electrical test equipment and measuring devices are connected to a circuit:**
c. Ohm meter

Ohm Meter

The ohm meter is an instrument used to determine resistance. A simple ohm meter (Figure 9) consists of a battery, a meter movement calibrated in ohms, and a variable resistor.

Ohm meters are connected to a component which is removed from the circuit as illustrated in Figure 9. The reason for removing the component is that measurement of current through the component determines the resistance. If the component remains in the circuit, and a parallel path exists in the circuit, the current will flow in the path of least resistance and give an erroneous reading.

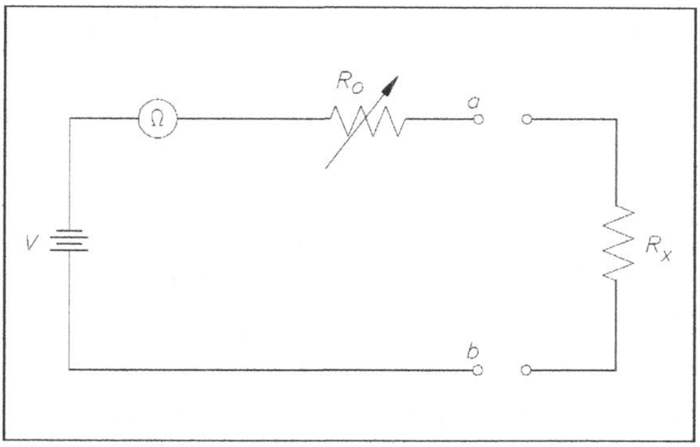

Figure 9 Simple Ohm Meter Circuit

R_o, in Figure 9, is an adjustable resistor whose purpose is to zero the ohm meter and correct for battery aging. It is also a current-limiting resistor which includes the meter resistance R_m. Zeroing the ohm meter is accomplished by shorting the ohm meter terminals ab and adjusting R_o to give full-scale deflection.

Equation (14-10) is the mathematical representation for determining full-scale deflection meter current.

$$I_m = \frac{V}{R_o} \tag{14-10}$$

When the unknown resistance R_x is connected across the ohm meter terminals, the current is measured by calculating the total series resistance and applying Equation (14-10). Equation (14-11) is the mathematical representation of this concept.

$$I = \frac{V}{R_o + R_x} \tag{14-11}$$

An easy way to determine ohm meter deflection is by use of a deflection factor (D). Deflection factor is the ratio of circuit current to meter current. Equation (14-12) is the mathematical representation of the deflection factor.

$$D = \frac{I}{I_m} = \frac{\frac{V}{R_o + R_x}}{\frac{V}{R_o}} = \frac{R_o}{R_o + R_x} \tag{14-12}$$

The current through the circuit can be determined by solving for I. Equation (14-13) is the mathematical representation of this relationship.

$$I = DI_m \tag{14-13}$$

To solve for R_x using Equations (14-10) through (14-13), the relationship between deflection factor and the meter resistance to the unknown resistance can be shown. Equation (14-14) is the mathematical representation of this relationship.

$$R_x = \frac{1 - D}{D} R_o \tag{14-14}$$

If half-scale deflection occurs, then $R_x = R_o$, so that the value of R_o is marked at mid-scale on the ohm meter face.

Example 1: An ohm meter has a meter movement with a 100 μA full-scale deflection. The open circuit voltage at terminals ab is 24 V. The ohm meter is zeroed and then an unknown resistance R_x is measured, which produces quarter-scale deflection. Find R_x.

Solution:

First find R_o.

$$R_o = \frac{V}{I_m}$$

$$= \frac{24}{1 \times 10^{-6}}$$

$$R_o = 2.4 \times 10^5 \Omega \text{ or } 240 K\Omega$$

Then solve for R_x:

$$R_x = \frac{1 - D}{D} R_o$$

$$= \left(\frac{1 - \frac{1}{4}}{\frac{1}{4}} \right) (240)$$

$$= (3)(240)$$

$$R_x = 720 \ K\Omega$$

Therefore, quarter scale deflection of this ohm meter face would read 720 KΩ.

Example 2: An ohm meter with R_o = 30 Ω, and full scale current I_m = 300 μA. Find I with: 1) 0 Ω, 2) 5 Ω, 3) 10 Ω, 4) 15 Ω, and 5) 1 MΩ resistors across the meter terminal.

Solution:

First, the deflection factor for each resistor must be found.

$$D = \frac{R_o}{R_o + R_x}$$

1. $R_x = 0\ \Omega$

   $D = \frac{30}{30} = 1$

2. $R_x = 5\ \Omega$

   $D = \frac{30}{30 + 5} = 0.86$

3. $R_x = 10\ \Omega$

   $D = \frac{30}{30 + 10} = 0.75$

4. $R_x = 15\ \Omega$

   $D = \frac{30}{30 + 15} = 0.67$

5. $R_x = 1\ M\Omega$

   $D = \frac{30}{1 \times 10^6} = 1 \times 10^{-6} = 0.000001$ approximately 0

Then find I by using:

$I = DI_m$

1. $R_x = 0\ \Omega$

   $I = (1)(300 \times 10^{-6}) = 300\ \mu A$ full-scale deflection

2. $R_x = 5\Omega$

   $I = (0.86)(300 \times 10^{-6}) = 258\ \mu A$

3. $R_x = 10\Omega$

   $I = (0.75)(300 \times 10^{-6}) = 225\ \mu A$

4. $R_x = 15\Omega$

   $I = (0.67)(300 \times 10^{-6}) = 201\ \mu A$

5. $R_x = 1\ M\Omega$

   $I = (0)(300 \times 10^{-6}) = 0\ \mu A$ zero deflection

NOTE: As the resistance was increased from 0 to 5Ω, meter current decreased by 42 μA. Similarly, when resistance was increased from 5 to 10Ω, the current decreased by 33 μA. Thus, an ammeter scale used to measure resistance is nonlinear (Figure 10). The ohm meter scale is a reversal of the ammeter and voltmeter scales. In other words, the zero resistance (R_x = 0) is at the right end of the scale and infinite resistance (R_x = 1 MΩ) is at the left end of the scale.

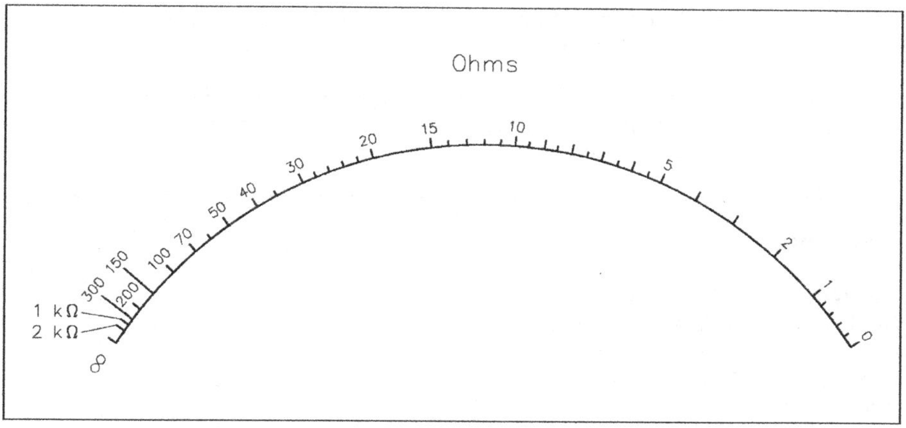

Figure 10 Ohm Meter Scale

Summary

Ohm meters are summarized below.

Ohm Meter Summary

- Measures circuit resistance
- Connected to a component removed from the circuit

WATTMETERS

Wattmeters are used to determine DC power or real AC power delivered to the load.

EO 1.2 STATE the electrical parameters measured by each of the following in-place measuring devices:
d. Wattmeter

EO 1.3 EXPLAIN how the following electrical test equipment and measuring devices are connected to a circuit:
d. Wattmeter

Wattmeter

The wattmeter is an instrument which measures DC power or true AC power. The wattmeter uses fixed coils to indicate current, while the movable coil indicates voltage (Figure 11). Coils L_{I1} and L_{I2} are the fixed coils in series with one another and serve as an ammeter. The two I terminals are connected in series with the load. The movable coil L_v, and its multiplier resistor R_s, are used as a voltmeter, with the V terminals connected in parallel with the load. The meter deflection is proportional to the VI, which is power.

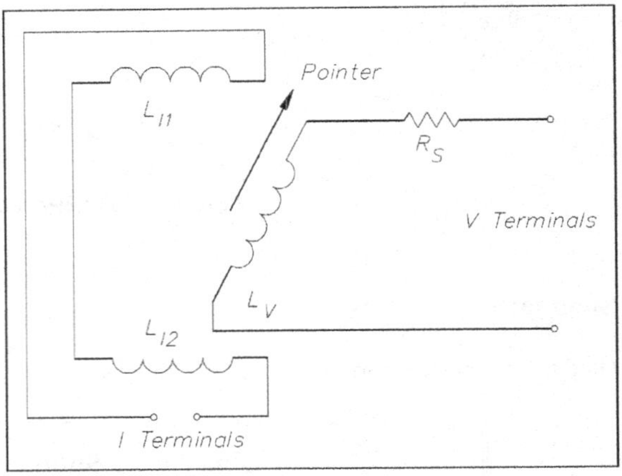

Figure 11 Wattmeter Schematic

Wattmeters are rated in terms of their maximum current, voltage, and power. All of these ratings must be observed to prevent damage to the meter.

Equation (14-15) is the mathematical representation of calculating power in a DC circuit.

$$P = VI \text{ or } P = I^2R \tag{14-15}$$

Equation (14-16) is the mathematical representation for calculating power in an AC circuit.

$$P = V_{Rms} I_{Rms} \cos\theta \text{ or } P = I^2R \quad (14\text{-}16)$$

Three-Phase Wattmeter

Total power in a 3ϕ circuit is the sum of the powers of the separate phases. The total power could be measured by placing a wattmeter in each phase (Figure 12); however, this method is not feasible since it is often impossible to break into the phases of a delta load. It also may not be feasible for the Y load, since the neutral point to which the wattmeters must be connected is not always accessible.

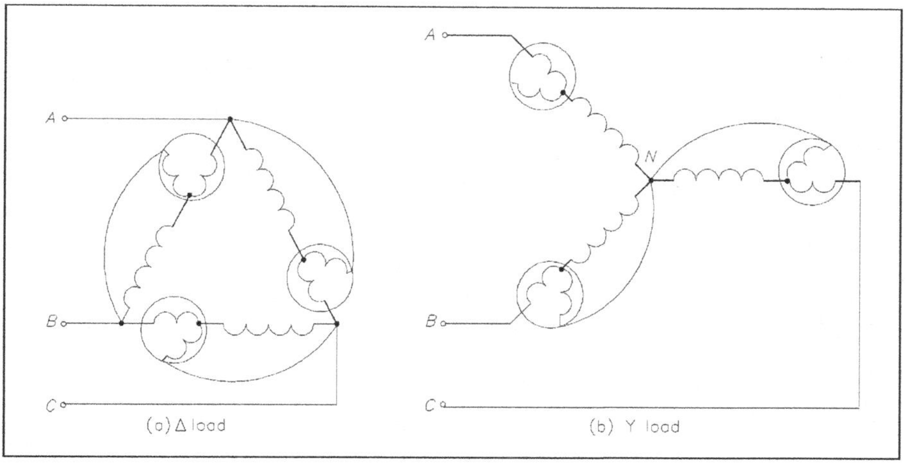

Figure 12 Wattmeters in Each Phase

Normally, only two wattmeters are used in making 3ϕ power measurements (Figure 13).

In balanced 3ϕ systems, with any power factor, total power is calculated by adding the A and B phase powers. Equation (14-17) is the mathematical representation for calculating total power (P_T).

$$P_T = W_A + W_B \quad (14\text{-}17)$$

where

W_A and W_B are the power readings in Phase A and Phase B

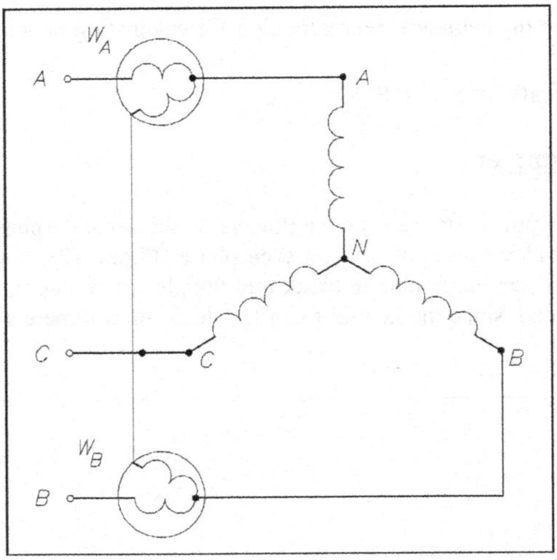

Figure 13 Two Wattmeters to Measure 3ϕ Power

Summary

Wattmeters are summarized below.

Wattmeter Summary

- Measures real power delivered to the load
- Single-phase AC or DC - voltage component (movable coil) connected in parallel with the load and the current component (fixed coil) connected in series with the load
- Three-phase AC - summation of Phase A and B powers

OTHER ELECTRICAL MEASURING DEVICES

Other measuring devices are used to aid operators in determining the electric plant conditions at a facility, such as the ampere-hour meter, power factor meter, ground detector, and synchroscope.

EO 1.2 STATE the electrical parameters measured by each of the following in-place measuring devices:

e. Ampere-hour meter
f. Power factor meter
g. Ground detector
h. Synchroscope

EO 1.3 EXPLAIN how the following electrical test equipment and measuring devices are connected to a circuit:

e. Ampere-hour meter
f. Power factor meter
g. Ground detector
h. Synchroscope

Ampere-Hour Meter

The ampere-hour meter registers ampere-hours and is an integrating meter similar to the watt-hour meter used to measure electricity usage in a home. Typical ampere-hour meters are digital indicators similar to the odometer used in automobiles. The ampere-hour meter is a direct current meter that will register in either direction depending on the direction of current flow. For example, starting from a given reading, it will register the amount of discharge of a battery; when the battery is placed on charge, it will operate in the opposite direction, returning once again to its starting point. When this point is reached, the battery has received a charge equal to the discharge, and the charge is stopped. It is normally desired to give a battery a 10% overcharge. This is accomplished by designing the ampere-hour meter to run 10% slow in the charge direction. These meters are subject to inaccuracies and cannot record the internal losses of a battery. They attempt to follow the charge and discharge, but inherently do not indicate the correct state of charge. Similar to an ammeter, the ampere-hour meter is connected in series. Although the ampere-hour meters were used quite extensively in the past, they have been largely superseded by the voltage-time method of control.

Power Factor Meter

A power factor meter is a type of electrodynamometer movement when it is made with two movable coils set at right angles to each other. The method of connection of this type of power factor meter, in a 3ϕ circuit, is shown in Figure 14. The two stationary coils, S and S^1, are connected in series in Phase B. Coils M and M^1 are mounted on a common shaft, which is free to move without restraint or control springs. These coils are connected with their series resistors from Phase B to Phase A and from Phase B to Phase C. At a power factor of unity, one potential coil current leads and one lags the current in Phase B by 30°; thus, the coils are balanced in the position shown in Figure 14. A change in power factor will cause the current of one potential coil to become more in phase and the other potential coil to be more out of phase with the current in Phase B, so that the moving element and pointer take a new position of balance to show the new power factor.

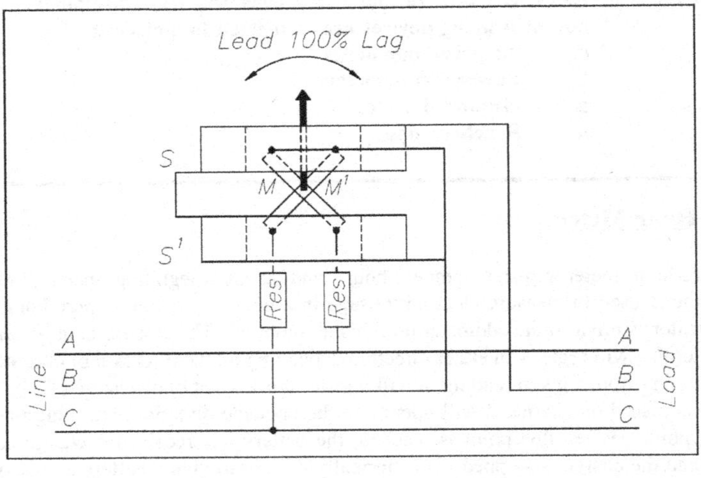

Figure 14 3ϕ Power Factor Meter Schematic

Ground Detector

The ground detector is an instrument which is used to detect conductor insulation resistance to ground. An ohm meter, or a series of lights, can be used to detect the insulation strength of an ungrounded distribution system. Most power distribution systems in use today are of the grounded variety; however, some ungrounded systems still exist.

In the ohm meter method (Figure 15), a DC voltage is applied to the conductor. If a leakage path exists between the conductor insulator and ground, a current will flow through the ground to the ohm meter proportional to the insulation resistance of the conductor.

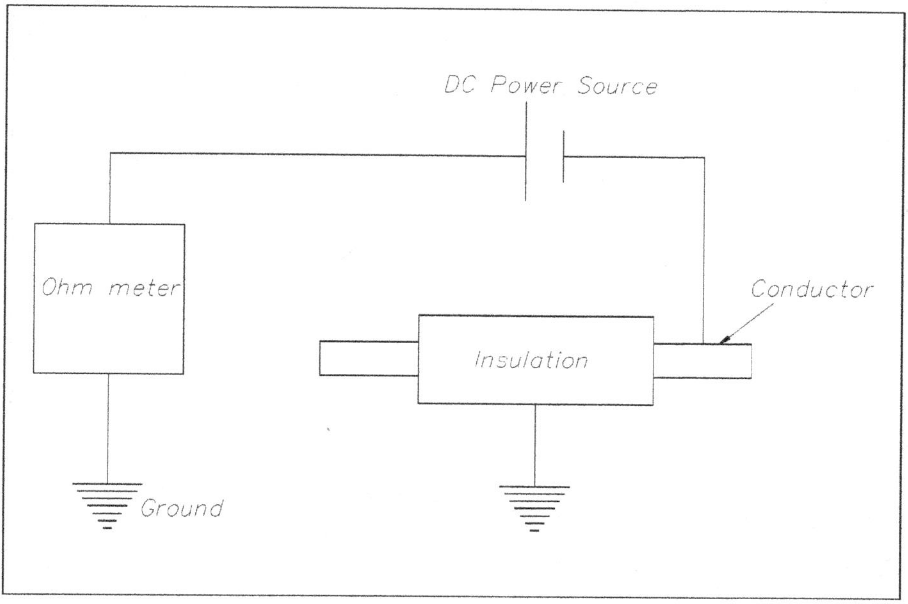

Figure 15 Simple Ohm Meter Ground Detector

In the ground detector lamp method (Figure 16), a set of three lamps connected through transformers to the system is used. To check for grounds, the switch is closed and the brilliance of the lamps is observed. If the lamps are equally bright, no ground exists and all the lamps receive the same voltage. If any one lamp is dark, and the other two lamps are brighter, the phase in which the darkened lamp is in is grounded. In this case, the primary winding of the transformer is shorted to ground and receives no voltage.

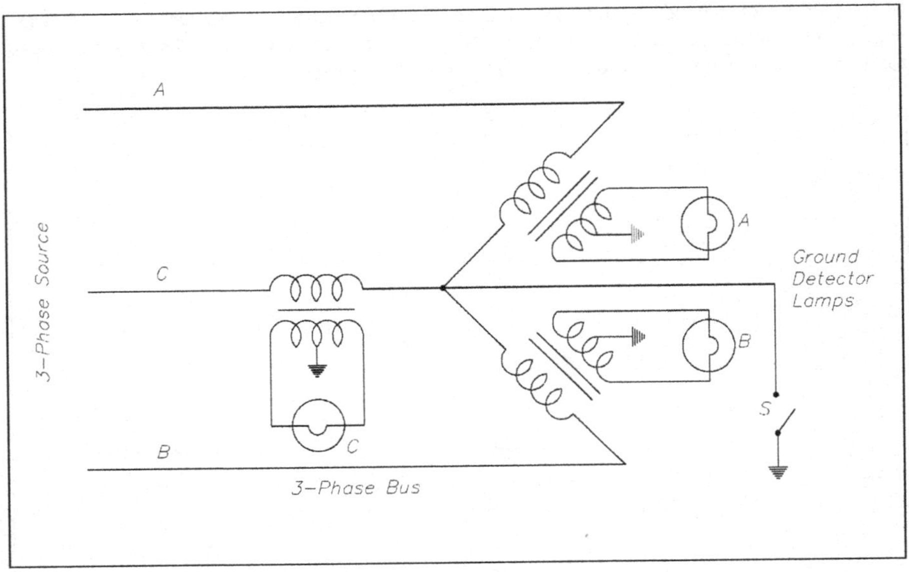

Figure 16 Ground Detector Lamp Circuit

Synchroscope

A synchroscope indicates when two AC generators are in the correct phase relation for connecting in parallel and shows whether the incoming generator is running faster or slower than the on-line generator. The synchroscope consists of a two-phase stator. The two stator windings are at right angles to one another, and by means of a phase-splitting network, the current in one phase leads the current of the other phase by 90°, thereby generating a rotating magnetic field. The stator windings are connected to the incoming generator, and a polarizing coil is connected to the running generator.

The rotating element is unrestrained and is free to rotate through 360°. It consists of two iron vanes mounted in opposite directions on a shaft, one at the top and one at the bottom, and magnetized by the polarizing coil.

If the frequencies of the incoming and running generators are different, the synchroscope will rotate at a speed corresponding to the difference. It is designed so that if incoming frequency is higher than running frequency, it will rotate in the clockwise direction; if incoming frequency is less than running frequency, it will rotate in the counterclockwise direction. When the synchroscope indicates 0° phase difference, the pointer is at the "12 o'clock" position and the two AC generators are in phase.

Summary

The important information contained in this chapter is summarized below.

Measuring Devices Summary

Ampere-hour Meter

- Measures current flow (either direction) through a given point
- Connected in series

Power Factor Meter

- Measures power factor between phases in a 3-phase circuit
- Connected in series with one phase

Ground Detector

- Measures conductor insulation
- Connected out of circuit to ground

Synchroscope

- Measures relationship between generator frequencies
- Connected by a two-phase stator at right angles

TEST EQUIPMENT

The multimeter can be used as an ammeter, an ohm meter, or a voltmeter. Meggers are used to measure insulation resistance.

EO 1.3 EXPLAIN how the following electrical test equipment and measuring devices are connected to a circuit:
i. Megger

EO 1.4 STATE the electrical parameters measured by each of the following test instruments:
a. Multimeter
b. Megger

Multimeter

The multimeter is a portable single instrument capable of measuring various electrical values including voltage, resistance, and current. The volt-ohm-milliammeter (VOM) is the most commonly used multimeter. The typical VOM has a meter movement with a full scale current of 50 μA, or a sensitivity of 20 KΩ/V, when used as a DC voltmeter. A single meter movement is used to measure current, AC and DC voltage, and resistance. Range switches are usually provided for scale selection (e.g., 0-1V, 0-10V, etc).

Megger

The megger is a portable instrument used to measure insulation resistance. The megger consists of a hand-driven DC generator and a direct reading ohm meter. A simplified circuit diagram of the instrument is shown in Figure 17.

The moving element of the ohm meter consists of two coils, A and B, which are rigidly mounted to a pivoted central shaft and are free to rotate over a C-shaped core (C on Figure 17). These coils are connected by means of flexible leads. The moving element may point in any meter position when the generator is not in operation.

As current provided by the hand-driven generator flows through Coil B, the coil will tend to set itself at right angles to the field of the permanent magnet. With the test terminals open, giving an infinite resistance, no current flows in Coil A. Thereby, Coil B will govern the motion of the rotating element, causing it to move to the extreme counter-clockwise position, which is marked as infinite resistance.

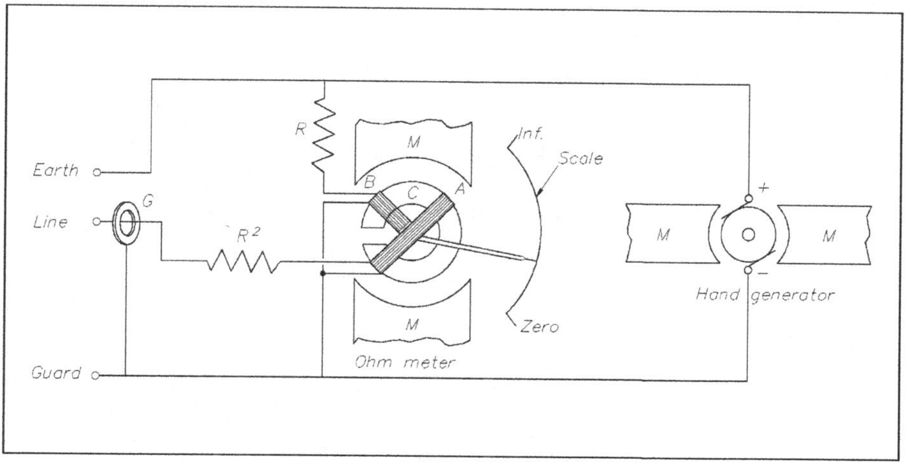

Figure 17 Simple Megger Circuit Diagram

Coil A is wound in a manner to produce a clockwise torque on the moving element. With the terminals marked "line" and "earth" shorted, giving a zero resistance, the current flow through the Coil A is sufficient to produce enough torque to overcome the torque of Coil B. The pointer then moves to the extreme clockwise position, which is marked as zero resistance. Resistance (R^1) will protect Coil A from excessive current flow in this condition.

When an unknown resistance is connected across the test terminals, line and earth, the opposing torques of Coils A and B balance each other so that the instrument pointer comes to rest at some point on the scale. The scale is calibrated such that the pointer directly indicates the value of resistance being measured.

Summary

Test equipment is summarized below.

Test Equipment Summary

- Multimeters measure current, voltage, and resistance.
- Meggers measure insulation resistance.
- Meggers are connected out of circuit.

Department of Energy
Fundamentals Handbook

ELECTRICAL SCIENCE
Module 15
Electrical Distribution Systems

TABLE OF CONTENTS

LIST OF FIGURES iii

LIST OF TABLES iv

REFERENCES v

OBJECTIVES vi

SYSTEM COMPONENTS AND PROTECTION DEVICES 1

Single (One-Line) Diagram 1
Commercial or Utility Power 1
Diesel Power 1
Failure-Free Power 3
Neutral Grounding 3
Voltage Class 3
Protective Relays 3
Overlapping Protective Zones 4
Fuses 5
Summary 6

CIRCUIT BREAKERS 7

Introduction 7
Low-Voltage Air Circuit Breakers 7
High-Voltage Circuit Breakers 11
Circuit Breaker Control 11
Summary 14

MOTOR CONTROLLERS 15

Motor Controllers 15
Manual Controllers 15
Magnetic Controller 16
Motor Controller Types and Operation 18
Summary 20

TABLE OF CONTENTS (Cont.)

WIRING SCHEMES AND GROUNDING 21

Introduction 21
Terminology 22
Single-Phase Power 22
Three-Phase Wiring Schemes 24
3-Wire, Three-Phase Delta System 24
4-Wire, Three-Phase Delta System 25
4-Wire, Three-Phase Wye System 26
Summary 27

LIST OF FIGURES

Figure 1	One-Line Distribution Diagram	2
Figure 2	Protective Relaying Zones	4
Figure 3	Types of Fuses	5
Figure 4	Molded Case Circuit Breaker	7
Figure 5	Cutaway View of Molded Case Circuit Breaker	8
Figure 6	Large Air Circuit Breaker	10
Figure 7	Simple Circuit Breaker Control Circuit - Breaker Open	12
Figure 8	Simple Circuit Breaker Control Circuit - Breaker Closed	13
Figure 9	Single Phase Manual Controller	16
Figure 10	Typical Three-Phase Magnetic Controller	17
Figure 11	Magnetic Contactor Assembly	18
Figure 12	LVP Controller	18
Figure 13	LVR Controller	19
Figure 14	LVRE Controller	19
Figure 15	Three-Phase to Single-Phase Connections	23
Figure 16	3-Wire Edison Scheme	24
Figure 17	3-Wire, Three-Phase Delta Scheme	25
Figure 18	4-Wire Delta System	26
Figure 19	4-Wire, Three Phase Wye System	27

LIST OF TABLES

NONE

REFERENCES

- Gussow, Milton, Schaum's Outline Series, Basic Electricity, McGraw-Hill.
- Academic Program for Nuclear Power Plant Personnel, Volume II, Columbia, MD: General Physics Corporation, Library of Congress Card #A 326517, 1982.
- Nasar and Unnewehr, Electromechanics and Electric Machines, John Wiley and Sons.
- Van Valkenburgh, Nooger, and Neville, Basic Electricity, Vol. 5, Hayden Book Company.
- Lister, Eugene C., Electric Circuits and Machines, 5th Edition, McGraw-Hill.
- Croft, Carr, Watt, and Summers, American Electricians Handbook, 10th Edition, McGraw-Hill.
- Mason, C. Russel, The Art and Science of Protective Relaying, John Wiley and Sons.
- Mileaf, Harry, Electricity One - Seven, Revised 2nd Edition, Hayden Book Company.
- Kidwell, Walter, Electrical Instruments and Measurements, McGraw-Hill.
- National Electrical Code, Batterymarch Park, Quincy, MA: National Fire Protection Association.

TERMINAL OBJECTIVE

1.0 Given the functional characteristics of an AC power source and the intended load, **DESCRIBE** the necessary components and the wiring scheme to provide a safe Electrical Distribution System.

ENABLING OBJECTIVES

1.1 **EXPLAIN** the following terms as they apply to Electrical Distribution Systems:
   a. Single (one-line) diagram
   b. Commercial or utility power
   c. Diesel power
   d. Failure-free power
   e. Neutral grounding
   f. Voltage class
   g. Protective relays
   h. Overlapping protective zones

1.2 **DESCRIBE** the protection provided by each of the following:
   a. Fuses
   b. Protective relays

1.3 **STATE** the purpose of circuit breakers.

1.4 Given a simple schematic of a circuit breaker control circuit, **DESCRIBE** the operation of that breaker during remote operation and automatic tripping.

1.5 **LIST** the three most widely-used protective features that may be incorporated into a circuit breaker control circuit.

1.6 **STATE** the function of motor controllers.

1.7 **STATE** three protective features (overloads) that may be incorporated into a motor controller.

1.8 Given a simplified drawing of a motor controller, **DESCRIBE** the operation of that motor controller.

ENABLING OBJECTIVES (Cont.)

1.9 **DEFINE** the following terms as they apply to wiring schemes used in power distribution systems:

a. Ampacity
b. Bond
c. Conductor
d. Ground
e. Ground voltage
f. Leg
g. Neutral
h. Phase voltage

1.10 **DESCRIBE** the two methods of connecting single-phase loads to a three-phase power source.

1.11 **DESCRIBE** the purpose of the following power distribution schemes.

a. 3-wire, single-phase Edison system
b. 3-wire, three-phase Delta system
c. 4-wire, three-phase Delta system
d. 4-wire, three-phase Wye system

Intentionally Left Blank

SYSTEM COMPONENTS AND PROTECTION DEVICES

Nuclear facilities rely on dependable electrical distribution systems to provide power to key vital equipment. Knowledge of the basic electrical power distribution system and its components will help the operator understand the importance of electrical power distribution systems.

EO 1.1 EXPLAIN the following terms as they apply to Electrical Distribution Systems:
- **a. Single (one-line) diagram**
- **b. Commercial or utility power**
- **c. Diesel power**
- **d. Failure-free power**
- **e. Neutral grounding**
- **f. Voltage class**
- **g. Protective relays**
- **h. Overlapping protective zones**

EO 1.2 DESCRIBE the protection provided by each of the following:
- **a. Fuses**
- **b. Protective relays**

Single (One-Line) Diagram

A *single*, or *one-line diagram* of a distribution system is a simple and easy-to-read diagram showing power supplies, loads, and major components in the distribution system (Figure 1).

Commercial or Utility Power

Commercial or *utility power* is electrical power that is provided by commercial generating systems to the facility.

Diesel Power

Diesel power is power generated by a diesel-driven generator. Diesel-driven generators are the most economical and practical source of "standby power."

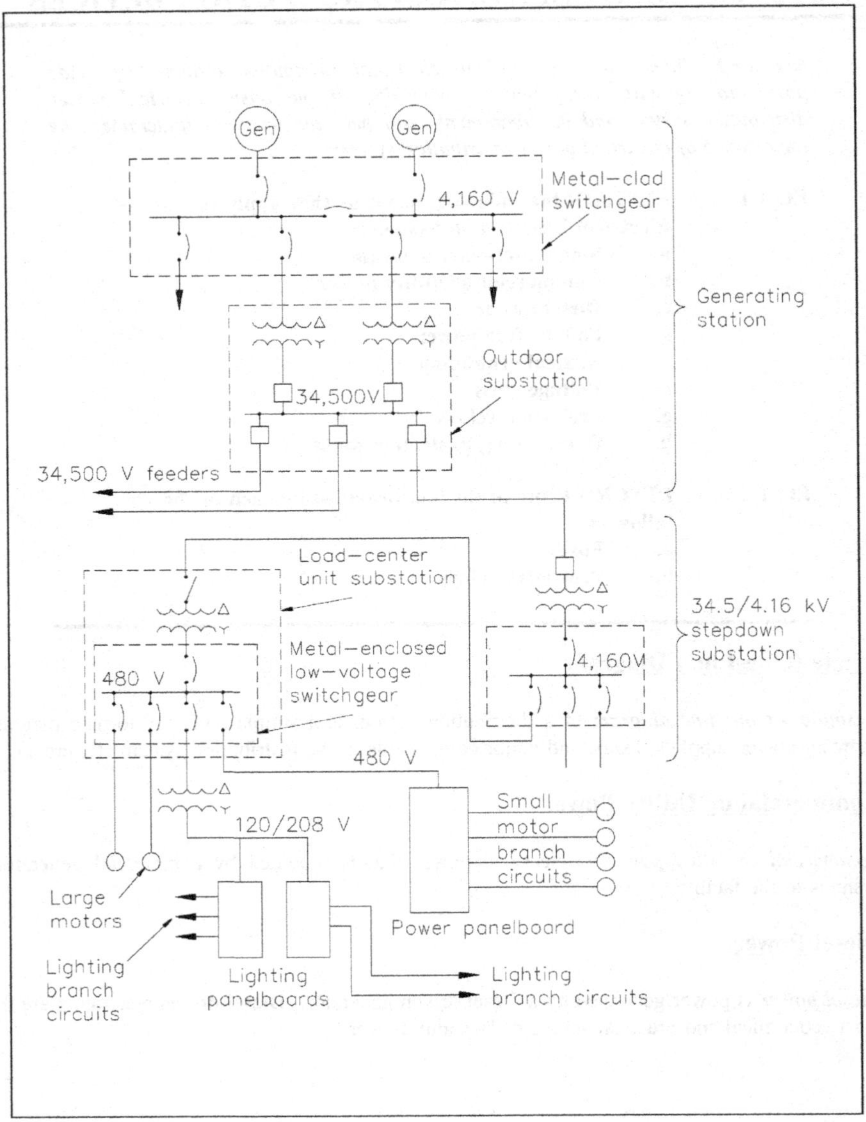

Figure 1 One-Line Distribution Diagram

Failure-Free Power

Failure-free power is accomplished by providing vital equipment with automatic switching between two or more power supplies so that interruption of power is minimized.

Neutral Grounding

Neutral grounding in electrical distribution systems helps prevent accidents to personnel and damage to property caused by: fire in case of lightning; a breakdown between primary and secondary windings of transformers; or accidental contact of high-voltage wires and low- voltage wires. If some point on the circuit is grounded (in this case neutral ground), lightning striking the wires will be conducted into the ground, and breakdown between the primary and secondary windings of a transformer will cause the primary transformer fuses to blow. Another advantage of neutral grounding is that it reduces the amount of insulation required for high-voltage transmission lines.

Voltage Class

Voltage in distribution systems is classified into three groups: high voltage, intermediate voltage, and low voltage. *High voltage* is voltage that is above 15,000 volts, *intermediate voltage* is voltage between 15,000 volts and 600 volts, and *low voltage* is voltage at 600 volts or less.

Protective Relays

Protective relays are designed to cause the prompt removal of any part of a power system that might cause damage or interfere with the effective and continuous operation of the rest of the system. Protective relays are aided in this task by circuit breakers that are capable of disconnecting faulty components or subsystems.

Protective relays can be used for types of protection other than short circuit or overcurrent. The relays can be designed to protect generating equipment and electrical circuits from any undesirable condition, such as undervoltage, underfrequency, or interlocking system lineups.

There are only two operating principles for protective relays: (1) electromagnetic attraction and (2) electromagnetic induction. Electromagnetic attraction relays operate by a plunger being drawn up into a solenoid or an armature that is attracted to the poles of an electromagnet. This type of relay can be actuated by either DC or AC systems. Electromagnetic induction relays operate on the induction motor principle whereby torque is developed by induction in a rotor. This type of relay can be used only in AC circuits.

Overlapping Protective Zones

A separate zone of protection is provided around each system element (Figure 2). Any failure that may occur within a given zone will cause the tripping or opening of all circuit breakers within that zone. For failures that occur within a region where two protective zones overlap, more breakers will be tripped than are necessary to disconnect the faulty component; however, if there were no overlap of protective zones, a fault in a region between the two zones would result in no protective action at all. Therefore, it is desirable for protective zone overlap to ensure the maximum system protection.

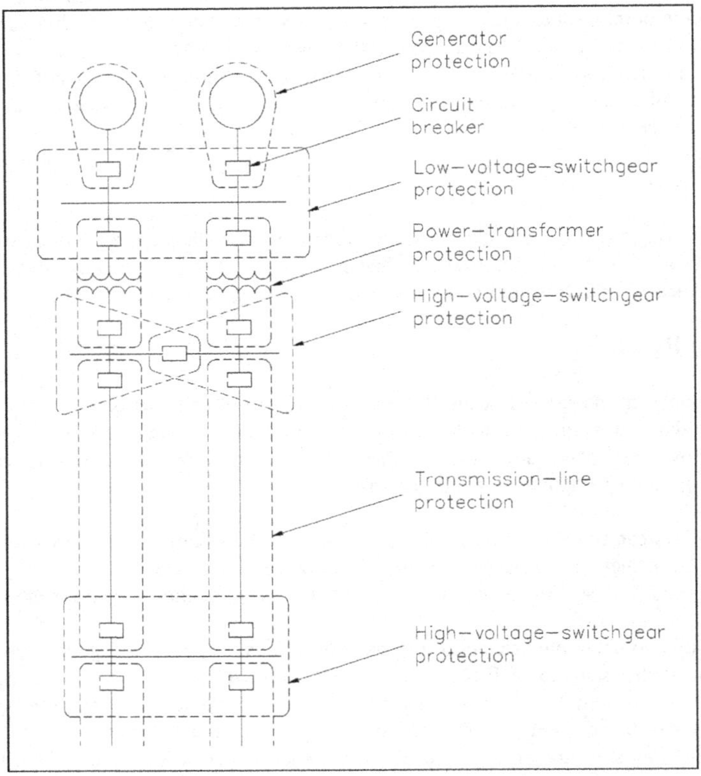

Figure 2 Protective Relaying Zones

Fuses

A fuse is a device that protects a circuit from an overcurrent condition only. It has a fusible link directly heated and destroyed by the current passing through it. A fuse contains a current-carrying element sized so that the heat generated by the flow of normal current through it does not cause it to melt the element; however, when an overcurrent or short-circuit current flows through the fuse, the fusible link will melt and open the circuit. There are several types of fuses in use (Figure 3).

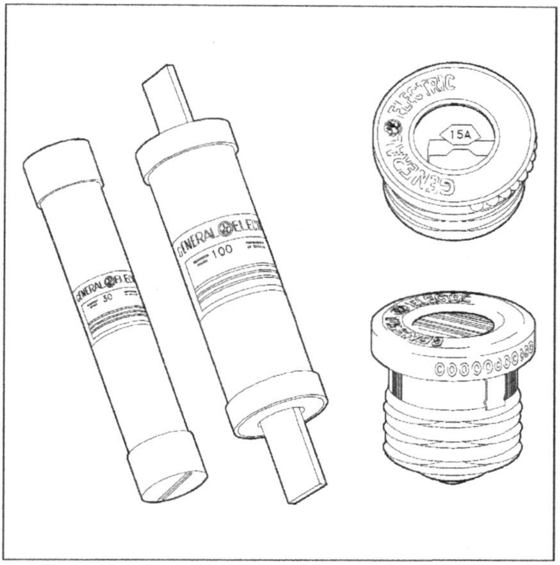

Figure 3 Types of Fuses

The plug fuse is a fuse that consists of a zinc or alloy strip, a fusible element enclosed in porcelain or pyrex housing, and a screw base. This type of fuse is normally used on circuits rated at 125 V or less to ground and has a maximum continuous current-carrying capacity of 30 amps.

The cartridge fuse is constructed with a zinc or alloy fusible element enclosed in a cylindrical fiber tube with the element ends attached to a metallic contact piece at the ends of the tube. This type of fuse is normally used on circuits rated at either 250 volts or 600 volts and has a maximum continuous current-carrying capacity of 600 amps.

Summary

The important information contained in this chapter is summarized below.

System Components and Protection Devices Summary

- Single (one-line) diagram - simple and easy to read diagram showing power supplies, loads, and major components in the distribution system
- Commercial or utility power - electric power supplied to the facility
- Diesel power - economical/practical source of standby power
- Failure-free power - power supplied to vital equipment with automatic switching so that interruption of power is minimized
- Neutral grounding - helps prevent accidents to personnel and damage to property by fire
- Voltage class - high voltage > 15,000 volts, intermediate voltage is 600-15,000 volts, low voltage ≤ 600 volts
- Protective relays - cause prompt removal of any part of a power system that suffers a short circuit
- Overlapping protective zones - created around each element of the power system to prevent element failure from interrupting the whole system operation
- Breakers - disconnect component from the power system
- Fuse - protects component from overcurrent

CIRCUIT BREAKERS

A circuit breaker is a device that is used to completely disconnect a circuit when any abnormal condition exists. The circuit breaker can be designed to actuate under any undesirable condition.

EO 1.3 **STATE the purpose of circuit breakers.**

EO 1.4 **Given a simple schematic of a circuit breaker control circuit, DESCRIBE the operation of that breaker during remote operation and automatic tripping.**

EO 1.5 **LIST the three most widely-used protective features that may be incorporated into a circuit breaker control circuit.**

Introduction

The purpose of a circuit breaker is to break the circuit and stop the current flow when the current exceeds a predetermined value without causing damage to the circuit or the circuit breaker. Circuit breakers are commonly used in place of fuses and sometimes eliminate the need for a switch. A circuit breaker differs from a fuse in that it "trips" to break the circuit and may be reset, while a fuse melts and must be replaced. Air circuit breakers (ACBs) are breakers where the interruption of the breaker contacts takes place in an air environment. Oil circuit breakers (OCBs) use oil to quench the arc when the breaker contacts open.

Low-Voltage Air Circuit Breakers

A low-voltage circuit breaker is one which is suited for circuits rated at 600 volts or lower. One of the most commonly used low-voltage air circuit breakers is the molded case circuit breaker (Figure 4).

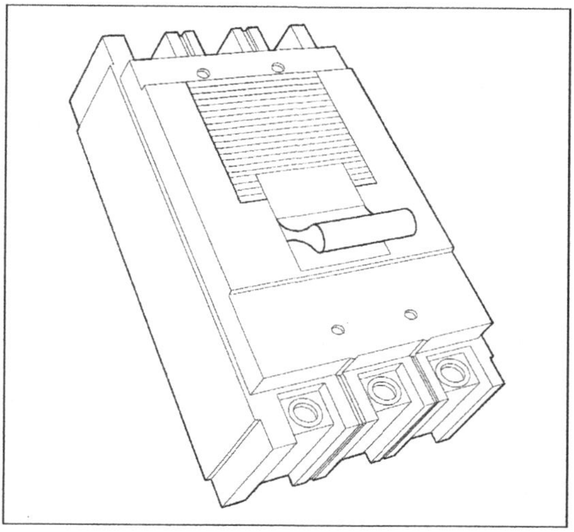

Figure 4 Molded Case Circuit Breaker

A cutaway view of the molded case circuit breaker is shown in Figure 5.

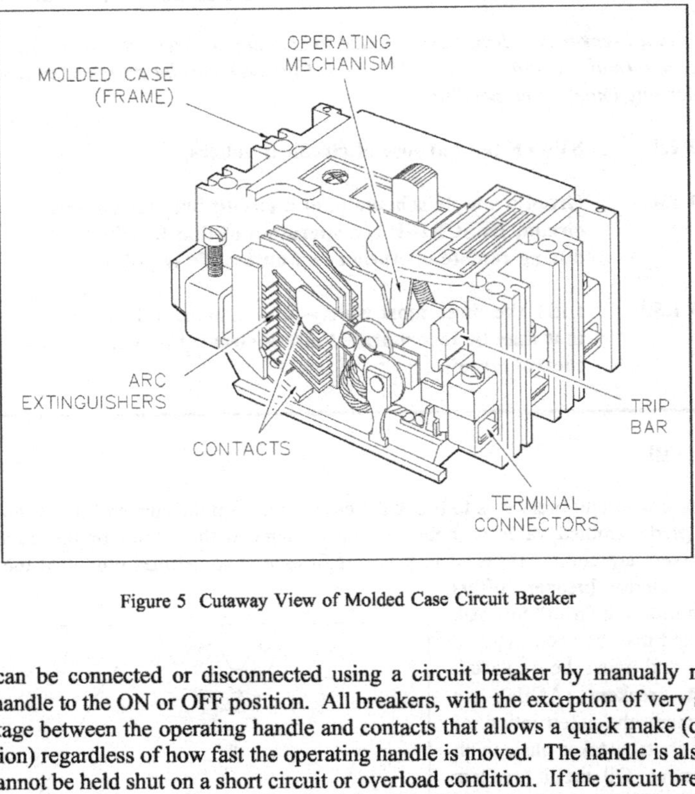

Figure 5 Cutaway View of Molded Case Circuit Breaker

A circuit can be connected or disconnected using a circuit breaker by manually moving the operating handle to the ON or OFF position. All breakers, with the exception of very small ones, have a linkage between the operating handle and contacts that allows a quick make (quick break contact action) regardless of how fast the operating handle is moved. The handle is also designed so that it cannot be held shut on a short circuit or overload condition. If the circuit breaker opens under one of these conditions, the handle will go to the trip-free position. The trip-free position is midway between the ON and OFF positions and cannot be re-shut until the handle is pushed to the OFF position and reset.

A circuit breaker will automatically trip when the current through it exceeds a pre-determined value. In lower current ratings, automatic tripping of the circuit breaker is accomplished by use of thermal tripping devices. Thermal trip elements consist of a bimetallic element that can be calibrated so that the heat from normal current through it does not cause it to deflect. An abnormally high current, which could be caused by a short circuit or overload condition, will cause the element to deflect and trip the linkage that holds the circuit breaker shut. The circuit breaker will then be opened by spring action. This bimetallic element, which is responsive to the heat produced by current flowing through it, has an inverse-time characteristic. If an extremely high current is developed, the circuit breaker will be tripped very rapidly.

For moderate overload currents, it will operate more slowly. Molded case breakers with much larger current ratings also have a magnetic trip element to supplement the thermal trip element. The magnetic unit utilizes the magnetic force that surrounds the conductor to operate the circuit breaker tripping linkage.

When the separable contacts of an air circuit breaker are opened, an arc develops between the two contacts. Different manufacturers use many designs and arrangements of contacts and their surrounding chambers. The most common design places the moving contacts inside of an arc chute. The construction of this arc chute allows the arc formed as the contacts open to draw out into the arc chute. When the arc is drawn into the arc chute, it is divided into small segments and quenched. This action extinguishes the arc rapidly, which minimizes the chance of a fire and also minimizes damage to the breaker contacts.

Molded case circuit breakers come in a wide range of sizes and current ratings. There are six frame sizes available: 100, 225, 400, 600, 800, and 2,000 amps. The size, contact rating, and current interrupting ratings are the same for all circuit breakers of a given frame size. The continuous current rating of a breaker is governed by the trip element rating. The range of voltage available is from 120 to 600 volts, and interrupting capacity ranges as high as 100,000 amps.

Much larger air circuit breakers are used in large commercial and industrial distribution systems. These circuit breakers are available in much higher continuous current and interrupting ratings than the molded case circuit breaker. Breakers of this type have current ratings as high as 4,000 amps, and interrupting ratings as high as 150,000 amps.

Most large air circuit breakers use a closing device, known as a "stored energy mechanism," for fast, positive closing action. Energy is stored by compressing large powerful coil springs that are attached to the contact assembly of a circuit breaker. Once these springs are compressed, the latch may be operated to release the springs, and spring pressure will shut the circuit breaker. Circuit breaker closing springs may be compressed manually or by means of a small electric motor. This type of circuit breaker can be classified as either a manually- or electrically-operated circuit breaker.

When a large air circuit breaker is closed, the operating mechanism is latched. As the circuit breaker is closed, a set of tripping springs, or coils, are compressed, and the circuit breaker may then be tripped by means of a trip latch. The trip latch mechanism may be operated either manually or remotely by means of a solenoid trip coil.

As previously stated, circuit breakers may be operated either manually or electrically. Electrically-operated circuit breakers are used when circuit breakers are to be operated at frequent intervals or when remote operation is required.

When the electrically-operated stored energy circuit breaker is tripped, the spring is recharged by the spring charging motor so that the breaker is ready for the next closing operation. The manually-operated circuit breaker closing springs are normally compressed by a hand crank just prior to operation of the breaker. Figure 6 shows a large air circuit breaker which is classified as a manually-operated stored energy circuit breaker. The closing springs are compressed by pulling downward on the large operating handle on the front of the breaker. Closing this circuit breaker is accomplished manually by depressing the small closing lever. Tripping this circuit breaker is done by means of the tripping lever, located at the bottom front of the breaker.

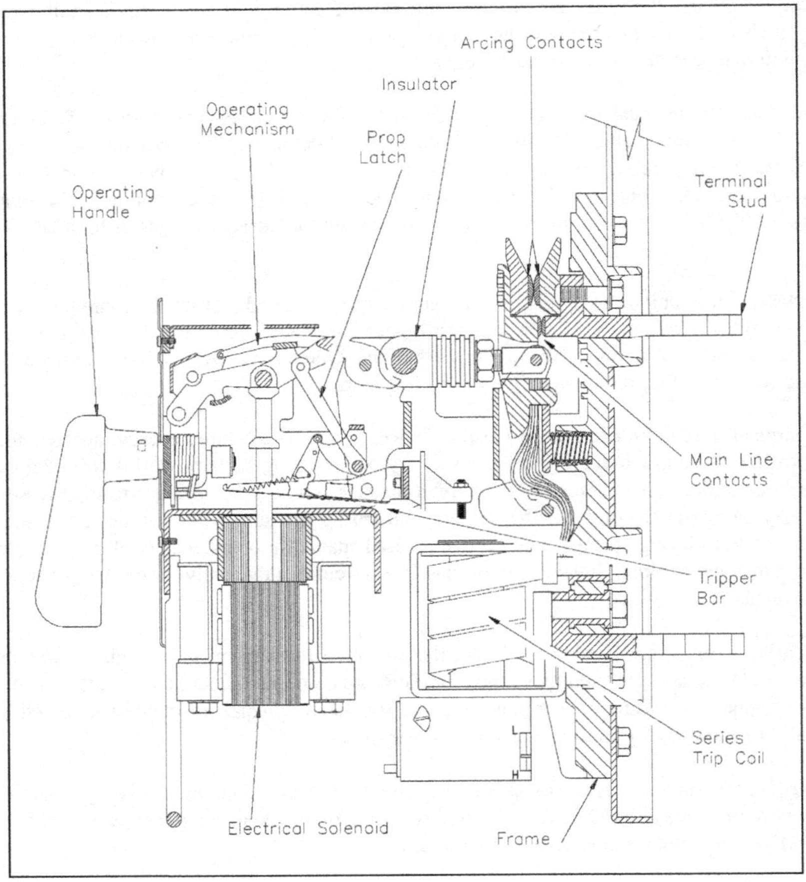

Figure 6 Large Air Circuit Breaker

High-Voltage Circuit Breakers

High-voltage circuit breakers (including breakers rated at intermediate voltage) are used for service on circuits with voltage ratings higher than 600 volts. Standard voltage ratings for these circuit breakers are from 4,160 to 765,000 volts and three-phase interrupting ratings of 50,000 to 50,000,000 kVA.

In the early stages of electrical system development, the major portion of high-voltage circuit breakers were oil circuit breakers. However, magnetic and compressed-air type air circuit breakers have been developed and are in use today.

The magnetic air circuit breaker is rated up to 750,000 kVA at 13,800 volts. This type of circuit breaker interrupts in air between two separable contacts with the aid of magnetic blowout coils. As the current-carrying contacts separate during a fault condition, the arc is drawn out horizontally and transferred to a set of arcing contacts. Simultaneously, the blowout coil provides a magnetic field to draw the arc upward into the arc chutes. The arc, aided by the blowout coil magnetic field and thermal effects, accelerates upward into the arc chute, where it is elongated and divided into many small segments.

The construction of this type of circuit breaker is similar to that of a large air circuit breaker used for low-voltage applications, except that they are all electrically operated.

Compressed-air circuit breakers, or air-blast circuit breakers, depend on a stream of compressed air directed toward the separable contacts of the breaker to interrupt the arc formed when the breaker is opened. Air-blast circuit breakers have recently been developed for use in extra high-voltage applications with standard ratings up to 765,000 volts.

Oil circuit breakers (OCBs) are circuit breakers that have their contacts immersed in oil. Current interruption takes place in oil which cools the arc developed and thereby quenches the arc. The poles of small oil circuit breakers can be placed in one oil tank; however, the large high-voltage circuit breakers have each pole in a separate oil tank. The oil tanks in oil circuit breakers are normally sealed. The electrical connections between the contacts and external circuits are made through porcelain bushings.

Circuit Breaker Control

As we have discussed, circuit breakers may be remotely operated. In order to operate the breakers from a remote location, there must be an electrical control circuit incorporated. Figure 7 shows a simple control circuit for a remotely-operated breaker.

Control power is supplied by an AC source and then rectified to DC. The major components of a simple control circuit are: the rectifier unit, the closing relay, the closing coil, the tripping coil, the auxiliary contacts, and the circuit breaker control switch.

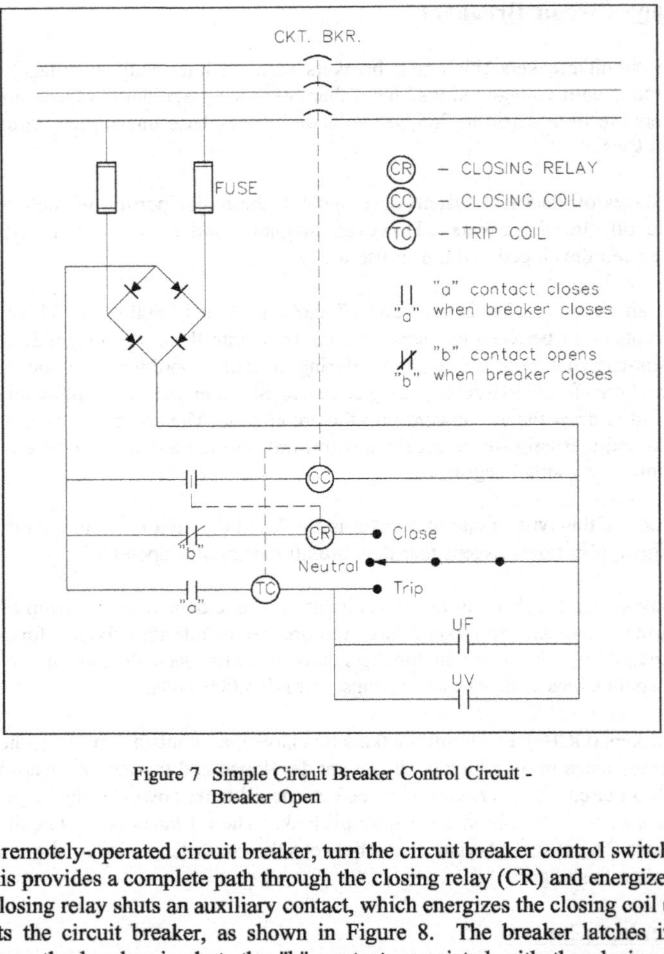

Figure 7 Simple Circuit Breaker Control Circuit - Breaker Open

To close the remotely-operated circuit breaker, turn the circuit breaker control switch to the close position. This provides a complete path through the closing relay (CR) and energizes the closing relay. The closing relay shuts an auxiliary contact, which energizes the closing coil (CC), which, in turn, shuts the circuit breaker, as shown in Figure 8. The breaker latches in the closed position. Once the breaker is shut, the "b" contact associated with the closing relay opens, de-energizing the closing relay and, thereby, the closing coil. When the breaker closes, the "a" contact also closes, which enables the trip circuit for manual or automatic trips of the breaker. The circuit breaker control switch may now be released and will automatically return to the neutral position.

To open the circuit breaker, turn the circuit breaker control switch to the trip position. This action energizes the trip coil (TC), which acts directly on the circuit breaker to release the latching mechanism that holds the circuit breaker closed.

When the latching mechanism is released, the circuit breaker will open, opening the "a" contact for the tripping coil and de-energizing the tripping coil. Also, when the circuit breaker opens, the "b" contact will close, thereby setting up the circuit breaker to be remotely closed using the closing relay, when desired. The circuit breaker control switch may now be released.

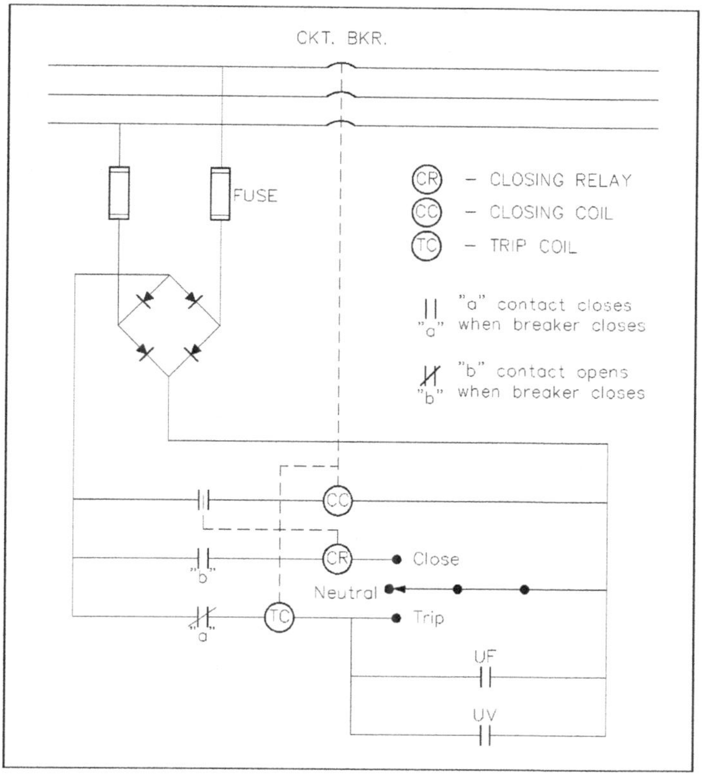

Figure 8 Simple Circuit Breaker Control Circuit - Breaker Closed

As you can see from Figure 7 or 8, the circuit breaker control circuit can be designed so that any one of a number of protective features may be incorporated. The three most commonly-used automatic trip features for a circuit breaker are overcurrent (as discussed previously), underfrequency, and undervoltage. If any one of the conditions exists while the circuit breaker is closed, it will close its associated contact and energize the tripping coil, which, in turn, will trip the circuit breaker.

Summary

The important information covered in this chapter is summarized below.

Circuit Breaker Summary

- The purpose of a circuit breaker is to provide a means for connecting and disconnecting circuits of relatively high capacities without causing damage to them.
- The three most commonly-used automatic trip features for a circuit breaker are overcurrent, underfrequency, and undervoltage.

MOTOR CONTROLLERS

Motor controllers range from a simple toggle switch to a complex system using solenoids, relays, and timers. The basic functions of a motor controller are to control and protect the operation of a motor.

EO 1.6 STATE the function of motor controllers.

EO 1.7 STATE three protective features (overloads) that may be incorporated into a motor controller.

EO 1.8 Given a simplified drawing of a motor controller, DESCRIBE the operation of that motor controller.

Motor Controllers

Motor controllers range from a simple toggle switch to a complex system using solenoids, relays, and timers. The basic functions of a motor controller are to control and protect the operation of a motor. This includes starting and stopping the motor, and protecting the motor from overcurrent, undervoltage, and overheating conditions that would cause damage to the motor. There are two basic categories of motor controllers: the manual controller and the magnetic controller.

Manual Controllers

A manual controller, illustrated by Figure 9, is a controller whose contact assembly is operated by mechanical linkage from a toggle-type handle or a pushbutton arrangement. The controller is operated by hand.

The manual controller is provided with thermal and direct-acting overload units to protect the motor from overload conditions. The manual controller is basically an "ON-OFF" switch with overload protection.

Manual controllers are normally used on small loads such as machine tools, fans, blowers, pumps, and compressors. These types of controllers are simple, and they provide quiet operation. The contacts are closed simply by moving the handle to the "ON" position or pushing the START button. They will remain closed until the handle is moved to the "OFF" position or the STOP button is pushed. The contacts will also open if the thermal overload trips.

Manual controllers do NOT provide low voltage protection or low voltage release. When power fails, the manual controller contacts remain closed, and the motor will restart when power is restored. This feature is highly desirable for small loads because operator action is not needed to restart the small loads in a facility; however, it is undesirable for larger loads because it could cause a hazard to equipment and personnel.

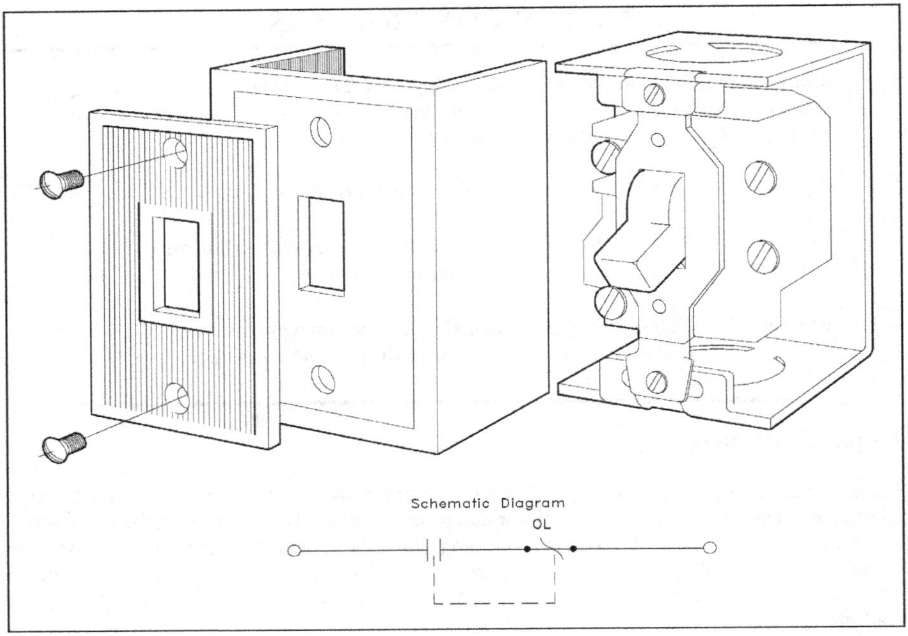

Figure 9 Single Phase Manual Controller

Magnetic Controller

A large percentage of controller applications require that the controller be operated from a remote location or operate automatically in response to control signals. As discussed, manual controllers cannot provide this type of control; therefore, magnetic controllers are necessary.

Basic operations using a magnetic controller, such as the closing of switches or contacts, are performed by magnetic contactors. A magnetic controller is one that will automatically perform all operations in the proper sequence after the closure of a master switch. The master switch (for example, float switch, pressure switch, or thermostat) is frequently operated automatically. But in some cases, such as pushbuttons, drum switches, or knife switches, the master switch is manually operated. Figure 10 shows a typical magnetic controller and its component parts.

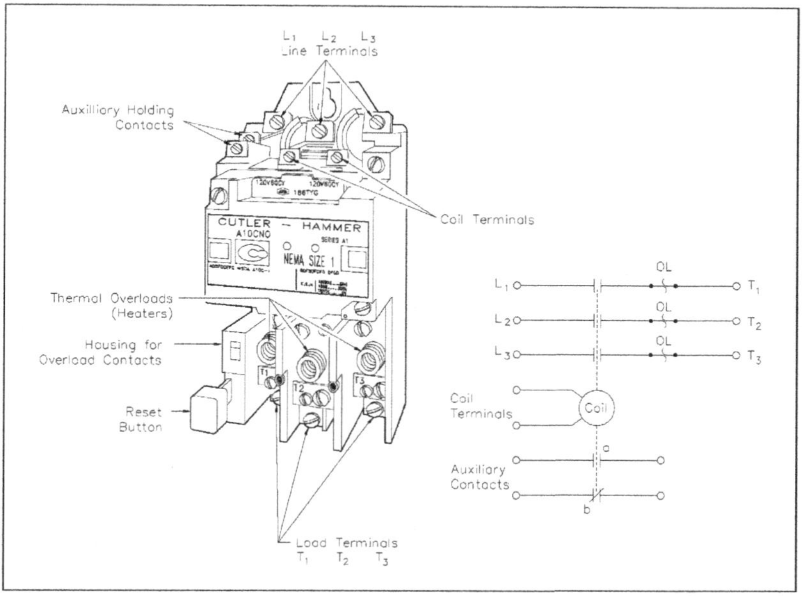

Figure 10 Typical Three-Phase Magnetic Controller

A magnetic contactor (Figure 11) is a device operated by an electromagnet.

The magnetic contactor consists of an electromagnet and a movable iron armature on which movable and stationary contacts are mounted. When there is no current flow through the electromagnetic coil, the armature is held away by a spring. When the coil is energized, the electromagnet attracts the armature and closes the electrical contacts.

Overload devices are incorporated into magnetic controllers. These overload devices protect the motor from overcurrent conditions that would be extremely harmful. There are many types and forms of overload devices. The following types of overload devices are commonly used in motor-control equipment.

- Fuses
- Thermal overloads
- Magnetic overloads

The thermal overload device is shown in Figure 10.

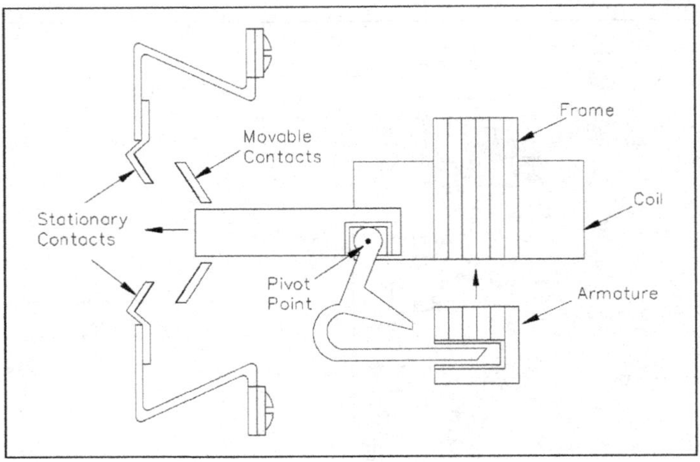

Figure 11 Magnetic Contactor Assembly

Motor Controller Types and Operation

Within the two basic categories of motor controllers, there are three major types of AC across-the-line controllers in use today. There are low-voltage protection (LVP), low-voltage release (LVR), and low-voltage release effect (LVRE) controllers.

The main purpose of an LVP controller is to de-energize the motor in a low voltage condition and keep it from re-starting automatically upon return of normal voltage (Figure 12).

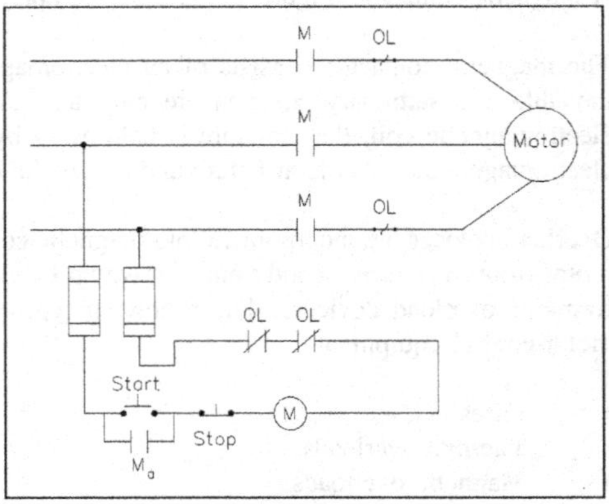

Figure 12 LVP Controller

LVP Controller Operation:

1. Push the START button, which energizes contactor coil M, closing the M and M_a contacts. When the START button is released, the circuit will remain complete, because the M_a contact remains closed, shunting the open start switch.

2. When a low voltage condition occurs, the M coil will drop out at some pre-determined value of voltage, and the M and M_a contacts will open. The START button must then be pushed to restart the motor.

3. Depressing the STOP button de-energizes the M coil, which then opens the M and M_a contacts.

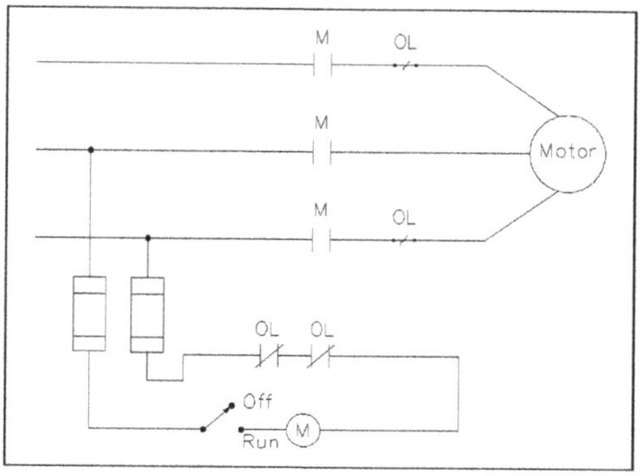

Figure 13 LVR Controller

The purpose of the LVR controller is to de-energize the motor in a low voltage condition and restart the motor when normal voltage is restored. This type of controller (Figure 13) is used primarily on small and/or critical loads (e.g., cooling water pumps required for safety-related equipment).

LVR Controller Operation:

1. Place the START switch in Run which energizes coil M, closing the M contacts and starting the motor.

2. When a low voltage condition occurs, the M coil drops out, opening the M contacts and de-energizing the motor. When normal voltage is restored, the M coil is again energized, closing the M contacts and restarting the motor.

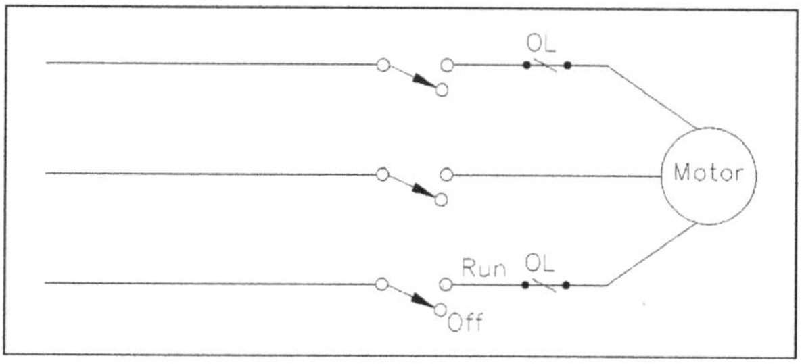

Figure 14 LVRE Controller

The LVRE controller maintains the motor across the line at all times. This type of controller is of the manual variety and is found mostly on small loads that must start automatically upon restoration of voltage (Figure 14). An LVRE controller may or may not contain overloads. If overloads are used, they will be placed in the lines to the load.

The motor controllers that have been discussed are very basic. There are many automatic control functions that can be incorporated into these types of controllers, but they are beyond the scope of this text.

Summary

The important information contained in this chapter is summarized below.

Motor Controllers Summary

- Motor controller - controls and protects the operation of a motor
- Controller's protective features - fuses, thermal overloads, and magnetic overloads
- LVP - de-energizes motor on low voltage and keeps it from automatically restarting
- LVR - de-energizes motor on low voltage and restarts when the voltage is restored to normal
- LVRE - maintains motor across the line at all times

WIRING SCHEMES AND GROUNDING

Nuclear facilities rely on standardized wiring schemes to provide both single-phase and three-phase power distribution systems and protective grounds to insure safe operation.

EO 1.9 DEFINE the following terms as they apply to wiring schemes used in power distribution systems:

a. Ampacity
b. Bond
c. Conductor
d. Ground
e. Ground voltage
f. Leg
g. Neutral
h. Phase voltage

EO 1.10 DESCRIBE the two methods of connecting single-phase loads to a three-phase power source.

EO 1.11 DESCRIBE the purpose of the following power distribution schemes.

a. 3-wire, single-phase Edison system
b. 3-wire, three-phase Delta system
c. 4-wire, three-phase Delta system
d. 4-wire, three-phase Wye system

Introduction

Many advisory boards exist to insure the standardization of electrical installations in accordance with accepted designs and safe practices. The Institute of Electrical and Electronics Engineers (IEEE) and the American National Standards Institute (ANSI) are two advisory boards that have published numerous standards. These standards are utilized by the Department of Energy and the nuclear industry. However, for a day-to-day practical guide for noncritical installations, the recognized guide is the National Electrical Code Handbook (NEC), published by the National Fire Protection Association and endorsed by ANSI. The NEC Handbook is the primary source of much of the material presented in this chapter and may serve as a ready reference for specific questions not covered in this fundamental discussion.

Terminology

To understand wiring schemes used in power distribution systems, you must be familiar with the following terms.

- *Ampacity* - the current in amperes that a conductor can carry continuously under the conditions of use without exceeding its temperature rating.
- *Bond* - the permanent joining of metallic parts or circuits assuring electrical continuity and capacity to safely conduct any current likely to be imposed.
- *Conductor* - any wire, cable, or substance capable of carrying an electrical current.
- *Ground* - a conducting connection, whether intentional or accidental, between a circuit or piece of equipment and the earth, or some body serving as earth; a place of zero electrical potential.
- *Ground voltage* - the voltage between any given conductor and any point at ground potential.
- *Leg* - a current-carrying conductor intended to deliver power to or from a load normally at an electrical potential other than ground.
- *Neutral* - a current-carrying conductor normally tied to ground so that the electrical potential is zero.
- *Phase voltage* - the greatest root mean square (effective) difference of potential between any two legs of the circuit.

Single-Phase Power

The source of single-phase (1ϕ) power in all facilities is by generation from a single-phase generator or by utilization of one phase of a three-phase (3ϕ) power source. Basically, each phase of the 3ϕ distribution system is a single-phase generator electrically spaced 120 degrees from the other two; therefore, a 3ϕ power source is convenient and practical to use as a source of single-phase power.

Single-phase loads can be connected to three-phase systems utilizing two methods. The diagram shown in Figure 15 illustrates these connections.

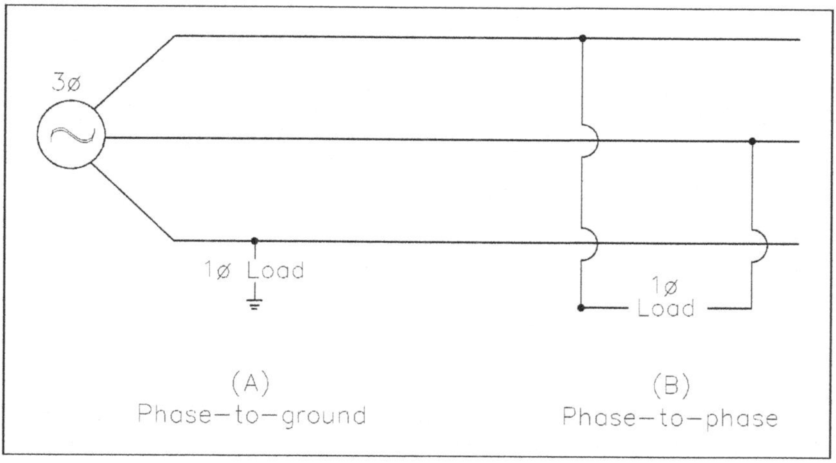

Figure 15 Three-Phase To Single-Phase Connections

The first scheme (Figure 15A) provides for the connection of the load from a phase leg to any ground point and is referred to as a phase-to-ground scheme. The remaining scheme (Figure 15B) connects the single-phase load between any two legs of the three-phase source and is referred to as a phase-to-phase connection. The choice of schemes, phase-to phase or phase-to-ground, allows several voltage options depending on whether the source three-phase system is a delta or wye configuration. This will be discussed in the three-phase segment of this chapter.

The only approved method of wiring single-phase power is the scheme commonly referred to as the 3-wire, single-phase Edison system. The illustration in Figure 16 depicts the use of a center-tapped transformer, with the center tap grounded, providing half voltage (120 V) connections on either side or full voltage (240 V) across both sides.

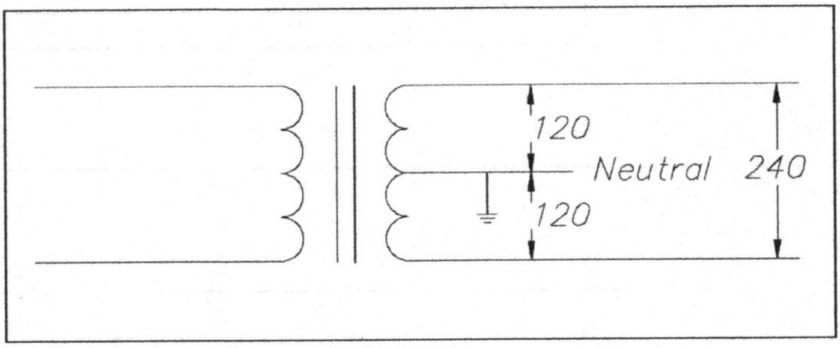

Figure 16 3-Wire Edison Scheme

The physical connections to the transformer secondary involve two insulated conductors and one bare conductor. If the conductor is a current-carrying leg or neutral leg, the conductor will be insulated. The remaining uninsulated conductor will serve as a safety ground and will be bonded to the ground point of the system. In all cases, 3 wires will be presented to the load terminals, and the safety ground will be bonded to each junction box, or device, in the distribution system. In the case of half voltage (120 V) use, the intended path of the current is from the supply leg through the load and back to the source on the neutral leg. No current would be carried on the ground unless a fault occurred in the system, in which case the current would flow safely to ground.

In the full voltage system (240 V), the insulated conductors are connected across the full winding of the transformer, and the uninsulated conductor is again bonded to the grounded center tap. In a balanced system, all currents will flow on the insulated conductors, and the grounded neutral will carry no current, acting only in a ground capacity. In the case of either an unbalanced load or a fault in the system, the bare conductor will carry current, but the potential will remain at zero volts because it is tied to the ground point. As in the case of the half voltage system, the uninsulated conductor will be bonded to each device in the system for safety.

Three-Phase Wiring Schemes

Unlike the single-phase wiring scheme that must make a provision for a neutral leg and separate ground, the three-phase system needs neither a separate neutral nor a ground to operate safely. However, to prevent any unsafe condition, all 3- and 4-wire, three-phase systems can include an effective ground path. As with the previous single-phase discussion, only the secondary side of the transformer and its connected load need to be studied.

3-Wire, Three-Phase Delta System

The simplest three-phase system is the 3-wire Delta configuration, normally used for transmission of power in the intermediate voltage class from approximately 15,000 volts to 600 volts. The diagram in Figure 17 depicts the two methods of connecting the Delta secondary.

The upper diagram depicts the ungrounded Delta, normally confined to protected environments such as fully enclosed ducts or overhead transmission lines that cannot be reached without extraordinary means. Each conductor's ground voltage is equal to the full phase voltage of the system.

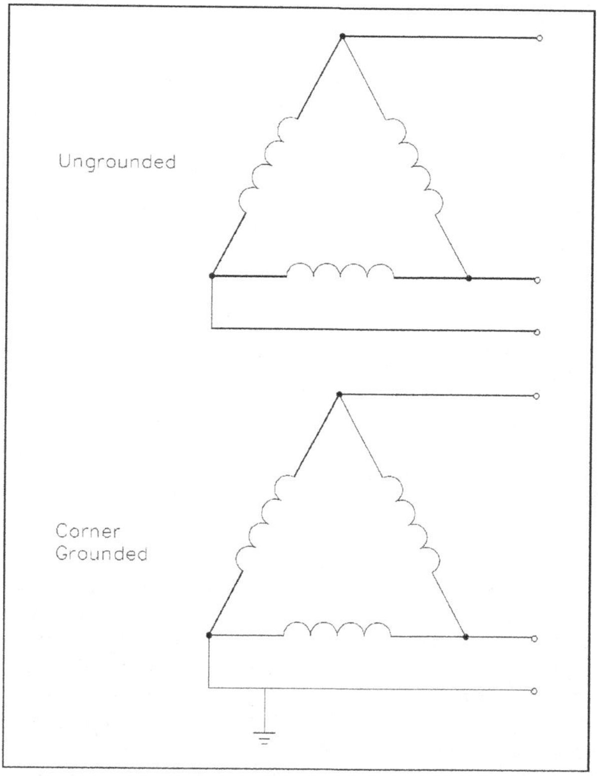

Figure 17 3-Wire, Three-Phase Delta Scheme

The lower diagram shows a ground point affixed to one corner of the Delta, which effectively lowers one phase's voltage reference to ground to zero, but retains a phase-to-phase voltage potential. The corner-grounded phase acts in much the same way as the grounded neutral of the single-phase Edison system, carrying current and maintaining ground potential.

The corner-grounded Delta system has an obvious economy in wiring costs, and the grounded phase can be used to physically protect the other two phases from accidental grounding or lightning strikes in outdoor settings. This system is rarely used for low voltage (under 600 V), however, because of the absence of a safety ground required by many facilities for circuits involving potential worker contact.

4-Wire, Three-Phase Delta System

The 4-wire, three-phase Delta system combines the ungrounded Delta discussed above for three-phase loads with the convenience of the Edison system for single-phase loads. As depicted in the example illustration in Figure 18, one side of the Delta has a grounded-neutral conductor connected to a center tap winding on one phase.

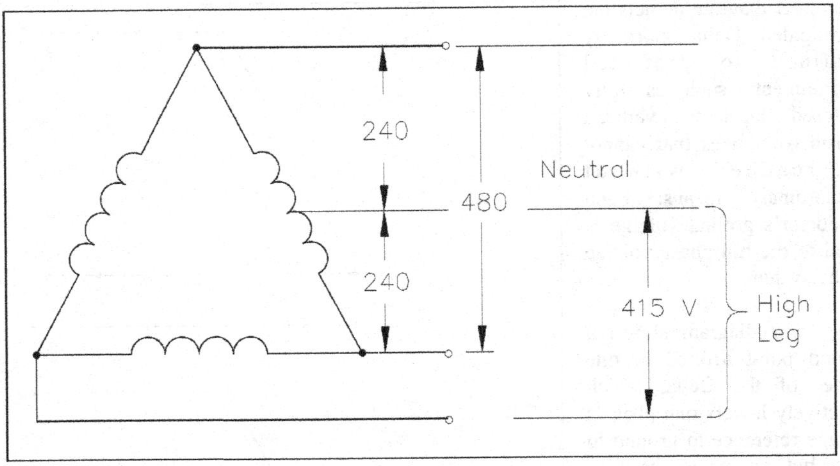

Figure 18 4-Wire Delta System

The single-phase voltage on each side of the half-tap is one-half the voltage available in the normal phase-to-phase relationship. This provides the same half- or full-voltage arrangement seen in the normal Edison scheme with a grounded neutral. Notice also that the legs coming from the corners of the Delta would have a normal ungrounded appearance if it were not for the center tap of one phase. Thus, at any given location in the system, either three-phase power at full voltage or single-phase power with half or full voltage is equally possible. However, there are several strict precautions that must be observed in the operation of this system. First, all loads must be carefully balanced on both the single-phase and three-phase legs. Second, because the voltage between one leg and the grounded neutral is considerably higher than the rest of the single-phase system, a measurement between the neutral and the phase must be taken to identify the "high leg," or "bastard voltage." Last, the "high leg" is never used as a single-phase source because no ground or grounded neutral exists for this circuit.

4-Wire, Three-Phase Wye System

Until now, the voltage, the phase voltage, and the ground voltage of the three-phase systems have been equal, with the one exception of one phase of the corner-grounded Delta. The Wye system has completely different voltage characteristics from the Delta system. In the Wye system, the ground voltage or voltage available from phase to ground is the phase voltage divided by 1.73.

In Figure 19, an example of the Wye system, or center-grounded Wye as it is commonly referred to, extends three current-carrying insulated conductors and an insulated grounded neutral to the loads. Depending on the selection of conductors, one of the following is available: a reduced-voltage single phase between a phase leg and the neutral; a full-voltage single-phase circuit between any two phase legs; or a full-voltage three-phase power. Again, some precautions must be taken when balancing the single-phase loads in the system. The full load ampacity of the neutral must be sized to 1.73 times the highest phase ampacity. This is done to avoid either an over-current condition if a fault is present or the operation of single-phase loads at reduced voltage if the loads become severely unbalanced by accidental interruption.

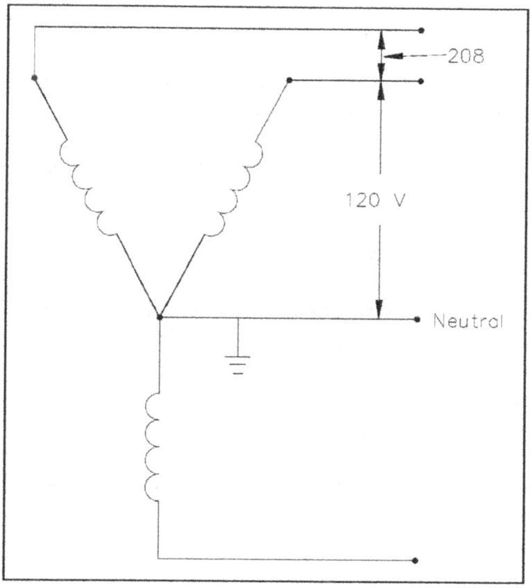

Figure 19 4-Wire, Three-Phase Wye System

As with all other grounded systems, bonds are established between the grounded neutral and all components of the system. This system is recognized as the safest possible multi-purpose distribution system for low voltage and is commonly seen in the 208/120-volt range in many facilities.

Summary

The important information in this chapter is summarized on the following page.

Wiring Schemes And Grounding Summary

Terminology

Ampacity - current-carrying capacity of a conductor in amperes

Bond - permanent joining of metallic parts or circuits assuring electrical continuity

Conductor - any wire, cable, or substance capable of carrying an electrical current

Ground - a conducting connection between a circuit or piece of equipment and the earth, or some body serving as earth

Ground voltage - the voltage between any given conductor and any point at ground potential

Leg - a current-carrying conductor intended to deliver power to or from a load

Neutral - a current-carrying conductor intended to deliver power to or from a load normally at an electrical potential other than ground

Phase voltage - the greatest root mean square (effective) difference of potential between any two legs of the circuit

Two methods to connect single-phase loads to a three-phase system are:

- Phase-to-phase
- Phase-to-ground

The purposes of the following wiring schemes are:

3-wire, single-phase Edison system - the only approved method of wiring single-phase power

3-wire, three-phase Delta system - normally used for transmission of power in the intermediate voltage class from approximately 15,000 volts to 600 volts

4-wire, three-phase Delta system - combines the ungrounded Delta for three-phase loads with the convenience of the Edison system for single-phase loads

4-wire, three-phase Wye system - the safest possible multi-purpose distribution system for low voltage

end of text.

CONCLUDING MATERIAL

Review activities:

DOE - ANL-W, BNL, EG&G Idaho,
EG&G Mound, EG&G Rocky Flats,
LLNL, LANL, MMES, ORAU, REECo,
WHC, WINCO, WEMCO, and WSRC.

Preparing activity:

DOE - NE-73
Project Number 6910-0017/4

Intentionally Left Blank

Made in the USA
Monee, IL
10 October 2026

65356098R00305